Hyo J. Eom

Electromagnetic Wave Theory for Boundary-Value Problems

Springer
Berlin
Heidelberg
New York
Hong Kong
London
Milan
Paris
Tokyo

Hyo J. Eom

Electromagnetic Wave Theory for Boundary-Value Problems

An Advanced Course on Analytical Methods

With 111 Figures

Springer

Professor Hyo J. Eom
Korea Advanced Institute of Science and Technology
Dept. of Electrical Engineering and Computer Science
373-1, Guseong-dong, Yuseong-gu
Daejeon, 305-701
Korea
e-mail: hjeom@ee.kaist.ac.kr

ISBN 3-540-21266-3 Springer-Verlag Berlin Heidelberg New York

Library of Congress Control Number: 2004102306

Springer-Verlag is a part of Springer Science + Business Media

springeronline.com

Printed in Germany

Typesetting: Camera ready by author
Cover-Design: medio Technologies AG
Printed on acid free paper 62/3111/kk - 5 4 3 2 1 SPIN: 11664246

To the memory of my parents

Preface

In this era of reliance on communication and information technologies, electromagnetic wave engineering has become increasingly important. Electromagnetic wave theory finds practical applications in mobile telecommunication and optical fiber transmission. This book is written as a text for the two-semester graduate course on electromagnetic wave theory offered at the Korea Advanced Institute of Science and Technology (KAIST), Daejeon, Korea. The prerequisites for this course are undergraduate courses in advanced calculus and in electromagnetism. In particular, some mathematical background in differential equations and complex variable analysis would be helpful.

To understand scattering, propagation, and radiation of electromagnetic waves, boundary-value problems must be solved using electromagnetic wave theory based on Maxwell's equations. The motivation for writing this text is to present solutions to canonical boundary-value problems, thereby helping students better understand electromagnetic wave theory in addition to enhancing their analytic skills. Selected topics are not only representative for electromagnetic boundary-value problems, but also instructive in radio wave communication applications. This text has the following unique attributes:

- Each topic deals with canonical boundary-value problems that can be solved using basic analytic skills in Fourier series, Fourier transform, and residue calculus.
- No numerical computations are presented.
- Necessary mathematical formulas and derivations are included in the text.
- The technique of Fourier transform and mode matching is utilized to derive rigorous solutions to selected problems in Sections 3.6, 4.4, 7.4, 10.3, 10.4, 10.5, 10.6, and 10.7.

Fundamental electromagnetic issues in communication engineering are presented in Chapters 1 through 11. Chapter 1 introduces Maxwell's equations and related basics including the constitutive relations, boundary conditions, powers, and potential concept. Chapter 2 deals with uniform plane waves, polarization, Gaussian beam, plane wave propagation across plane boundaries, and transmission line theory. Chapter 3 analyzes electromagnetic waves that propagate along a metallic rectangular waveguide, dielectric slab waveguide, circular waveguide, and shielded stripline. Chapter 4 investigates resonance in rectangular, circular, and spherical cavities as well as wave coupling using a groove guide coupler. Chapter 5 discusses wave propagation in anisotropic media including uniaxial and ferrite materials. Chapter 6 introduces some theorems that are useful for scattering computation. Relevant scattering examples are introduced to illustrate these theorems. Chapter 7 considers wave scattering from a circular cylinder, sphere, step, slit, and circular aperture. Chapter 8 introduces Sturm-Liouville problems and free-space Green's functions. Chapter 9 illustrates applications of Green's functions dealing with radiation from currents in waveguides. Chapter 10 presents antenna fundamentals and antenna radiation in terms of the dipole antenna, loop antenna, aperture antenna, groove-backed antenna, slit array antenna, slotted coaxial line antenna, and flanged coaxial line antenna. Chapter 11 considers half-space radiation problems using dipole and line sources.

The selected topics can be covered in two semesters. The following sections are covered in the first semester:
Chapter 1, Sections 1.1 through 1.6
Chapter 2, Sections 2.1 through 2.6
Chapter 3, Sections 3.1 through 3.5
Chapter 4, Sections 4.1 through 4.2
Chapter 6, Sections 6.1 through 6.6
Chapter 7, Section 7.1
Chapter 8, Sections 8.1 through 8.4
Chapter 9, Section 9.1
Chapter 10, Sections 10.1 through 10.2
The remaining topics are more advanced and, therefore, are recommended for second semester course material.

Throughout the text, the $\exp(-i\omega t)$ time factor is suppressed and the International System of Units (SI) is implicitly assumed. Vectors are denoted with overlines, such as $\overline{A}$, $\overline{B}$, etc.

My sincere appreciation goes to KAIST for providing me with a superb academic environment in which I dared to write this text. The many valuable comments, criticisms, and creative ideas received from students at KAIST have been very much appreciated. My thanks go to Professors Jong K. Park, Yong H. Cho, Sang W. Nam, Young K. Cho, and Sang Y. Shin for their reading of the manuscript and helpful comments. Excellent assistance from my graduate students in drawing figures is greatly acknowledged. I gratefully acknowledge support from the Radio Education Research Center of Information and Communications University, Daejeon, Korea. I also wish to express my gratitude for the patience and support of my family during the writing of this text. Finally, any comments regarding the text would be gratefully received.

Daejeon, Korea *Hyo J. Eom*

Contents

Notations

$\overline{A}$	magnetic vector potential
$\overline{B}$	magnetic flux density
$\overline{D}$	electric flux density
$\overline{E}$	electric field intensity
$\overline{F}$	electric vector potential
$g(\overline{r};\overline{r}')$	Green's function
$\overline{H}$	magnetic field intensity
$H_m^{(1)}(\cdot)$	Hankel function of the first kind of order m
$H_m^{(2)}(\cdot)$	Hankel function of the second kind of order m
i	imaginary unit ($i = \sqrt{-1}$)
$I_m(\cdot)$	modified Bessel function of the first kind of order m
$\mathrm{Im}\,(\cdot)$	imaginary part of $(\cdot)$
$\overline{J}$	electric current density
$\overline{J}_c$	conduction current density
$\overline{J}_i$	impressed electric current density
$J_m(\cdot)$	Bessel function of the first kind of order m
$\hat{J}_m(\cdot)$	spherical Bessel function of the first kind of order m
$j_m(\cdot)$	spherical Bessel function of the first kind of order m
$\overline{J}_s$	surface electric current density
k	wavenumber
$\overline{k}$	wave vector
$K_m(\cdot)$	modified Bessel function of the second kind of order m
$\overline{M}$	magnetic current density, magnetic polarization vector
$\overline{M}_s$	surface magnetic current density
$\hat{n}$	normal unit vector
$N_m(\cdot)$	Bessel function of the second kind of order m
$\overline{P}$	electric polarization vector
P_{av}	time-average power
$P_n^m(\cdot)$	associated Legendre function of the first kind
PEC	perfect electric conductor

PMC	perfect magnetic conductor
$Q_n^m(\cdot)$	associated Legendre function of the second kind
Re $(\cdot)$	real part of $(\cdot)$
Res $(\cdot)$	residue of $(\cdot)$
$\overline{S}_{av}$	time-average Poynting vector
TE	transverse electric
TEM	transverse electromagnetic
TM	transverse magnetic
δ	skin depth
$\delta(\cdot)$	delta function
δ_{mn}	Kronecker delta
ϵ	permittivity
ϵ_0	permittivity of free space (vacuum)
$\bar{\bar{\epsilon}}$	tensor permittivity
ε_m	$\varepsilon_0 = 2$ and $\varepsilon_1 = \varepsilon_2 = \cdots = 1$
η	intrinsic impedance of medium
θ	angles
λ	wavelength
μ	permeability
μ_0	permeability of free space (vacuum)
$\bar{\bar{\mu}}$	tensor permeability
ρ_e	electric charge density
ρ_m	magnetic charge density
σ	conductivity
ϕ	angles
ϕ_e	electric scalar potential
ϕ_m	magnetic scalar potential
χ_e	electric susceptibility
χ_m	magnetic susceptibility
ω	angular frequency
∇	del operator
∇^2	Laplacian operator
$(\cdot)^*$	complex conjugate of $(\cdot)$

1

Electromagnetic Basics

1.1 Maxwell's Equations

Most basic laws in electromagnetic theory were established in the 19th century by a variety of talented scientists, including Oersted, Ampère, and Faraday. Later, James Clerk Maxwell unified these basic laws into fundamental equations, which macroscopically state the relations between electromagnetic fields and their sources (current and charge). These fundamental equations are called Maxwell's equations. Let us first consider Maxwell's equations in a differential equation form.

1.1.1 Maxwell's Equations in Differential Equation Form

Maxwell's equations are written in terms of time-varying quantities as

$$\nabla \times \overline{E}(\overline{r},t) = -\frac{\partial \overline{B}(\overline{r},t)}{\partial t} \qquad \text{(Faraday's law)} \tag{1.1}$$

$$\nabla \times \overline{H}(\overline{r},t) = \frac{\partial \overline{D}(\overline{r},t)}{\partial t} + \overline{J}(\overline{r},t) \qquad \text{(Ampère's law)} \tag{1.2}$$

$$\nabla \cdot \overline{D}(\overline{r},t) = \rho_e(\overline{r},t) \qquad \text{(Gauss' law)} \tag{1.3}$$

$$\nabla \cdot \overline{B}(\overline{r},t) = 0 \qquad \text{(Magnetic Gauss' law)} \tag{1.4}$$

where the time-varying quantities, functions of position vector $\overline{r}$ and time t, are

$\overline{E}(\overline{r},t)$ = electric field intensity or electric field (volts/meter)
$\overline{H}(\overline{r},t)$ = magnetic field intensity or magnetic field (amperes/meter)
$\overline{D}(\overline{r},t)$ = electric flux density (coulombs/meter^2)
$\overline{B}(\overline{r},t)$ = magnetic flux density (webers/meter^2)
$\overline{J}(\overline{r},t)$ = electric current density (amperes/meter^2)

$\rho_e(\overline{r},t)$ = electric charge density (coulombs/meter^3) .

Maxwell revised Ampère's law by adding the term $\frac{\partial \overline{D}(\overline{r},t)}{\partial t}$, which is called the electric displacement current density. He noted that the addition of the electric displacement current density was necessary to make Maxwell's equations a consistent set. He also predicted the existence of electromagnetic waves, which was later experimentally verified by Heinrich Rudolf Hertz. Maxwell's equations are thus considered empirical in that they were formulated and verified by experimental observations. Maxwell's equations are instrumental in understanding macroscopic electromagnetic phenomena encountered in wave scattering, radiation, and propagation.

Maxwell's equations (1.1) through (1.4) are in an asymmetric form due to the absence of magnetic sources. However, from a mathematical viewpoint, it is convenient to transform (1.1) through (1.4) into a symmetric form, which can be achieved by introducing the fictitious magnetic sources $\overline{M}(\overline{r},t)$ and $\rho_m(\overline{r},t)$. Consequently, Maxwell's equations can be written in a symmetric form as

$$\nabla \times \overline{E}(\overline{r},t) = -\frac{\partial \overline{B}(\overline{r},t)}{\partial t} - \overline{M}(\overline{r},t) \tag{1.5}$$

$$\nabla \times \overline{H}(\overline{r},t) = \frac{\partial \overline{D}(\overline{r},t)}{\partial t} + \overline{J}(\overline{r},t) \tag{1.6}$$

$$\nabla \cdot \overline{D}(\overline{r},t) = \rho_e(\overline{r},t) \tag{1.7}$$

$$\nabla \cdot \overline{B}(\overline{r},t) = \rho_m(\overline{r},t) \tag{1.8}$$

where
$\overline{M}(\overline{r},t)$ = magnetic current density (volts/meter^2)
$\rho_m(\overline{r},t)$ = magnetic charge density (webers/meter^3) .

The hypothetical assumption of $\overline{M}$ and ρ_m is purely for mathematical convenience. This means that as long as the field produced by the equivalent sources $\overline{M}$ and ρ_m is identical with the field produced by the real sources $\overline{J}$ and ρ_e, the assumption of $\overline{M}$ and ρ_m is justified.

1.1.2 Continuity Equations

The assumption of charge conservation stipulates that the current and charge densities be conserved since the sum of them can be neither destroyed nor created. Charge conservation is mathematically stated in terms of the continuity equations given as

$$\nabla \cdot \overline{J}(\overline{r},t) + \frac{\partial \rho_e(\overline{r},t)}{\partial t} = 0 \tag{1.9}$$

$$\nabla \cdot \overline{M}(\overline{r},t) + \frac{\partial \rho_m(\overline{r},t)}{\partial t} = 0\,. \tag{1.10}$$

The continuity equations state that the creation of the current $\overline{J}$ (or $\overline{M}$) should be at the expense of a decrease in the charge density ρ_e (or ρ_m).

It is important to note that Gauss' laws (1.7) and (1.8) are derivable from (1.5), (1.6), (1.9), and (1.10). For instance, applying the divergence to (1.5) gives

$$\underbrace{\nabla \cdot \left[\nabla \times \overline{E}(\overline{r},t)\right]}_{0} = -\frac{\partial}{\partial t}\left[\nabla \cdot \overline{B}(\overline{r},t)\right] - \nabla \cdot \overline{M}(\overline{r},t)\,. \tag{1.11}$$

Substituting (1.10) into (1.11) yields

$$\frac{\partial}{\partial t}\left[\nabla \cdot \overline{B}(\overline{r},t)\right] = \frac{\partial \rho_m(\overline{r},t)}{\partial t}\,. \tag{1.12}$$

Equation (1.12) is integrated with respect to time, and setting the integration constant equal to zero produces (1.8). Similarly, applying the divergence to (1.6) results in (1.7). This implies that the two Gauss' laws are dependent equations, whereas the remaining two Maxwell's curl equations, (1.5) and (1.6), are independent. In general, (1.5) and (1.6) must be solved to determine unique and complete solutions to electromagnetic boundary-value problems.

1.1.3 Maxwell's Equations in Integral Form

It is sometimes useful to transform Maxwell's equations into an integral form. In view of Fig. 1.1, Stokes' theorem is given by

$$\int_S \nabla \times \overline{A} \cdot d\overline{s} = \oint_C \overline{A} \cdot d\overline{l} \tag{1.13}$$

where $d\overline{s}$ is the differential surface vector and $d\overline{l}$ is the differential line vector encircling the surface S. Applying Stokes' theorem to (1.5) yields

$$\oint_C \overline{E}(\overline{r},t) \cdot d\overline{l} = -\int_S \left[\frac{\partial \overline{B}(\overline{r},t)}{\partial t} + \overline{M}(\overline{r},t)\right] \cdot d\overline{s}\,. \tag{1.14}$$

Similarly, applying Stokes' theorem to (1.6) yields

$$\oint_C \overline{H}(\overline{r},t) \cdot d\overline{l} = \int_S \left[\frac{\partial \overline{D}(\overline{r},t)}{\partial t} + \overline{J}(\overline{r},t)\right] \cdot d\overline{s}\,. \tag{1.15}$$

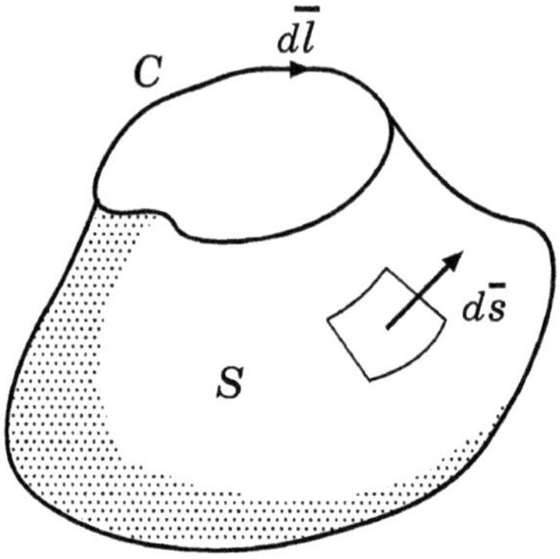

Fig. 1.1. Surface S encircled by line C.

In view of Fig. 1.2, the divergence theorem gives

$$\int_V \nabla \cdot \overline{A}\, dv = \oint_S \overline{A} \cdot d\overline{s} \tag{1.16}$$

where S is the closed surface surrounding the volume V, and $d\overline{s}$ is the differential surface vector pointing outward from the surface. Applying the divergence theorems to (1.7) and (1.8), respectively, gives

$$\oint_S \overline{D}(\overline{r}, t) \cdot d\overline{s} = \int_V \rho_e(\overline{r}, t)\, dv \tag{1.17}$$

$$\oint_S \overline{B}(\overline{r}, t) \cdot d\overline{s} = \int_V \rho_m(\overline{r}, t)\, dv\ . \tag{1.18}$$

Equations (1.14), (1.15), (1.17), and (1.18) constitute Maxwell's equations in an integral form.

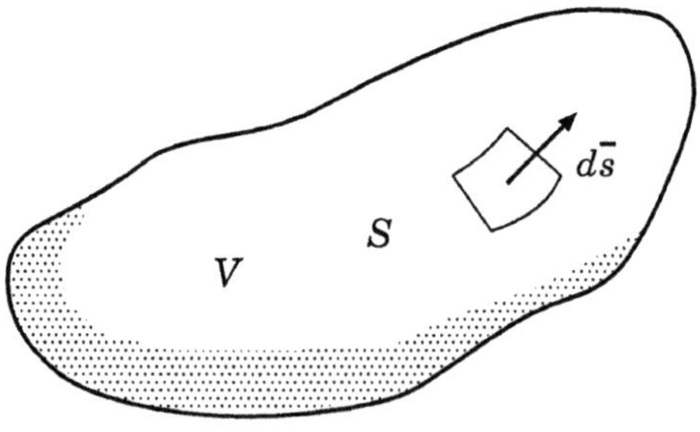

Fig. 1.2. Volume V surrounded by surface S.

1.1.4 Maxwell's Equations in Time-Harmonic Form

In most wireless communication channels, information is transmitted by a sinusoidal (time-harmonic) electromagnetic wave with a carrier frequency. Let us assume that a time-varying (instantaneous) field $\overline{E}(\overline{r},t)$ with a carrier angular frequency ω is represented as

$$\overline{E}(\overline{r},t) = \overline{E}_0(\overline{r})\cos(\omega t + \phi) \tag{1.19}$$

where $\overline{E}_0(\overline{r})$ is the amplitude vector of $\overline{E}(\overline{r},t)$, and ϕ is its phase. The time-varying field $\overline{E}(\overline{r},t)$ is a real function of the position vector $\overline{r}$ and time t. To effectively represent the time-varying field $\overline{E}(\overline{r},t)$ in a time-harmonic case, the phasor field $\overline{E}(\overline{r}) = \overline{E}_0(\overline{r})e^{-i\phi}$ is introduced. Let us use the $e^{-i\omega t}$ time convention for the phasor field. The relation between the time-varying field $\overline{E}(\overline{r},t)$ and the corresponding phasor field $\overline{E}(\overline{r})$ is

$$\overline{E}(\overline{r},t) = \mathrm{Re}\left[\overline{E}(\overline{r})e^{-i\omega t}\right] \tag{1.20}$$

where the notation $\mathrm{Re}\,[\cdot]$ denotes the real part of $[\cdot]$. The phasor field $\overline{E}(\overline{r})$ suppresses the redundant time factor $e^{-i\omega t}$, thereby substantially simplifying the mathematical expressions.

In a time-harmonic case, Maxwell's equations using phasors can be obtained from (1.5) through (1.8) with the time-varying forms replaced by the corresponding phasors and with $\dfrac{\partial}{\partial t}$ replaced by $(-i\omega)$. Consequently, Maxwell's equations and continuity equations are represented in terms of the phasors as

$$\nabla \times \overline{E}(\overline{r}) = i\omega\overline{B}(\overline{r}) - \overline{M}(\overline{r}) \tag{1.21}$$

$$\nabla \times \overline{H}(\overline{r}) = -i\omega\overline{D}(\overline{r}) + \overline{J}(\overline{r}) \tag{1.22}$$

$$\nabla \cdot \overline{D}(\overline{r}) = \rho_e(\overline{r}) \tag{1.23}$$

$$\nabla \cdot \overline{B}(\overline{r}) = \rho_m(\overline{r}) \tag{1.24}$$

$$\nabla \cdot \overline{J}(\overline{r}) - i\omega\rho_e(\overline{r}) = 0 \tag{1.25}$$

$$\nabla \cdot \overline{M}(\overline{r}) - i\omega\rho_m(\overline{r}) = 0\,. \tag{1.26}$$

In the remainder of this book, the discussion is limited to time-harmonic cases, unless otherwise specified. The short notation $\overline{E}$ is used throughout for the phasor $\overline{E}(\overline{r})$.

1.2 Constitutive Relations

To determine the electromagnetic fields $\overline{E}$ and $\overline{H}$, two independent curl Maxwell's equations, (1.21) and (1.22), must be solved. This implies that $\overline{D}$ and $\overline{B}$ must be a priori related to $\overline{E}$ and $\overline{H}$. The relationships amongst $\overline{E}$, $\overline{B}$, $\overline{D}$, $\overline{H}$, and $\overline{J}$ are given by medium characteristics, referred to as constitutive relations. These relations can be constructed from the electric charge properties of the material medium.

1.2.1 Material Media

Consider a material medium that is electrically neutral and free from the influence of external fields. When the fields $\overline{E}$ and $\overline{H}$ are externally applied to the material medium, the electric balance within the material medium breaks down and the charges (atoms and electrons) are redistributed, thereby creating net electric and magnetic dipoles within the molecules. Here, the material field vectors $\overline{P}$ and $\overline{M}$ are introduced to account for the electric and magnetic dipole moment distributions respectively caused by the applied fields $\overline{E}$ and $\overline{H}$. The resulting total electric and magnetic flux densities, $\overline{D}$ and $\overline{B}$, are the sum of the contributions from the applied fields and dipole moments

$$\overline{D} = \epsilon_0 \overline{E} + \overline{P} \tag{1.27}$$

$$\overline{B} = \mu_0 \left(\overline{H} + \overline{M}\right) \tag{1.28}$$

where
$\overline{P} = \epsilon_0 \chi_e \overline{E}$ = electric polarization vector (coulombs/meter^2)
$\overline{M} = \chi_m \overline{H}$ = magnetic polarization vector (amperes/meter)
ϵ_0 = dielectric constant (permittivity) of free space $\approx \dfrac{10^{-9}}{36\pi}$ (farads/meter)
μ_0 = permeability of free space $= 4\pi \times 10^{-7}$ (henrys/meter)
χ_e = electric susceptibility
χ_m = magnetic susceptibility .

Introducing the material medium permittivity $\epsilon = \epsilon_0(1+\chi_e) \equiv \epsilon_0\epsilon_r$ and permeability $\mu = \mu_0(1+\chi_m) \equiv \mu_0\mu_r$ produces

$$\overline{D} = \epsilon \overline{E} \tag{1.29}$$

$$\overline{B} = \mu \overline{H} \tag{1.30}$$

where ϵ_r and μ_r are the relative permittivity and permeability of the material medium, respectively. In practice, the values of ϵ and μ vary with the measuring frequency.

1.2.2 Conduction Current and Loss Tangent

When an electric field is applied to a conductive and lossy medium, a conduction current is generated within the lossy medium, dissipating the applied energy into heat. According to Ohm's law, the conduction current density $\overline{J}_c$ is proportional to the applied electric field $\overline{E}$

$$\overline{J}_c = \sigma \overline{E} \tag{1.31}$$

where the proportionality constant σ is called the conductivity of the lossy medium. The expressions (1.29), (1.30), and (1.31) are the constitutive relations. The constitutive relations characterize the medium-electromagnetic field interdependence through the medium parameters, ϵ, μ, and σ. The medium parameters, ϵ, μ, and σ, are associated with the capacitance (C), inductance (L), and conductance (G), respectively, of an equivalent electric circuit. Let us decompose the electric current term $\overline{J}$ into the impressed and conduction terms

$$\overline{J} = \overline{J}_i + \overline{J}_c \,. \tag{1.32}$$

The impressed term $\overline{J}_i$ is the current source that is externally applied to the medium. Since most practical media, except for some magnetic materials, are magnetically lossless, the magnetic conduction current $\overline{M}_c$ is zero and the permeability μ becomes a real scalar. The magnetic current $\overline{M}$ can, therefore, be regarded as an impressed one such as $\overline{M} = \overline{M}_i$.

In view of the constitutive relations, Maxwell's equations become

$$\nabla \times \overline{E} = i\omega\mu\overline{H} - \overline{M_i} \tag{1.33}$$

$$\nabla \times \overline{H} = -i\omega\epsilon_c\overline{E} + \overline{J}_i \tag{1.34}$$

$$\nabla \cdot \underbrace{(\epsilon_c \overline{E})}_{\overline{D}} = \rho_e \tag{1.35}$$

$$\nabla \cdot \underbrace{(\mu \overline{H})}_{\overline{B}} = \rho_m \tag{1.36}$$

where the complex permittivity ϵ_c is

$$\epsilon_c \equiv \epsilon \left(1 + i\frac{\sigma}{\omega\epsilon}\right) \tag{1.37}$$

and ρ_e is the electric charge density associated with $\overline{J}_i$ $\left(\nabla \cdot \overline{J}_i = i\omega\rho_e\right)$.

The loss tangent is defined as $\dfrac{\sigma}{\omega\epsilon}$ and signifies the degree of energy loss caused by the conduction current.

- When $\frac{\sigma}{\omega\epsilon} \gg 1$, the material is classified as a good conductor.
- When $\frac{\sigma}{\omega\epsilon} \ll 1$, the material is a good dielectric.

1.3 Boundary Conditions

The field behavior across a boundary between two media follows certain rules, usually known as boundary conditions, which can be derived from Maxwell's equations. The boundary conditions are used to solve boundary-value problems. This section discusses the conditions imposed on the boundary between two adjacent dielectric media.

1.3.1 Boundary Conditions Between Two Media

Consider an electromagnetic field near the boundary between media 1 and 2, as shown in Fig. 1.3. Rewriting (1.33) in an integral form yields

$$\oint_C \overline{E} \cdot d\overline{l} = \int_S i\omega\mu\overline{H} \cdot d\overline{s} - \int_S \overline{M_i} \cdot d\overline{s} \tag{1.38}$$

where the contour C encircles the surface S occupying media 1 and 2. Equation (1.38) is divided by Δx and we let $\Delta y \to 0$. The right-hand side of (1.38) then becomes $-\overline{M}_s$, where $\overline{M}_s (= \overline{M}_i \Delta y)$ is the surface magnetic current density present at the boundary. The left-hand side of (1.38) reduces to $\hat{n} \times (\overline{E}_1 - \overline{E}_2)$, where $\hat{n}$ is a normal unit vector. Note that $\hat{n} \times \overline{E}_1$ and $\hat{n} \times \overline{E}_2$ are the tangential electric fields in media 1 and 2 at the boundary, respectively. Hence, (1.38) can be rewritten as

$$\hat{n} \times \left(\overline{E}_1 - \overline{E}_2\right) = -\overline{M}_s . \tag{1.39}$$

The boundary condition (1.39) states that the tangential electric field is discontinuous across the boundary due to $-\overline{M}_s$. Thus, if $\overline{M}_s = 0$ at the boundary, the tangential electric field must be continuous.

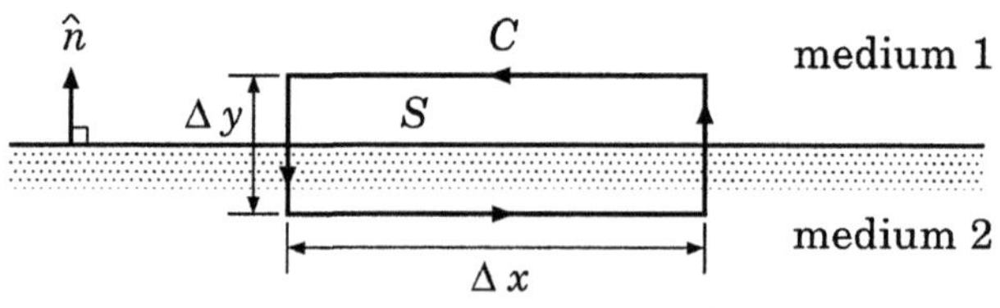

Fig. 1.3. Infinitesimally small surface S occupying media 1 and 2.

The boundary condition for the tangential magnetic field is also obtained from (1.34). Rewriting (1.34), similarly, yields

$$\oint_C \overline{H} \cdot d\overline{l} = -\int_S i\omega\epsilon_c \overline{E} \cdot d\overline{s} + \int_S \overline{J}_i \cdot d\overline{s} \, . \tag{1.40}$$

Thus, when $\Delta y \to 0$, (1.40) becomes

$$\hat{n} \times \left(\overline{H}_1 - \overline{H}_2\right) = \overline{J}_s \tag{1.41}$$

where $\overline{J}_s$ is the surface electric current density. The boundary condition (1.41) states that the tangential magnetic field is discontinuous across the boundary due to $\overline{J}_s$ situated at the boundary. In the case $\overline{J}_s = 0$, the tangential magnetic field across the boundary must be continuous.

Considering the small pill box shown in Fig. 1.4, Equation (1.35) is transformed into

$$\oint_S \overline{D} \cdot d\overline{s} = \int_V \rho_e \, dv \, . \tag{1.42}$$

Letting $\Delta x \to 0$ results in

$$\hat{n} \cdot \left(\overline{D}_1 - \overline{D}_2\right) = q_{es} \tag{1.43}$$

where q_{es} is the surface electric charge density at the boundary. Note that $\hat{n} \cdot \overline{D}_1$ denotes the normal component of $\overline{D}_1$ at the boundary. The boundary condition (1.43) indicates that the normal component of $\overline{D}$ is discontinuous across the boundary by the amount q_{es}.

Similarly, it is possible to obtain the boundary condition for the normal component of $\overline{B}$. Equation (1.36) produces

$$\oint_S \overline{B} \cdot d\overline{s} = \int_V \rho_m \, dv \, . \tag{1.44}$$

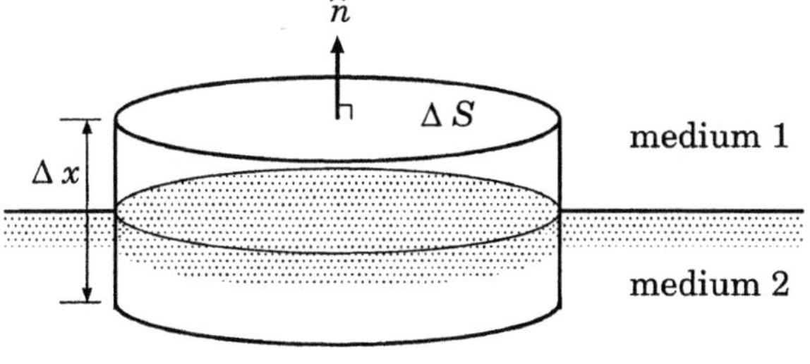

Fig. 1.4. Small volume V surrounded by surface S across media 1 and 2.

As $\Delta x \to 0$, (1.44) yields

$$\hat{n} \cdot \left(\overline{B}_1 - \overline{B}_2\right) = q_{ms} \tag{1.45}$$

which states that the normal component of $\overline{B}$ across the boundary is discontinuous by the amount q_{ms}, which is the surface magnetic charge density at the boundary.

Note that the boundary conditions (1.39) and (1.41) are an independent set, while (1.43) and (1.45) are dependent on (1.39) and (1.41). The tangential field continuities, (1.39) and (1.41), must be used to obtain a unique solution to Maxwell's equations.

1.3.2 Boundary Conditions with Perfect Electric Conductor

It is of practical importance to examine the boundary conditions between a dielectric medium and a perfect electric conductor (PEC). Assume that medium 2 is perfectly conducting (PEC: infinite conductivity, $\sigma_2 = \infty$) and no magnetic surface current is present at the boundary ($\overline{M}_s = 0$). When an electromagnetic wave impinges on the PEC, the field in medium 2 is null ($\overline{E}_2 = \overline{H}_2 = 0$) and an electric current $\overline{J}_{ind}$ is induced on the surface. When a surface electric current density $\overline{J}_s$ is placed on the PEC surface, $\overline{J}_s$ is short-circuited due to the image current density induced on the surface. Since the field inside the perfect conductor vanishes ($\overline{E}_2 = \overline{H}_2 = 0$), the boundary conditions become

$$\hat{n} \times \overline{E}_1 = 0 \tag{1.46}$$

$$\hat{n} \times \overline{H}_1 = \overline{J}_{ind} \tag{1.47}$$

$$\hat{n} \cdot \overline{D}_1 = q_{ind} \tag{1.48}$$

$$\hat{n} \cdot \overline{B}_1 = 0 \tag{1.49}$$

where $\overline{J}_{ind}$ and q_{ind} denote the electric current density and charge density induced on the PEC surface, respectively.

1.4 Poynting Vector

The energy and power calculation in a time-harmonic case is usually performed in terms of the time-average Poynting vector. Here, the Poynting vector formulation is discussed commencing from Maxwell's equations.

1.4.1 Complex Poynting Vector

The two Maxwell's curl equations (1.33) and (1.34) are rewritten as

$$\nabla \times \overline{E} = i\omega\mu\overline{H} - \overline{M}_i \tag{1.50}$$

$$\nabla \times \overline{H}^* = i\omega\epsilon_c^*\overline{E}^* + \overline{J}_i^*$$

$$= (\sigma + i\omega\epsilon)\overline{E}^* + \overline{J}_i^* \tag{1.51}$$

where the symbol $(\cdot)^*$ denotes the complex conjugate of $(\cdot)$. Taking the dot products of (1.50) and (1.51) with $\overline{H}^*$ and $\overline{E}$, respectively, and then subtracting yields

$$\overline{E} \cdot \left(\nabla \times \overline{H}^*\right) - \overline{H}^* \cdot (\nabla \times \overline{E})$$

$$= \overline{E} \cdot \overline{J}_i^* + \sigma\overline{E} \cdot \overline{E}^* + i\omega\epsilon\overline{E} \cdot \overline{E}^* - i\omega\mu\overline{H} \cdot \overline{H}^* + \overline{H}^* \cdot \overline{M}_i\,. \tag{1.52}$$

In view of the vector identity

$$\nabla \cdot \left(\overline{E} \times \overline{H}^*\right) = \overline{H}^* \cdot (\nabla \times \overline{E}) - \overline{E} \cdot \left(\nabla \times \overline{H}^*\right) \tag{1.53}$$

the following result is obtained:

$$-\nabla \cdot \left(\frac{1}{2}\overline{E} \times \overline{H}^*\right) = \frac{1}{2}\overline{E} \cdot \overline{J}_i^* + \frac{1}{2}\overline{H}^* \cdot \overline{M}_i + \frac{1}{2}\sigma\overline{E} \cdot \overline{E}^*$$

$$+2i\omega\left(\frac{1}{4}\epsilon\overline{E} \cdot \overline{E}^* - \frac{1}{4}\mu\overline{H} \cdot \overline{H}^*\right)\,. \tag{1.54}$$

A physical interpretation of (1.54) would be more evident from its integral representation. Consider a volume V that contains the current sources $\overline{J}_i$ and $\overline{M}_i$, as shown in Fig. 1.5. Applying a volume integral to (1.54) then produces

$$-\frac{1}{2}\int_V \left(\overline{E} \cdot \overline{J}_i^* + \overline{H}^* \cdot \overline{M}_i\right) dv$$

$$= \oint_S \frac{1}{2}\left(\overline{E} \times \overline{H}^*\right) \cdot d\overline{s} + \frac{1}{2}\int_V \sigma|\overline{E}|^2\, dv$$

$$+2i\omega\int_V \left(\frac{1}{4}\epsilon|\overline{E}|^2 - \frac{1}{4}\mu|\overline{H}|^2\right) dv\,. \tag{1.55}$$

Rewriting (1.55) yields

$$P_i = P_c + P_d + i2\omega(W_e - W_m) \tag{1.56}$$

where

$$P_i = -\frac{1}{2}\int_V \left(\overline{E}\cdot\overline{J}_i^* + \overline{H}^*\cdot\overline{M}_i\right) dv \tag{1.57}$$

$$P_c = \oint_S \frac{1}{2}\left(\overline{E}\times\overline{H}^*\right)\cdot d\overline{s} \tag{1.58}$$

$$P_d = \frac{1}{2}\int_V \sigma|\overline{E}|^2\, dv \tag{1.59}$$

$$W_e = \int_V \frac{1}{4}\epsilon|\overline{E}|^2\, dv \tag{1.60}$$

$$W_m = \int_V \frac{1}{4}\mu|\overline{H}|^2\, dv\ . \tag{1.61}$$

Note that P_i is the complex power supplied by the sources $\overline{J}_i^*$ and $\overline{M}_i$ within the volume V, P_d is the real power dissipation due to the medium conductivity σ, and W_e and W_m are the time-average electric and magnetic energy, respectively. If relation (1.56) is reinterpreted as a law of complex power conservation, P_c can be regarded as the complex power flow leaving the surface S that surrounds the volume V. The term $\left(\overline{E}\times\overline{H}^*\right)$ is called the complex Poynting vector.

1.4.2 Time-Average Poynting Vector

It is also useful to obtain the time-average Poynting vector, which accounts for the flow of the time-average power density. Taking the real part of (1.56) yields a law of real power conservation. Therefore, the time-average Poynting vector $\overline{S}_{av}$ is given by

$$\overline{S}_{av} = \frac{1}{2}\mathrm{Re}\left(\overline{E}\times\overline{H}^*\right)\ . \tag{1.62}$$

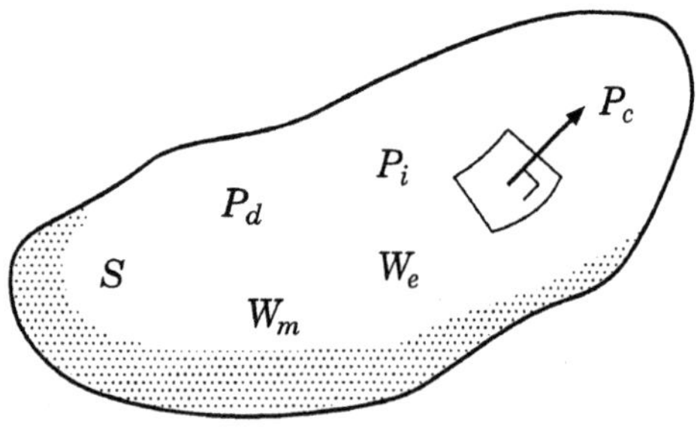

Fig. 1.5. Surface S surrounding volume V containing sources $\overline{J}_i$ and $\overline{M}_i$.

Alternatively, the time-average Poynting vector $\overline{S}_{av}$ can be obtained by averaging the instantaneous (time-varying) Poynting vector over a time period T with the time-varying fields $\overline{E}(\overline{r},t)$ and $\overline{H}(\overline{r},t)$ as

$$\begin{aligned}\overline{S}_{av} &= \frac{1}{T}\int_0^T \overline{E}(\overline{r},t) \times \overline{H}(\overline{r},t)\, dt \\ &= \frac{1}{T}\int_0^T \text{Re}\,\left(\overline{E}e^{-i\omega t}\right) \times \text{Re}\,\left(\overline{H}e^{-i\omega t}\right)\, dt \\ &= \frac{1}{4T}\int_0^T \left(\overline{E}e^{-i\omega t} + \overline{E}^* e^{i\omega t}\right) \times \left(\overline{H}e^{-i\omega t} + \overline{H}^* e^{i\omega t}\right) dt \\ &= \frac{1}{4T}\int_0^T \left(\overline{E} \times \overline{H}^* + \overline{E}^* \times \overline{H}\right) dt \\ &= \frac{1}{2}\text{Re}\,\left(\overline{E} \times \overline{H}^*\right) . \end{aligned} \tag{1.63}$$

1.5 Vector and Scalar Potentials

In electromagnetic scattering and radiation problems, Maxwell's equations must be solved to determine the electromagnetic fields $\overline{E}$ and $\overline{H}$. It is convenient to solve Maxwell's equations using a potential concept. The fields $\overline{E}$ and $\overline{H}$ are considered to be the responses to the sources $\overline{J}$ and $\overline{M}$. This section introduces the magnetic and electric vector potentials to relate $\overline{J}$ and $\overline{M}$ with $\overline{E}$ and $\overline{H}$. Also, the Helmholtz equations for the magnetic and electric vector potentials are derived.

1.5.1 Potential Representations

If Maxwell's equations are assumed to be linear, the superposition principle applies, thereby enabling $\overline{E}$ and $\overline{H}$ to be decomposed into

$$\overline{E} = \overline{E}_e + \overline{E}_m \tag{1.64}$$

$$\overline{H} = \overline{H}_e + \overline{H}_m \tag{1.65}$$

where $\overline{E}_e$ and $\overline{H}_e$ are the responses due to $\overline{J}$, and $\overline{E}_m$ and $\overline{H}_m$ are the responses due to $\overline{M}$. First, Maxwell's equations associated with $\overline{E}_e$ and $\overline{H}_e$ excited by $\overline{J}$ alone ($\overline{M} = \rho_m = 0$ and $\overline{J} \neq 0$) are considered.

$$\nabla \times \overline{E}_e = i\omega\mu\overline{H}_e \tag{1.66}$$

$$\nabla \times \overline{H}_e = -i\omega\epsilon\overline{E}_e + \overline{J} \tag{1.67}$$

$$\nabla \cdot \left(\epsilon\overline{E}_e\right) = \rho_e \tag{1.68}$$

$$\nabla \cdot \left(\mu\overline{H}_e\right) = 0 \,. \tag{1.69}$$

Since $\nabla \cdot \left(\mu\overline{H}_e\right) = 0$ and the divergence of the curl of any vector function is zero, $\overline{A}$ is introduced as

$$\overline{H}_e = \frac{1}{\mu}\nabla \times \overline{A} \tag{1.70}$$

where $\overline{A}$ is referred to as the magnetic vector potential. Substituting $\overline{H}_e$ into (1.66) gives

$$\nabla \times \left(\overline{E}_e - i\omega\overline{A}\right) = 0 \,. \tag{1.71}$$

Since the curl of the gradient of any scalar function is zero, an electric scalar potential ϕ_e exists, such as

$$\overline{E}_e - i\omega\overline{A} = -\nabla\phi_e \,. \tag{1.72}$$

Substituting $\overline{H}_e$ from (1.70) and $\overline{E}_e$ from (1.72) into (1.67) yields

$$-\nabla \times \nabla \times \overline{A} + \mu\epsilon\omega^2\overline{A} + i\omega\mu\epsilon\nabla\phi_e = -\mu\overline{J} \,. \tag{1.73}$$

Substituting $\overline{E}_e$ from (1.72) into (1.68) gives

$$\nabla^2\phi_e - i\omega\nabla \cdot \overline{A} = -\frac{\rho_e}{\epsilon} \,. \tag{1.74}$$

For the unique determination of $\overline{A}$, the condition $\nabla \cdot \overline{A}$ must be specified in addition to the condition $\nabla \times \overline{A}$. To that end, the Lorentz condition is chosen

$$\nabla \cdot \overline{A} = i\omega\mu\epsilon\phi_e \,. \tag{1.75}$$

In view of (1.75) and the vector identity $\nabla \times \nabla \times \overline{A} = -\nabla^2\overline{A} + \nabla(\nabla \cdot \overline{A})$, (1.73) and (1.74) are simplified to

$$\left(\nabla^2 + k^2\right)\overline{A} = -\mu\overline{J} \tag{1.76}$$

$$\left(\nabla^2 + k^2\right)\phi_e = -\frac{\rho_e}{\epsilon} \tag{1.77}$$

where k $(= \omega\sqrt{\mu\epsilon})$ is the wavenumber. Equations (1.76) and (1.77) are called the nonhomogeneous Helmholtz equations.

Once $\overline{A}$ and ϕ_e are determined from (1.76) and (1.75), respectively, $\overline{H}_e$ and $\overline{E}_e$ can be straightforwardly obtained from (1.70) and (1.72), respectively. The result is

$$\overline{E}_e = i\omega\overline{A} + \frac{i}{\omega\mu\epsilon}\nabla\left(\nabla\cdot\overline{A}\right) \tag{1.78}$$

$$\overline{H}_e = \frac{1}{\mu}\nabla\times\overline{A}\,. \tag{1.79}$$

Similarly, it is possible to determine $\overline{E}_m$ and $\overline{H}_m$ resulting from only the magnetic source ($\overline{M}\neq 0$ and $\overline{J}=\rho_e=0$). Consider Maxwell's equations associated with $\overline{M}$

$$\nabla\times\overline{E}_m = i\omega\mu\overline{H}_m - \overline{M} \tag{1.80}$$

$$\nabla\times\overline{H}_m = -i\omega\epsilon\overline{E}_m \tag{1.81}$$

$$\nabla\cdot\left(\epsilon\overline{E}_m\right) = 0 \tag{1.82}$$

$$\nabla\cdot\left(\mu\overline{H}_m\right) = \rho_m\,. \tag{1.83}$$

Since $\nabla\cdot\left(\epsilon\overline{E}_m\right)=0$, the electric vector potential $\overline{F}$ is introduced as

$$\overline{E}_m = -\frac{1}{\epsilon}\nabla\times\overline{F}\,. \tag{1.84}$$

Substituting $\overline{E}_m$ into (1.81) gives

$$\nabla\times\left(\overline{H}_m - i\omega\overline{F}\right) = 0\,. \tag{1.85}$$

The magnetic scalar potential ϕ_m exists as

$$\overline{H}_m - i\omega\overline{F} = -\nabla\phi_m\,. \tag{1.86}$$

Substituting $\overline{E}_m$ from (1.84) and $\overline{H}_m$ from (1.86) into (1.80) yields

$$-\nabla\times\nabla\times\overline{F} + \mu\epsilon\omega^2\overline{F} + i\omega\mu\epsilon\nabla\phi_m = -\epsilon\overline{M}\,. \tag{1.87}$$

Substituting $\overline{H}_m$ from (1.86) into (1.83) gives

$$\nabla^2\phi_m - i\omega\nabla\cdot\overline{F} = -\frac{\rho_m}{\mu}\,. \tag{1.88}$$

Choosing the condition

$$\nabla\cdot\overline{F} = i\omega\mu\epsilon\phi_m \tag{1.89}$$

results in

$$\left(\nabla^2 + k^2\right) \overline{F} = -\epsilon \overline{M} \tag{1.90}$$

$$\left(\nabla^2 + k^2\right) \phi_m = -\frac{\rho_m}{\mu} \tag{1.91}$$

where k $(= \omega\sqrt{\mu\epsilon})$ is the wavenumber. The fields are therefore given as

$$\overline{E}_m = -\frac{1}{\epsilon}\nabla \times \overline{F} \tag{1.92}$$

$$\overline{H}_m = i\omega\overline{F} + \frac{i}{\omega\mu\epsilon}\nabla\left(\nabla \cdot \overline{F}\right) . \tag{1.93}$$

Collecting (1.78), (1.79), (1.92), and (1.93) produces

$$\begin{aligned}\overline{E} &= \overline{E}_e + \overline{E}_m \\ &= i\omega\overline{A} + \frac{i}{\omega\mu\epsilon}\nabla\left(\nabla \cdot \overline{A}\right) - \frac{1}{\epsilon}\nabla \times \overline{F}\end{aligned} \tag{1.94}$$

$$\begin{aligned}\overline{H} &= \overline{H}_e + \overline{H}_m \\ &= \frac{1}{\mu}\nabla \times \overline{A} + i\omega\overline{F} + \frac{i}{\omega\mu\epsilon}\nabla\left(\nabla \cdot \overline{F}\right)\end{aligned} \tag{1.95}$$

where the fields $\overline{E}$ and $\overline{H}$ can be completely described in terms of $\overline{A}$ and $\overline{F}$. To determine $\overline{A}$ and $\overline{F}$, we must solve the nonhomogeneous Helmholtz equations (1.76) and (1.90), which are subject to the boundary conditions.

1.5.2 Free-Space Solution

The evaluation of (1.76) and (1.90) in free space, however, is trivial. Consider the Helmholtz equation (1.76) where $\overline{J}$ is located in free space, as shown in Fig. 1.6. Here, a Green's function, $g(\overline{r};\overline{r}')$, which is the response to the delta source $\delta(\overline{r} - \overline{r}')$, is introduced. The governing equation is

$$\left(\nabla^2 + k^2\right) g(\overline{r};\overline{r}') = -\delta(\overline{r} - \overline{r}') . \tag{1.96}$$

Noting that

$$\overline{J}(\overline{r}) = \int_{V'} \delta(\overline{r} - \overline{r}')\overline{J}(\overline{r}')\, dv' \tag{1.97}$$

Equation (1.96) is multiplied by $\mu\overline{J}(\overline{r}')$ and integrated to obtain

$$\left(\nabla^2 + k^2\right) \int_{V'} g(\overline{r};\overline{r}')\mu\overline{J}(\overline{r}')\, dv' = -\int_{V'} \delta(\overline{r} - \overline{r}')\mu\overline{J}(\overline{r}')\, dv'$$

$$= -\mu\overline{J}(\overline{r})$$

$$= \left(\nabla^2 + k^2\right)\overline{A}(\overline{r})\ . \tag{1.98}$$

Hence

$$\overline{A}(\overline{r}) = \mu \int_{V'} g(\overline{r};\overline{r}')\overline{J}(\overline{r}')\, dv'\ . \tag{1.99}$$

The free-space Green's functions in one-, two-, and three-dimensional cases are available in Sections 8.2, 8.3, and 8.4, respectively. For a three-dimensional boundary-value problem, for instance, the Green's function is

$$g(\overline{r};\overline{r}') = \frac{1}{4\pi}\frac{e^{ik|\overline{r}-\overline{r}'|}}{|\overline{r}-\overline{r}'|}\ . \tag{1.100}$$

Therefore, the solutions to (1.76) and (1.90) are given by

$$\overline{A}(\overline{r}) = \frac{\mu}{4\pi}\int_{V'} \overline{J}(\overline{r}')\frac{e^{ik|\overline{r}-\overline{r}'|}}{|\overline{r}-\overline{r}'|}\, dv' \tag{1.101}$$

$$\overline{F}(\overline{r}) = \frac{\epsilon}{4\pi}\int_{V'} \overline{M}(\overline{r}')\frac{e^{ik|\overline{r}-\overline{r}'|}}{|\overline{r}-\overline{r}'|}\, dv'\ . \tag{1.102}$$

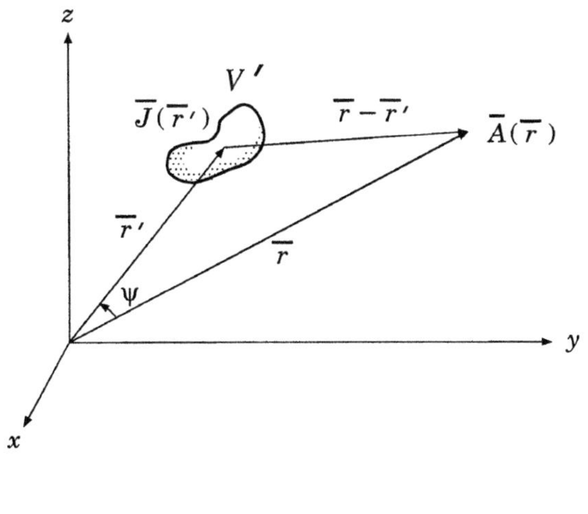

Fig. 1.6. Vector potential and current density in free space.

1.6 Static Fields

Consider Maxwell's equations when the time variation becomes zero $\left(\frac{\partial}{\partial t}=-i\omega \rightarrow 0\right)$ and, therefore, the fields become static ($k \rightarrow 0$). For simplicity, it is assumed that $\overline{M} = 0$ and $\rho_m = 0$. In a static case, the continuity equation (1.9) becomes $\nabla \cdot \overline{J} = 0$; thus, $\overline{J}$ and ρ_e behave as independent sources to each other. The static fields due to $\overline{J}$ and ρ_e are then given by

$$\nabla \times \overline{E} = 0 \tag{1.103}$$

$$\nabla \times \overline{H} = \overline{J} \tag{1.104}$$

$$\nabla \cdot \overline{D} = \rho_e \tag{1.105}$$

$$\nabla \cdot \overline{B} = 0 \, . \tag{1.106}$$

Equations (1.103) through (1.106) illustrate that $\overline{E}$ and $\overline{H}$ are decoupled, where $\overline{E}$ is determined by the static charge ρ_e, while $\overline{H}$ is determined by the static current $\overline{J}$. Let us first consider electrostatics where the electrostatic fields are generated by the static charge ρ_e.

1.6.1 Electrostatic Fields

The electrostatic fields are determined by

$$\nabla \times \overline{E} = 0 \tag{1.107}$$

$$\nabla \cdot \overline{D} = \rho_e \, . \tag{1.108}$$

The boundary conditions are

$$\hat{n} \times \left(\overline{E}_1 - \overline{E}_2\right) = 0 \tag{1.109}$$

$$\hat{n} \cdot \left(\overline{D}_1 - \overline{D}_2\right) = q_{es} \, . \tag{1.110}$$

The Helmholtz equation (1.77) reduces to Poisson's equation

$$\nabla^2 \phi_e = -\frac{\rho_e}{\epsilon} \, . \tag{1.111}$$

For a free-space problem, the solution to (1.111) is

$$\phi_e(\overline{r}) = \frac{1}{4\pi\epsilon} \int_{V'} \frac{\rho_e(\overline{r}')}{|\overline{r} - \overline{r}'|} \, dv' \, . \tag{1.112}$$

Once ϕ_e is determined from (1.111), $\overline{E}$ is given by

$$\overline{E} = -\nabla\phi_e \, . \tag{1.113}$$

Let us re-express the boundary conditions (1.109) and (1.110) in terms of electric scalar potentials as

$$-\frac{\partial\phi_e^1}{\partial t} + \frac{\partial\phi_e^2}{\partial t} = 0 \tag{1.114}$$

$$-\epsilon_1\frac{\partial\phi_e^1}{\partial n} + \epsilon_2\frac{\partial\phi_e^2}{\partial n} = q_{es} \tag{1.115}$$

where the notations $\frac{\partial}{\partial t}$ and $\frac{\partial}{\partial n}$ denote the rates of change in the tangential and normal directions to the boundary, respectively. Note that ϕ_e^1 and ϕ_e^2 are the electric scalar potentials in regions 1 and 2 along the boundary, respectively. Since the electrostatic fields are conservative, the electric scalar potentials are continuous across the boundary yielding

$$-\phi_e^1 + \phi_e^2 = 0 \tag{1.116}$$

$$-\epsilon_1\frac{\partial\phi_e^1}{\partial n} + \epsilon_2\frac{\partial\phi_e^2}{\partial n} = q_{es} \, . \tag{1.117}$$

Equations (1.116) and (1.117) are used to solve (1.111) in electrostatic boundary-value problems.

1.6.2 Magnetostatic Fields

The magnetostatic fields are determined by

$$\nabla \times \overline{H} = \overline{J} \tag{1.118}$$

$$\nabla \cdot \overline{B} = 0 \tag{1.119}$$

where the boundary conditions are

$$\hat{n} \times \left(\overline{H}_1 - \overline{H}_2\right) = \overline{J}_s \tag{1.120}$$

$$\hat{n} \cdot \left(\overline{B}_1 - \overline{B}_2\right) = 0 \, . \tag{1.121}$$

The Helmholtz equation (1.76) reduces to Poisson's equation

$$\nabla^2\overline{A} = -\mu\overline{J} \, . \tag{1.122}$$

For a free-space problem, the solution to (1.122) is trivial as

$$\overline{A}(\overline{r}) = \frac{\mu}{4\pi}\int_{V'} \frac{\overline{J}(\overline{r}')}{|\overline{r}-\overline{r}'|}\, dv' \,. \tag{1.123}$$

Once $\overline{A}$ is determined from (1.122), $\overline{H}$ is given by

$$\overline{H} = \frac{1}{\mu}\nabla \times \overline{A}\,. \tag{1.124}$$

Problems for Chapter 1

1. Starting with time-varying Maxwell's equations (1.1) and (1.2), derive the wave equation for $\overline{E}(\overline{r}, t)$. Assume that the medium is inhomogeneous with $\epsilon(\overline{r})$, $\mu(\overline{r})$, and $\sigma(\overline{r})$.
2. Maxwell's equations (1.33) through (1.36) are represented in terms of the field quantities $\overline{E}$ and $\overline{H}$. In the low-frequency limit, it is also possible to re-interpret Maxwell's equations in terms of the circuit parameters v (voltage) and i (current). Starting from Maxwell's equations, derive Kirchhoff's voltage and current laws.
3. The perfect magnetic conductor (PMC) is a fictitious medium and the tangential magnetic field on the PMC surface is assumed to be zero. Construct the boundary conditions across the boundary between the PMC and dielectric medium.
4. Two perfectly conducting radial plates are at the electrostatic potentials $\phi_e = 0$ and V, respectively, as shown in Fig. 1.7. Determine an electrostatic potential distribution for $0 < \phi < \phi_0$.

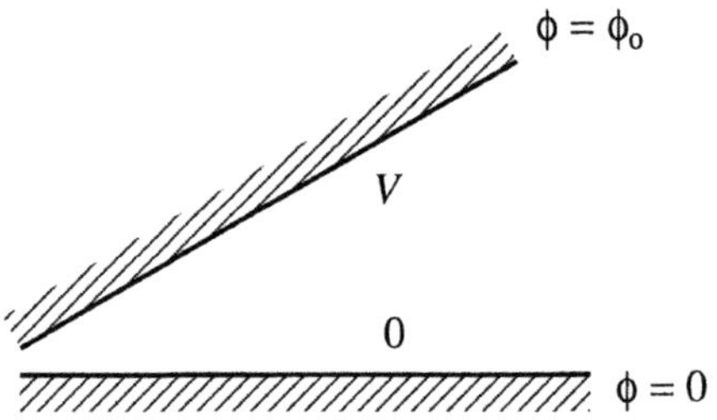

Fig. 1.7. Two perfectly conducting radial plates.

2

Plane Wave Propagation

2.1 Uniform Plane Wave

Consider electromagnetic wave propagation in a source-free, lossless, unbounded medium. A simple propagating wave, which adequately represents wave propagation often encountered in real situations, is a uniform plane wave. This section introduces the fundamentals of a uniform plane wave and analyzes its propagation characteristics.

2.1.1 Propagation in Lossless Medium

This subsection investigates the characteristics of a uniform plane wave starting from Maxwell's equations. Maxwell's equations (1.33) through (1.36) for a source-free, homogeneous, lossless region $\left(\overline{J}_i = \overline{M}_i = \rho_e = \rho_m = 0 \text{ and } \sigma = 0\right)$ are

$$\nabla \times \overline{E} = i\omega\mu\overline{H} \tag{2.1}$$

$$\nabla \times \overline{H} = -i\omega\epsilon\overline{E} \tag{2.2}$$

$$\nabla \cdot \overline{E} = 0 \tag{2.3}$$

$$\nabla \cdot \overline{H} = 0 \, . \tag{2.4}$$

Let us derive an equation for $\overline{E}$ that governs wave propagation. Taking the curl of (2.1) gives

$$\nabla \times \nabla \times \overline{E} = i\omega\mu\nabla \times \overline{H} \, . \tag{2.5}$$

Substituting (2.2) into (2.5) yields

$$\nabla \times \nabla \times \overline{E} = \omega^2 \mu \epsilon \overline{E} . \tag{2.6}$$

The vector identity gives

$$\nabla \times \nabla \times \overline{E} \equiv -\nabla^2 \overline{E} + \nabla \underbrace{(\nabla \cdot \overline{E})}_{0} = -\nabla^2 \overline{E} . \tag{2.7}$$

Substituting (2.7) into (2.6) yields the Helmholtz equation

$$\left(\nabla^2 + \omega^2 \mu \epsilon\right) \overline{E}(\overline{r}) = 0 . \tag{2.8}$$

The Helmholtz equation governs the behavior of the wave $\overline{E}(\overline{r})$. It would be instructive to consider a one-dimensional (z) wave propagation characteristic by letting $\frac{\partial}{\partial x} = \frac{\partial}{\partial y} = 0$ and $\overline{E}(\overline{r}) = \hat{x} E_x(z)$ in (2.8). The one-dimensional Helmholtz equation is given by

$$\left(\frac{d^2}{dz^2} + \omega^2 \mu \epsilon\right) E_x(z) = 0 \tag{2.9}$$

where the general solution is written in terms of the wavenumber $k = \omega\sqrt{\mu\epsilon}$ as

$$E_x(z) = E_x^+ e^{ikz} + E_x^- e^{-ikz} . \tag{2.10}$$

Let us investigate the behavior of $E_x^+ e^{ikz}$ in the time domain. Transforming $E_x^+ e^{ikz}$ into a time-varying form gives

$$E_x^+(z,t) = E_x^+ \cos(\omega t - kz) . \tag{2.11}$$

In view of Fig. 2.1, it is seen that $E_x^+(z,t)$ represents the one-dimensional wave that propagates along the $+z$-direction with a velocity u

$$u = \frac{dz}{dt} = \frac{\omega}{k} = \frac{1}{\sqrt{\mu\epsilon}} . \tag{2.12}$$

Substituting $\overline{E} = \hat{x} E_x^+ e^{ikz}$ into (2.1) and taking its time-varying form yields

$$H_y^+(z,t) = \frac{E_x^+(z,t)}{\eta} \tag{2.13}$$

where $\eta = \sqrt{\frac{\mu}{\epsilon}}$ is the intrinsic impedance of the medium and 120π (ohms) in air. A graphical representation for the propagating wave is illustrated in Fig. 2.2. Since a constant phase front of the wave, $E_x^+(z,t)$ and $H_y^+(z,t)$, constitutes the x-y plane, the wave is called a uniform plane wave.

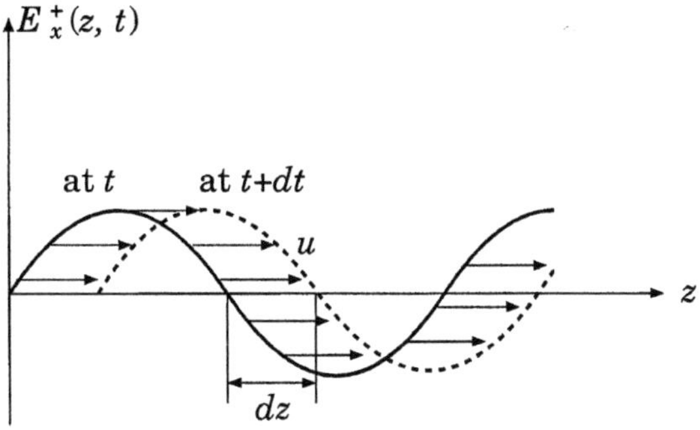

Fig. 2.1. Traveling wave at velocity u along $+z$-direction.

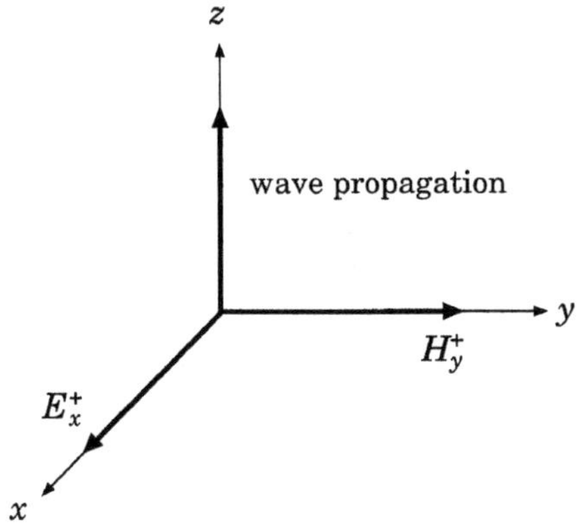

Fig. 2.2. Uniform plane wave propagating along $+z$ direction.

Let us consider a three-dimensional case using (2.8). The plane wave solution to (2.8) is assumed to be

$$\overline{E}(\overline{r}) = \underbrace{(\hat{x}E_x + \hat{y}E_y + \hat{z}E_z)}_{\overline{E}_0} \underbrace{\exp\left[i(k_x x + k_y y + k_z z)\right]}_{\exp(i\overline{k}\cdot\overline{r})} \tag{2.14}$$

where the wave vector is $\overline{k} = \hat{x}k_x + \hat{y}k_y + \hat{z}k_z$ and the position vector is $\overline{r} = \hat{x}x + \hat{y}y + \hat{z}z$. This represents a uniform plane wave that propagates in the $\overline{k}$ direction. Substituting (2.14) into (2.8) yields the dispersion relation

$$k_x^2 + k_y^2 + k_z^2 \equiv k^2 = \omega^2\mu\epsilon \tag{2.15}$$

where the wavenumber is $k = |\overline{k}| = \dfrac{2\pi}{\lambda}$ and λ is the wavelength. Substituting (2.14) into (2.3) gives

$$\nabla \cdot \overline{E} = i\underbrace{(k_x E_x + k_y E_y + k_z E_z)}_{\overline{k}\cdot\overline{E}_0} \exp(i\overline{k}\cdot\overline{r}) = 0 \qquad (2.16)$$

which results in

$$\overline{k}\cdot\overline{E}(\overline{r}) = 0\,. \qquad (2.17)$$

Equation (2.17) implies that the wave vector $\overline{k}$ and the electric field vector $\overline{E}(\overline{r})$ are perpendicular to each other. Furthermore, (2.14) and (2.1) yields

$$\overline{H}(\overline{r}) = \frac{1}{\omega\mu}\overline{k}\times\overline{E}(\overline{r}) \qquad (2.18)$$

indicating that the three vectors, $\overline{E}(\overline{r})$, $\overline{H}(\overline{r})$, and $\overline{k}$, are mutually perpendicular. A uniform plane wave is, therefore, referred to as a TEM (transverse electromagnetic) wave since there are no electromagnetic field components in the $\overline{k}$ direction. For a lossless medium, the time-average Poynting vector $\overline{S}_{av}$ is

$$\begin{aligned}\overline{S}_{av} &= \frac{1}{2}\,\mathrm{Re}\,\left(\overline{E}\times\overline{H}^*\right)\\ &= \frac{1}{2}|\overline{E}_0|^2\frac{\overline{k}}{\omega\mu}\end{aligned} \qquad (2.19)$$

which indicates that the real power flows along the $\overline{k}$ direction.

2.1.2 Propagation in Lossy Medium

Consider a lossy medium where the complex medium permittivity ϵ_c is

$$\epsilon_c = \epsilon\left(1 + i\frac{\sigma}{\omega\epsilon}\right)\,. \qquad (2.20)$$

Then, the solution to the Helmholtz equation

$$\left(\frac{d^2}{dz^2} + \omega^2\mu\epsilon_c\right)E_x(z) = 0 \qquad (2.21)$$

is

$$E_x(z) = E_x^+ e^{ikz} \qquad (2.22)$$

where $k = \omega\sqrt{\mu\epsilon_c}$. Expression (2.22) represents a uniform plane wave that propagates in a lossy medium with complex medium permittivity ϵ_c. Consider two different lossy media – good dielectric and good conducting media.

- For the good dielectric medium, $\frac{\sigma}{\omega\epsilon} \ll 1$

$$\begin{aligned} k &= \omega\sqrt{\mu\epsilon\left(1+i\frac{\sigma}{\omega\epsilon}\right)} \\ &\approx \omega\sqrt{\mu\epsilon}\left(1+i\frac{\sigma}{2\omega\epsilon}\right) . \end{aligned} \tag{2.23}$$

- For the good conductor, $\frac{\sigma}{\omega\epsilon} \gg 1$

$$\begin{aligned} k &= \omega\sqrt{\mu\epsilon\left(1+i\frac{\sigma}{\omega\epsilon}\right)} \\ &\approx \sqrt{i\omega\mu\sigma} \\ &= (1+i)\sqrt{\frac{\omega\mu\sigma}{2}} . \end{aligned} \tag{2.24}$$

Then, a wave that propagates within the good conductor is written as

$$E_x(z) = E_x^+ \exp\left[(i-1)\sqrt{\frac{\omega\mu\sigma}{2}}z\right] \tag{2.25}$$

which shows the wave decays exponentially as it propagates in the lossy medium. As the wave propagates over the distance $z = \sqrt{\frac{2}{\omega\mu\sigma}}$, the field amplitude decreases by a factor of $e^{-1} \approx 0.368$. This propagating distance is known as the skin depth δ of a good conductor

$$\delta = \sqrt{\frac{2}{\omega\mu\sigma}} . \tag{2.26}$$

The rate of wave attenuation depends upon the size of the skin depth δ. For a good conductor such as copper ($\sigma = 5.80 \times 10^7$ (siemens/m)) at 100 (MHz), the skin depth is $\delta = 6.6 \times 10^{-3}$ (mm), which is indeed very small compared with the wavelength. The field that exists within the interior of good conductors is, therefore, considered approximately as a null field at 100 (MHz). This means that good conductors like copper can be regarded as PEC (perfect electric conductors) at 100 (MHz) and even higher frequencies. Let us find the magnetic field within good conductors. Substituting $E_x(z)$ into

$$\nabla \times \overline{E} = i\omega\mu\overline{H} \tag{2.27}$$

gives

$$H_y(z) = \frac{1}{i\omega\mu}\frac{\partial E_x(z)}{\partial z}$$
$$= \frac{k}{\omega\mu}E_x^+ e^{ikz} \,. \tag{2.28}$$

The wave impedance Z in good conductors is given by

$$Z \equiv \frac{E_x(z)}{H_y(z)}$$
$$= \frac{\omega\mu}{k} = (1-i)\sqrt{\frac{\omega\mu}{2\sigma}} \tag{2.29}$$

indicating that the phase difference between electric and magnetic fields in good conductors is $\frac{\pi}{4}$ (radians).

2.2 Polarization

As an electromagnetic wave propagates, the tip of its electric field traces a certain curve on a plane perpendicular to the direction of wave propagation. The trace of the electric field tip is described in terms of the polarization. Let us consider the one-dimensional plane wave propagation of (2.14) with the assumption $k_x = k_y = 0$. Assume that the electric field phasor is given by

$$\overline{E}(z) = \overline{E}_0 e^{ikz}$$
$$= (\hat{x}E_x + \hat{y}E_y)\, e^{ikz}$$
$$\equiv [\hat{x}E_1 \exp(-i\theta_x) + \hat{y}E_2 \exp(-i\theta_y)]\, e^{ikz} \tag{2.30}$$

where E_1 and E_2 are the magnitudes of E_x and E_y, respectively. Then, the time-varying form of $\overline{E}(z)$ is

$$\overline{E}(z,t) = \mathrm{Re}\ \left[\overline{E}(z)e^{-i\omega t}\right]$$
$$\equiv \hat{x}E_x(z,t) + \hat{y}E_y(z,t) \tag{2.31}$$

where

$$E_x(z,t) = E_1\cos(\omega t - kz + \theta_x) \tag{2.32}$$

$$E_y(z,t) = E_2\cos(\omega t - kz + \theta_y) \,. \tag{2.33}$$

Let

$$u \equiv \frac{E_x(z,t)}{E_1}$$

$$= \cos(\omega t - kz + \theta_x) \tag{2.34}$$

$$v \equiv \frac{E_y(z,t)}{E_2}$$

$$= \cos(\omega t - kz + \theta_y)\ . \tag{2.35}$$

Then, the relation

$$v = \cos(\omega t - kz + \theta_x + \theta_y - \theta_x)$$

$$= \cos(\omega t - kz + \theta_x)\cos(\theta_y - \theta_x) - \sin(\omega t - kz + \theta_x)\sin(\theta_y - \theta_x)$$

$$= u\cos(\theta_y - \theta_x) \pm \sqrt{1 - u^2}\sin(\theta_y - \theta_x) \tag{2.36}$$

results in

$$u^2 + v^2 - 2uv\cos(\theta_y - \theta_x) = \sin^2(\theta_y - \theta_x)\ . \tag{2.37}$$

It is possible to transform (2.37) into the equation of an ellipse using ξ-η coordinates based on the relation

$$E_x(z,t) = E_\xi \cos\psi - E_\eta \sin\psi \tag{2.38}$$

$$E_y(z,t) = E_\xi \sin\psi + E_\eta \cos\psi\ . \tag{2.39}$$

The corresponding polarization ellipse is shown in Fig. 2.3. Substituting $E_x(z,t)$ and $E_y(z,t)$ into (2.37) yields

$$AE_\xi^2 + BE_\eta^2 + CE_\xi E_\eta = \sin^2(\theta_y - \theta_x) \tag{2.40}$$

where

$$A = \frac{\cos^2\psi}{E_1^2} + \frac{\sin^2\psi}{E_2^2} - \cos(\theta_y - \theta_x)\frac{\sin 2\psi}{E_1 E_2} \tag{2.41}$$

$$B = \frac{\sin^2\psi}{E_1^2} + \frac{\cos^2\psi}{E_2^2} + \cos(\theta_y - \theta_x)\frac{\sin 2\psi}{E_1 E_2} \tag{2.42}$$

$$C = -\frac{\sin 2\psi}{E_1^2} + \frac{\sin 2\psi}{E_2^2} - 2\cos(\theta_y - \theta_x)\frac{\cos 2\psi}{E_1 E_2}\ . \tag{2.43}$$

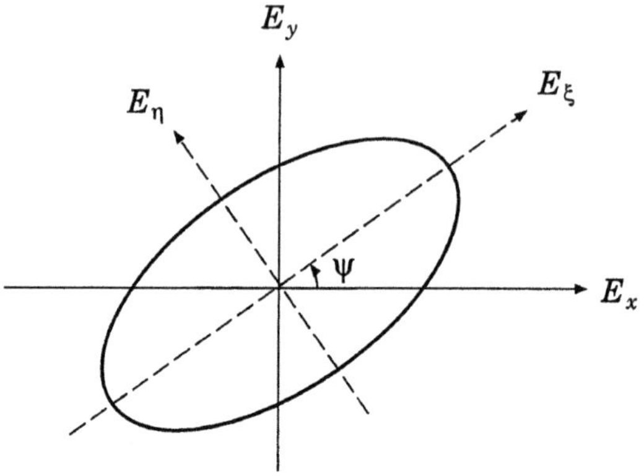

Fig. 2.3. Polarization ellipse.

- **Elliptic polarization**
  To obtain the equation of an ellipse, it is necessary to eliminate the cross term $E_\xi E_\eta$ ($C = 0$) by choosing the rotation angle ψ appropriately. The condition ($C = 0$) gives

$$\tan 2\psi = \tan 2\alpha \cos(\theta_y - \theta_x) \tag{2.44}$$

  where $\tan\alpha = \dfrac{E_2}{E_1}$. Thus, we obtain

$$AE_\xi^2 + BE_\eta^2 = \sin^2(\theta_y - \theta_x) \tag{2.45}$$

  which represents elliptic polarization where the tip of the electric field traces a tilted ellipse as the wave propagates along the z-direction.

- **Linear polarization**
  When $\theta_y - \theta_x = 0$ or π, then $\psi = \alpha + \dfrac{n\pi}{2}$ or $-\alpha + \dfrac{n\pi}{2}$, where n is an integer. For instance, when $\theta_y - \theta_x = 0$ and $\psi = \alpha$, A is shown to be 0 and (2.45) reduces to

$$E_\eta = 0 \tag{2.46}$$

  where the tip of the electric field traces a line on the E_ξ-axis and the wave is said to have linear polarization. Its graphical representation is shown in Fig. 2.4.

- **Circular polarization**
  When $\theta_y - \theta_x = \pm\dfrac{\pi}{2}$ and $E_1 = E_2$, (2.45) becomes

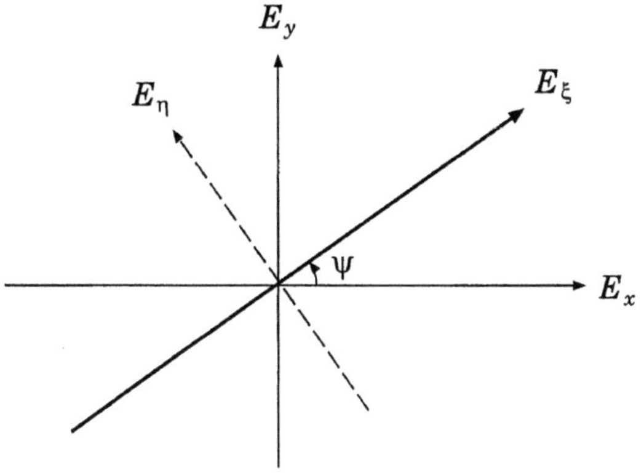

Fig. 2.4. Linear polarization.

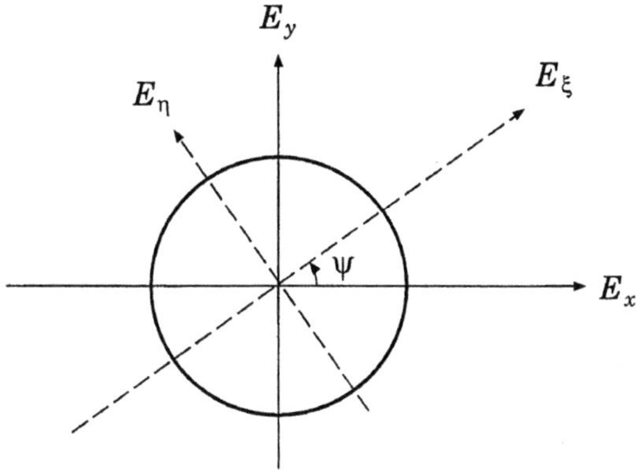

Fig. 2.5. Circular polarization.

$$\left(\frac{E_\xi}{E_1}\right)^2 + \left(\frac{E_\eta}{E_1}\right)^2 = 1 \tag{2.47}$$

which represents a circle on the E_ξ-E_η plane; thus, the wave has circular polarization. Its graphical representation is shown in Fig. 2.5. When $\theta_y - \theta_x = -\frac{\pi}{2}$, the polarization is called right-hand circular polarization, since the electric field behaves as a right-handed screw that advances in the $+z$-direction. Similarly, when $\theta_y - \theta_x = \frac{\pi}{2}$, (2.47) represents left-hand circular polarization.

- **Unpolarized**
  The intensity of natural sun light is completely random in any direction perpendicular to the direction of wave propagation. Therefore, the trace of the electric field tip is random, and thus irregular. This type of wave is called completely unpolarized.

2.3 Gaussian Beam

Consider the problem of laser beam propagation when a laser beam illuminates a large-sized target that is not far away from the source. The amplitude shape of a laser beam is commonly Gaussian due to the Gaussian field distribution across a laser source aperture. This section considers the representations of a Gaussian beam that emanates from line and point sources. The cases of line and point sources correspond to two- and three-dimensional problems, respectively. Note that some relevant discussion is available in [1] and [2, pp. 160-165].

2.3.1 Line Source

Consider a laser beam $\overline{E}(\overline{r}) = \hat{y}E(x,z)$ that propagates along the $+z$-direction from the source aperture $E(x,0)$, as shown in Fig. 2.6. The line source is infinitely long in the y-direction and the field is assumed to be uniform with respect to y $\left(\frac{\partial}{\partial y} = 0\right)$. The Gaussian beam $E(x,z)$ for $z > 0$ is expressed in terms of the inverse Fourier transform as

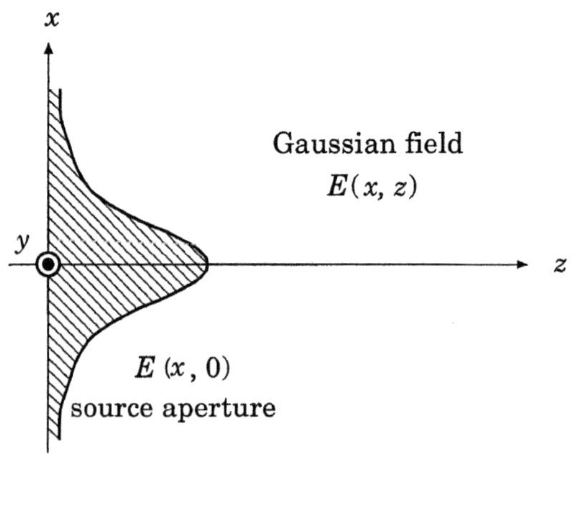

Fig. 2.6. Gaussian wave propagating along z-direction.

$$E(x,z) = \frac{1}{2\pi}\int_{-\infty}^{\infty} \tilde{E}(\zeta,z)e^{i\zeta x}\,d\zeta\,. \tag{2.48}$$

The expression $E(x,z)$ should satisfy the Helmholtz equation. Substituting $E(x,z)$ into (2.8) yields

$$\left(\frac{d^2}{dz^2} + \kappa^2\right)\tilde{E}(\zeta,z) = 0 \tag{2.49}$$

where $\kappa = \sqrt{k^2 - \zeta^2}$ and $k = \omega\sqrt{\mu\epsilon}$. The solution for the beam $\tilde{E}(\zeta,z)$ is assumed to propagate along the $+z$-direction. Therefore

$$\tilde{E}(\zeta,z) = \tilde{E}_0(\zeta)e^{i\kappa z}\,. \tag{2.50}$$

Substituting $\tilde{E}(\zeta,z)$ into $E(x,z)$ gives

$$E(x,z) = \frac{1}{2\pi}\int_{-\infty}^{\infty} \tilde{E}_0(\zeta)e^{i\zeta x + i\kappa z}\,d\zeta \tag{2.51}$$

where

$$\tilde{E}_0(\zeta) = \int_{-\infty}^{\infty} E(x,0)e^{-i\zeta x}\,dx\,. \tag{2.52}$$

The aim is to express $E(x,z)$ in a mathematically tractable form based on approximations. In practice, the source aperture $E(x,0)$ is modeled as approximately Gaussian in amplitude as

$$E(x,0) = \exp\left(-\frac{x^2}{\beta_0^2} + i\phi\right) \tag{2.53}$$

where β_0 is the beamwidth. To estimate the phase variation ϕ at the source aperture, let us consider the line source placed at $z = -z_0$, as shown in Fig. 2.7. The circular cylindrical phase front arriving at $z = 0$ from $z = -z_0$ is

$$e^{i\phi} = \exp\left(ik\sqrt{x^2 + z_0^2}\right)\,. \tag{2.54}$$

Assuming $z_0 \gg |x|$, then

$$e^{i\phi} \approx \exp\left(ikz_0 + \frac{ikx^2}{2z_0}\right)\,. \tag{2.55}$$

Therefore, the source aperture $E(x,0)$ can be rewritten as

$$E(x,0) = A\exp\left(-\frac{x^2}{\beta^2}\right) \tag{2.56}$$

where $A = e^{ikz_0}$ and $\dfrac{1}{\beta^2} = \dfrac{1}{\beta_0^2} - \dfrac{ik}{2z_0}$. Substituting $E(x,0)$ into (2.52) and using the integral formula for $\mathrm{Re}\,(p) > 0$

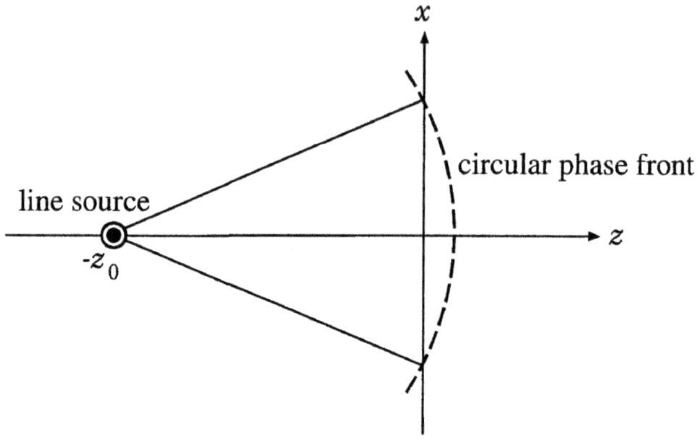

Fig. 2.7. Line source placed at $z = -z_0$.

$$\int_{-\infty}^{\infty} e^{-pu^2+qu}\,du = \sqrt{\frac{\pi}{p}}\exp\left(\frac{q^2}{4p}\right) \tag{2.57}$$

gives

$$\tilde{E}_0(\zeta) = \sqrt{\pi}A\beta\exp\left(-\frac{\beta^2\zeta^2}{4}\right). \tag{2.58}$$

Hence, the Gaussian beam is written as

$$E(x,z) = \frac{1}{2\pi}\int_{-\infty}^{\infty}\sqrt{\pi}A\beta\exp\left(-\frac{\beta^2\zeta^2}{4}+i\zeta x+i\kappa z\right)d\zeta\,. \tag{2.59}$$

The expression (2.59) is still difficult to analytically evaluate unless further approximation is made. Assume that the Gaussian beam is strongly collimated in the z-direction so that its transverse wave vector component (ζ) is much smaller than the longitudinal one (κ); thus

$$\begin{aligned}\kappa &= \sqrt{k^2-\zeta^2}\\ &\approx k-\frac{\zeta^2}{2k}\,.\end{aligned} \tag{2.60}$$

Substituting (2.60) into $E(x,z)$ and performing integration yields

$$E(x,z) = \frac{A}{\sqrt{1+iZ}}\exp\left[ikz-\frac{x^2}{\beta^2(1+iZ)}\right] \tag{2.61}$$

where $Z = \dfrac{2z}{\beta^2 k}$. Note that $E(x,z)$ represents an approximate form for the Gaussian beam that propagates along the $+z$-direction. When β_0 and $z_0 \to \infty$, $E(x,z)$ reduces to the uniform plane wave Ae^{ikz} as it should.

2.3.2 Point Source

Gaussian beam propagation from a point source can be similarly analyzed as in the line source case. Consider the laser beam $\overline{E}(\overline{r}) = \hat{y}E(x,y,z)$ that propagates along the $+z$-direction from the source aperture $E(x,y,0)$. The Gaussian beam $E(x,y,z)$ for $z > 0$ is expressed in terms of the inverse Fourier transform

$$E(x,y,z) = \frac{1}{(2\pi)^2}\int_{-\infty}^{\infty}\int_{-\infty}^{\infty}\tilde{E}(\zeta,\eta,z)e^{i(\zeta x+\eta y)}\,d\zeta\,d\eta\,. \tag{2.62}$$

The solution for the beam $E(x,y,z)$ propagating along the $+z$ direction is

$$E(x,y,z) = \frac{1}{(2\pi)^2}\int_{-\infty}^{\infty}\int_{-\infty}^{\infty}\tilde{E}_0(\zeta,\eta)e^{i(\zeta x+\eta y)+i\kappa z}\,d\zeta\,d\eta \tag{2.63}$$

where $\kappa = \sqrt{k^2-\zeta^2-\eta^2}$ and

$$\tilde{E}_0(\zeta,\eta) = \int_{-\infty}^{\infty}\int_{-\infty}^{\infty}E(x,y,0)e^{-i(\zeta x+\eta y)}\,dx\,dy\,. \tag{2.64}$$

The source aperture $E(x,y,0)$ is modeled as approximately Gaussian in amplitude with the beamwidth β_0, thereby yielding

$$E(x,y,0) = \exp\left[-\frac{(x^2+y^2)}{\beta_0^2}+i\phi\right]. \tag{2.65}$$

Assume that the point source placed at $z = -z_0$ produces quadratic phase variation ϕ over the source aperture at $z = 0$. The source aperture is, therefore, given by

$$E(x,y,0) = A\exp\left[-\frac{(x^2+y^2)}{\beta^2}\right] \tag{2.66}$$

where $A = e^{ikz_0}$ and $\dfrac{1}{\beta^2} = \dfrac{1}{\beta_0^2} - \dfrac{ik}{2z_0}$. Substituting $E(x,y,0)$ into (2.64) gives

$$\tilde{E}_0(\zeta,\eta) = \pi A\beta^2\exp\left[-\frac{(\zeta^2+\eta^2)\beta^2}{4}\right]. \tag{2.67}$$

Assume that the Gaussian beam is strongly collimated in the z-direction to yield

$$\begin{aligned}\kappa &= \sqrt{k^2-\zeta^2-\eta^2}\\ &\approx k - \frac{(\zeta^2+\eta^2)}{2k}\,.\end{aligned} \tag{2.68}$$

Substituting $\tilde{E}_0(\zeta, \eta)$ and (2.68) into (2.63) finally gives the approximate Gaussian beam expression for a point source

$$E(x, y, z) = \frac{A}{(1+iZ)} \exp\left[ikz - \frac{x^2+y^2}{\beta^2(1+iZ)}\right] \tag{2.69}$$

where $Z = \dfrac{2z}{\beta^2 k}$. When β_0 and $z_0 \to \infty$, $E(x, y, z)$ reduces to the uniform plane wave Ae^{ikz}.

2.4 Reflection at Plane Boundary

When a uniform plane wave impinges on a plane boundary, the reflection and transmission of an incident wave occurs at the boundary. The reflected and transmitted waves naturally take the form of uniform plane waves. This section investigates the reflection and transmission across a plane boundary separating two lossless and isotropic media. TE and TM waves are analyzed to obtain their reflection and transmission coefficients based on the boundary conditions.

2.4.1 TE Wave (Perpendicular Polarization)

Consider a TE wave that is incident on a plane boundary between two lossless, dielectric media, as shown in Fig. 2.8. For simplicity, it is assumed that the incident electric field has only a y-component and the incident wave vector lies on the x-z plane (plane of incidence). The incident wave, therefore, is referred to as a TE (transverse electric to the plane of incidence) wave. The electric field vector is perpendicular to the plane of incidence and the TE wave thus

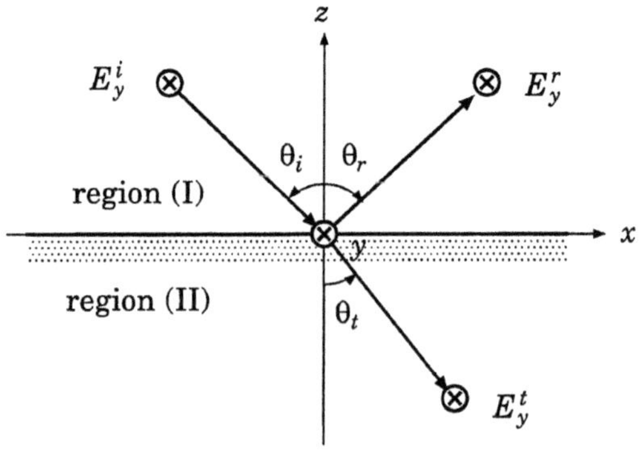

Fig. 2.8. TE wave incident on plane boundary.

corresponds to the case of perpendicular polarization. The wavenumbers in regions (I) and (II) are k_1 $(= \omega\sqrt{\mu_1\epsilon_1})$ and k_2 $(= \omega\sqrt{\mu_2\epsilon_2})$, respectively.

The total electric field in region (I) consists of the incident and reflected waves as

$$E_y^i(x,z) = \exp(ik_{xi}x - ik_{zi}z) \tag{2.70}$$

$$E_y^r(x,z) = R_{TE}\exp(ik_{xr}x + ik_{zr}z) \tag{2.71}$$

where R_{TE} is the reflection coefficient for a TE wave. The phase terms are given by $k_{xi} = k_1\sin\theta_i$, $k_{zi} = k_1\cos\theta_i$, $k_{xr} = k_1\sin\theta_r$, and $k_{zr} = k_1\cos\theta_r$.

The transmitted wave in region (II) is

$$E_y^t(x,z) = T_{TE}\exp(ik_{xt}x - ik_{zt}z) \tag{2.72}$$

where $k_{xt} = k_2\sin\theta_t$, $k_{zt} = k_2\cos\theta_t$, and T_{TE} is the transmission coefficient for a TE wave.

Let us enforce the boundary conditions for the tangential field continuity at $z = 0$. The tangential E_y continuity at $z = 0$

$$E_y^i(x,0) + E_y^r(x,0) = E_y^t(x,0) \tag{2.73}$$

requires

$$\exp(ik_{xi}x) + R_{TE}\exp(ik_{xr}x) = T_{TE}\exp(ik_{xt}x)\ . \tag{2.74}$$

Equation (2.74) must be satisfied irrespective of x, thereby yielding the phase-matching condition

$$k_{xi} = k_{xr} = k_{xt} \tag{2.75}$$

which results in $\sin\theta_i = \sin\theta_r$ (law of reflection) and $k_1\sin\theta_i = k_2\sin\theta_t$ (Snell's law) simultaneously. Hence, (2.74) becomes

$$1 + R_{TE} = T_{TE}\ . \tag{2.76}$$

The accompanying H_x and H_z components can be obtained by substituting (2.70) through (2.72) into Faraday's law $\nabla \times \overline{E} = i\omega\mu\overline{H}$. Since $H_x = \dfrac{i}{\omega\mu}\dfrac{\partial E_y}{\partial z}$, the tangential H_x continuity at $z = 0$

$$H_x^i(x,0) + H_x^r(x,0) = H_x^t(x,0) \tag{2.77}$$

gives

$$\frac{k_{zi}}{\mu_1}(R_{TE} - 1) = -\frac{k_{zt}}{\mu_2}T_{TE}\ . \tag{2.78}$$

Solving (2.76) and (2.78) for the reflection and transmission coefficients gives

$$R_{TE} = \frac{Z_2 - Z_1}{Z_2 + Z_1} \tag{2.79}$$

$$T_{TE} = \frac{2Z_2}{Z_2 + Z_1} \tag{2.80}$$

where $Z_2 = \dfrac{\eta_2}{\cos\theta_t}$, $Z_1 = \dfrac{\eta_1}{\cos\theta_i}$, $\eta_2 = \sqrt{\dfrac{\mu_2}{\epsilon_2}}$, and $\eta_1 = \sqrt{\dfrac{\mu_1}{\epsilon_1}}$.

Let us evaluate the time-average power densities entering and leaving the boundary at $z = 0$ for lossless media, where both Z_1 and Z_2 are real. The incident time-average power density is given in terms of the Poynting vector as

$$\begin{aligned} S_i &= \frac{1}{2}\mathrm{Re}\left(\overline{E}^i \times \overline{H}^{i*}\right) \cdot (-\hat{z}) \\ &= \frac{1}{2Z_1} \,. \end{aligned} \tag{2.81}$$

Similarly, the reflected and transmitted power densities are

$$\begin{aligned} S_r &= \frac{1}{2}\mathrm{Re}\left(\overline{E}^r \times \overline{H}^{r*}\right) \cdot \hat{z} \\ &= \frac{R_{TE}^2}{2Z_1} \end{aligned} \tag{2.82}$$

$$\begin{aligned} S_t &= \frac{1}{2}\mathrm{Re}\left(\overline{E}^t \times \overline{H}^{t*}\right) \cdot (-\hat{z}) \\ &= \frac{T_{TE}^2}{2Z_2} \,. \end{aligned} \tag{2.83}$$

A law of power conservation ($S_r + S_t = S_i$) across the boundary can be easily proved as

$$\begin{aligned} S_r + S_t &= \frac{1}{2}\left(\frac{R_{TE}^2}{Z_1} + \frac{T_{TE}^2}{Z_2}\right) \\ &= \frac{1}{2Z_1} = S_i \,. \end{aligned} \tag{2.84}$$

Let us consider the special case when $\epsilon_1 > \epsilon_2$ and $\mu_2 = \mu_1$. The transmitted angle is

$$\begin{aligned} \cos\theta_t &= \sqrt{1 - \sin^2\theta_t} \\ &= \sqrt{1 - \left(\frac{\epsilon_1}{\epsilon_2}\right)\sin^2\theta_i} \,. \end{aligned} \tag{2.85}$$

When the incident angle θ_i is greater than the critical angle $\theta_c = \sin^{-1}\sqrt{\left(\frac{\epsilon_2}{\epsilon_1}\right)}$, the term $\cos\theta_t$ (thus Z_2) becomes purely imaginary, thereby yielding $|R_{TE}| = 1$, $S_r = S_i$, and $S_t = 0$. This means that the total internal reflection occurs when a wave impinges on a less dense medium ($\epsilon_1 > \epsilon_2$) at an incident angle greater than the critical angle ($\theta_i > \theta_c$). The condition $S_t = 0$ should not be construed as a null field in region (II). The transmitted field in region (II) is evanescent, and it decays exponentially away from the boundary at $z = 0$.

2.4.2 TM Wave (Parallel Polarization)

Consider a TM (transverse magnetic to the plane of incidence) wave that impinges on the plane boundary at $z = 0$, as shown in Fig. 2.9. A TM wave has no magnetic field components in the plane of incidence. The electric field vector is parallel to the plane of incidence and the TM wave thus corresponds to the case of parallel polarization. In regions (I) ($k_1 = \omega\sqrt{\mu_1\epsilon_1}$) and (II) ($k_2 = \omega\sqrt{\mu_2\epsilon_2}$) the magnetic fields are

$$H_y^i(x,z) = \exp(ik_{xi}x - ik_{zi}z) \tag{2.86}$$

$$H_y^r(x,z) = R_{TM}\exp(ik_{xr}x + ik_{zr}z) \tag{2.87}$$

$$H_y^t(x,z) = T_{TM}\exp(ik_{xt}x - ik_{zt}z) \tag{2.88}$$

where $k_{xi} = k_1\sin\theta_i$, $k_{zi} = k_1\cos\theta_i$, $k_{xr} = k_1\sin\theta_r$, $k_{xt} = k_2\sin\theta_t$, $k_{zr} = k_1\cos\theta_r$, and $k_{zt} = k_2\cos\theta_t$.

Since the tangential $H_y(x,0)$ must be continuous, the continuity condition

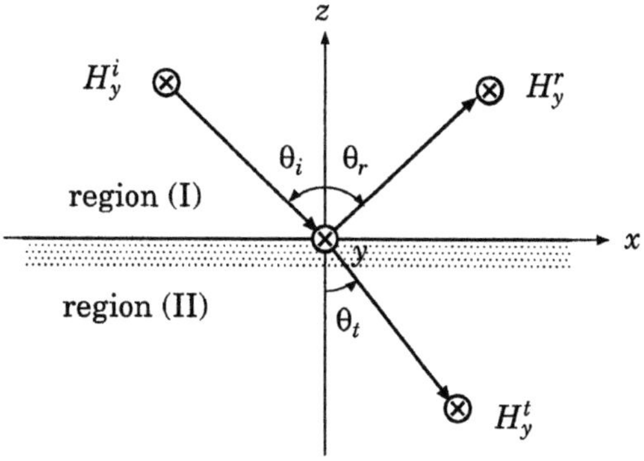

Fig. 2.9. TM wave incident on plane boundary.

$$H_y^i(x,0) + H_y^r(x,0) = H_y^t(x,0) \tag{2.89}$$

results in

$$1 + R_{TM} = T_{TM} \tag{2.90}$$

with the conditions $\sin\theta_i = \sin\theta_r$ (law of reflection) and $k_1 \sin\theta_i = k_2 \sin\theta_t$ (Snell's law).

Ampère's law, $\nabla \times \overline{H} = -i\omega\epsilon\overline{E}$, produces the tangential electric field, $E_x = -\frac{i}{\omega\epsilon}\frac{\partial H_y}{\partial z}$. Hence, $E_x(x,0)$ continuity

$$E_x^i(x,0) + E_x^r(x,0) = E_x^t(x,0) \tag{2.91}$$

yields

$$\frac{k_{zi}}{\epsilon_1}(R_{TM} - 1) = -\frac{k_{zt}}{\epsilon_2}T_{TM} \,. \tag{2.92}$$

Solving (2.90) and (2.92) for the reflection and transmission coefficients for a TM wave gives

$$R_{TM} = \frac{Y_2 - Y_1}{Y_2 + Y_1} \tag{2.93}$$

$$T_{TM} = \frac{2Y_2}{Y_2 + Y_1} \tag{2.94}$$

where $Y_2 = \frac{1}{\eta_2 \cos\theta_t}$, $Y_1 = \frac{1}{\eta_1 \cos\theta_i}$, $\eta_2 = \sqrt{\frac{\mu_2}{\epsilon_2}}$, and $\eta_1 = \sqrt{\frac{\mu_1}{\epsilon_1}}$. Note that R_{TM} and T_{TM} are also obtained from (2.79) and (2.80) based on the duality theorem, which permits interchanging $\eta_1 \to \frac{1}{\eta_1}$ and $\eta_2 \to \frac{1}{\eta_2}$.

Let us consider the special case when the reflection coefficient R_{TM} becomes zero for nonmagnetic media ($\mu_1 = \mu_2 = \mu_0$ and $\epsilon_1 \neq \epsilon_2$). The condition $Y_2 - Y_1 = 0$ gives

$$\sqrt{\epsilon_1}\cos\theta_t = \sqrt{\epsilon_2}\cos\theta_i \,. \tag{2.95}$$

Solving (2.95) and Snell's law

$$\sqrt{\epsilon_1}\sin\theta_i = \sqrt{\epsilon_2}\sin\theta_t \tag{2.96}$$

for θ_i gives

$$\begin{aligned} \theta_i &= \sin^{-1}\sqrt{\frac{\epsilon_2}{\epsilon_1 + \epsilon_2}} \\ &= \tan^{-1}\sqrt{\frac{\epsilon_2}{\epsilon_1}} \end{aligned} \tag{2.97}$$

which indicates that the total transmission occurs when a TM wave impinges on the half space at the incident angle θ_i. The angle (2.97) is referred to as the Brewster angle.

Let us consider the other special case when $\epsilon_1 > \epsilon_2$ and $\mu_2 = \mu_1 = \mu_0$. Like the previous TE wave case, the total internal reflection occurs when a TM wave impinges on a less dense medium ($\epsilon_1 > \epsilon_2$) at an incident angle greater than the critical angle $\theta_c = \sin^{-1}\sqrt{\left(\frac{\epsilon_2}{\epsilon_1}\right)}$.

2.5 Infinitely Long Transmission Lines

Transmission lines are commonly used to transfer electromagnetic energy from one location to another. Parallel wires, parallel plates, and coaxial lines are all typical examples of transmission lines. Transmission lines consist of two conductors that guide TEM (transverse electromagnetic) waves in a low-frequency regime where their field components are all transverse to the wave propagation direction.

2.5.1 Coaxial Line

This section investigates the behavior of wave propagation using an infinitely long coaxial line, as shown in Fig. 2.10. For a practical coaxial line with an inner conductor radius a and outer conductor radius b, the size is usually small $(b - a) \ll \lambda$ and its annular interior region is filled with dielectric material of permittivity ϵ and permeability μ. For simplicity, an ideal coaxial line is assumed where the dielectric material is lossless and the inner and outer conductors are PEC material. An input voltage source is applied between the inner and outer conductors at $z = 0$ to send a TEM wave along the z-direction. The TEM wave consists of E_ρ and H_ϕ components that propagate along the z-direction with the propagation constant k $(= \omega\sqrt{\mu\epsilon})$

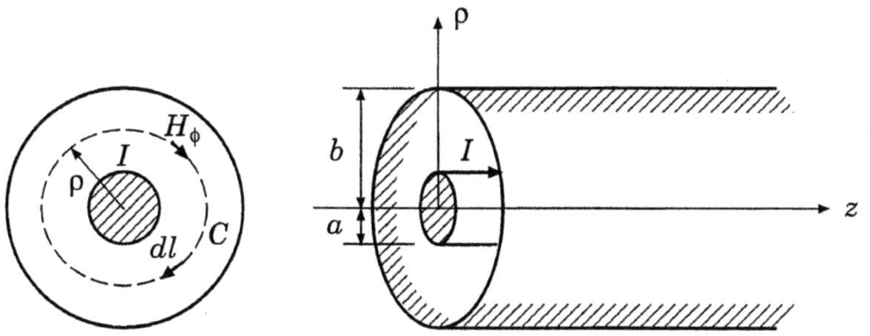

Fig. 2.10. Coaxial line lying along z-axis.

$$E_\rho(\rho, z) = E_\rho(\rho)e^{ikz} \tag{2.98}$$

$$H_\phi(\rho, z) = H_\phi(\rho)e^{ikz} \,. \tag{2.99}$$

Ampère's law

$$\nabla \times \overline{H} = -i\omega\epsilon\overline{E} + \overline{J} \tag{2.100}$$

reduces to Ampère's circuital law

$$\nabla \times \overline{H} = \overline{J} \tag{2.101}$$

for a TEM wave since the electric current density $\overline{J}$ that flows through the conductors is such that $\left|i\omega\epsilon\overline{E}\right| \ll \left|\overline{J}\right|$. Applying Stokes' theorem to Ampère's circuital law gives

$$\oint_C H_\phi(\rho, z)\, dl = I(\rho, z) \,. \tag{2.102}$$

Substituting the current $I(\rho, z) = I_0^+ e^{ikz}$ into (2.102) gives

$$H_\phi(\rho, z) = \frac{I_0^+}{2\pi\rho} e^{ikz} \,. \tag{2.103}$$

Faraday's law

$$\nabla \times \overline{E} = i\omega\mu\overline{H} \tag{2.104}$$

is rewritten as

$$\frac{\partial E_\rho(\rho, z)}{\partial z} = i\omega\mu H_\phi(\rho, z) \tag{2.105}$$

which results in

$$\begin{aligned} E_\rho(\rho, z) &= \sqrt{\frac{\mu}{\epsilon}} H_\phi(\rho, z) \\ &= \sqrt{\frac{\mu}{\epsilon}} \frac{I_0^+}{2\pi\rho} e^{ikz} \,. \end{aligned} \tag{2.106}$$

The time-average power carried by a coaxial line is

$$\begin{aligned} P_{av} &= \frac{1}{2}\mathrm{Re}\left(2\pi \int_a^b E_\rho H_\phi^* \rho\, d\rho\right) \\ &= \frac{|I_0^+|^2}{4\pi} \sqrt{\frac{\mu}{\epsilon}} \ln\left(\frac{b}{a}\right) \,. \end{aligned} \tag{2.107}$$

2.5.2 Voltage and Current

As a TEM wave propagates on a coaxial line, the direction of E_ρ and H_ϕ remains unchanged as if they were scalars. It is thus expedient to represent the wave propagation on transmission lines in terms of scalars such as the voltage and current. Let us reinterpret the transmission line problem by introducing $I^+(z)$ and $V^+(z)$ where $V^+(z)$ and $I^+(z)$ are the incident (forward) voltage and current that propagate along the $+z$-direction, as shown in Fig. 2.11. The incident waves, $I^+(z)$ and $V^+(z)$, are

$$I^+(z) \equiv I_0^+ e^{ikz} \tag{2.108}$$

$$\begin{aligned} V^+(z) &\equiv \int_a^b E_\rho(\rho, z)\, d\rho \\ &= \underbrace{\frac{I_0^+}{2\pi}\sqrt{\frac{\mu}{\epsilon}}\ln\left(\frac{b}{a}\right)}_{V_0^+} e^{ikz} \\ &\equiv V_0^+ e^{ikz} . \end{aligned} \tag{2.109}$$

The time-average power, which is delivered by the voltage and current, is

$$\begin{aligned} P_{av} &= \frac{1}{2}\text{Re}\ \left[V^+(z) I^{+*}(z)\right] \\ &= \frac{|I_0^+|^2}{4\pi}\sqrt{\frac{\mu}{\epsilon}}\ln\left(\frac{b}{a}\right) \end{aligned} \tag{2.110}$$

which is identical with (2.107). The characteristic impedance of a transmission line Z_0 is defined as the ratio of the voltage $V^+(z)$ to the current $I^+(z)$. The characteristic impedance of a coaxial line is, therefore

$$\begin{aligned} Z_0 &\equiv \frac{V_0^+}{I_0^+} \\ &= \frac{1}{2\pi}\sqrt{\frac{\mu}{\epsilon}}\ln\left(\frac{b}{a}\right) . \end{aligned} \tag{2.111}$$

Consider the reflected waves, $V^-(z)$ and $I^-(z)$, as shown in Fig. 2.11. Assuming

$$I^-(z) = I_0^- e^{-ikz} \tag{2.112}$$

and substituting $I^-(z)$ into Ampère's circuital law (2.102) gives

$$H_\phi(\rho, z) = \frac{I_0^-}{2\pi\rho} e^{-ikz} . \tag{2.113}$$

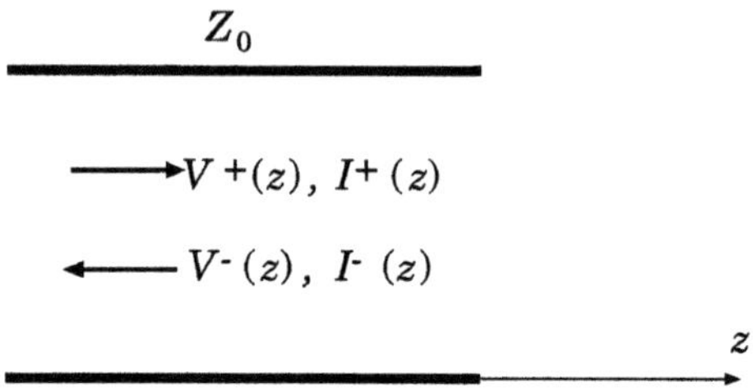

Fig. 2.11. Incident $[V^+(z),\, I^+(z)]$ and reflected $[V^-(z),\, I^-(z)]$ waves.

Substituting $H_\phi(\rho, z)$ into Faraday's law (2.105) yields

$$E_\rho(\rho, z) = -\sqrt{\frac{\mu}{\epsilon}} \frac{I_0^-}{2\pi\rho} e^{-ikz} . \tag{2.114}$$

Let us define the reflected voltage $V^-(z)$ as

$$\begin{aligned} V^-(z) &\equiv \int_a^b E_\rho(\rho, z)\, d\rho \\ &= \underbrace{-\frac{I_0^-}{2\pi}\sqrt{\frac{\mu}{\epsilon}} \ln\left(\frac{b}{a}\right)}_{V_0^-} e^{-ikz} \\ &\equiv V_0^- e^{-ikz} . \end{aligned} \tag{2.115}$$

Hence, the ratio yields

$$\frac{V^-(z)}{I^-(z)} = \frac{V_0^-}{I_0^-} = -Z_0 . \tag{2.116}$$

2.6 Terminated Transmission Lines

Consider a terminated, lossless transmission line that is excited by the source V_s with internal impedance Z_s at $z = 0$, as shown in Fig. 2.12. Termination with a load Z_l at $z = l$ generates reflected waves, thereby resulting in standing waves due to the interference between incident and reflected waves. The total standing wave voltage $V(z)$ consists of the incident $V^+(z)$ and reflected $V^-(z)$ components

$$\begin{aligned} V(z) &= V^+(z) + V^-(z) \\ &= V_0^+ e^{ikz} + V_0^- e^{-ikz} . \end{aligned} \tag{2.117}$$

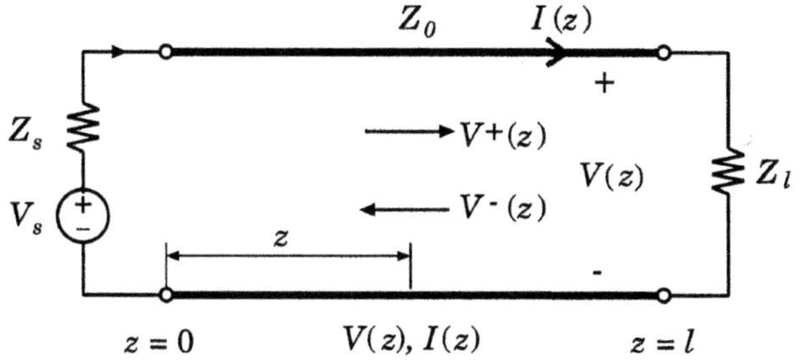

Fig. 2.12. Transmission line terminated with load Z_l.

Similarly, the current $I(z)$ is

$$I(z) = I^+(z) + I^-(z)$$

$$= I_0^+ e^{ikz} + I_0^- e^{-ikz} . \tag{2.118}$$

Since $I_0^+ = \frac{V_0^+}{Z_0}$, and $I_0^- = -\frac{V_0^-}{Z_0}$,

$$I(z) = \frac{1}{Z_0}\left(V_0^+ e^{ikz} - V_0^- e^{-ikz}\right) . \tag{2.119}$$

It is possible to solve (2.117) and (2.119) for the two unknown voltage amplitudes V_0^+ and V_0^-. Two boundary conditions must be enforced to determine V_0^+ and V_0^-. Applying Kirchhoff's voltage law at $z = l$ and 0 results in

$$Z_l = \frac{V(l)}{I(l)} \tag{2.120}$$

$$V_s = I(0)Z_s + V(0) . \tag{2.121}$$

Solving (2.120) and (2.121) for V_0^+ and V_0^- gives

$$V_0^+ = \frac{Z_0}{(Z_0 + Z_s) + \Gamma_l (Z_0 - Z_s)e^{2ikl}} V_s \tag{2.122}$$

$$V_0^- = \Gamma_l e^{2ikl} \underbrace{\frac{Z_0}{(Z_0 + Z_s) + \Gamma_l (Z_0 - Z_s)e^{2ikl}} V_s}_{V_0^+}$$

$$\equiv \Gamma_l e^{2ikl} V_0^+ \tag{2.123}$$

where $\Gamma_l \left(= \frac{Z_l - Z_0}{Z_l + Z_0}\right)$ is the voltage reflection coefficient at the load $z = l$.

2.6.1 Reflection Coefficient, Impedance, and Power

The voltage reflection coefficient $\Gamma(z)$ at z is

$$\Gamma(z) \equiv \frac{V^-(z)}{V^+(z)}$$

$$= \underbrace{\frac{Z_l - Z_0}{Z_l + Z_0}}_{\Gamma_l} e^{2ik(l-z)} = \Gamma_l e^{2ik(l-z)} \,. \tag{2.124}$$

The impedance $Z(z)$ at z is defined as

$$Z(z) \equiv \frac{V(z)}{I(z)} = Z_0 \frac{1+\Gamma(z)}{1-\Gamma(z)} \,. \tag{2.125}$$

Let us evaluate the time-average power delivery to the load P_{av} when the load voltage and current are

$$V(l) = V_0^+ e^{ikl} + V_0^- e^{-ikl}$$

$$= V_0^+ e^{ikl}(1+\Gamma_l) \tag{2.126}$$

$$I(l) = \frac{V_0^+}{Z_0} e^{ikl}(1-\Gamma_l) \,. \tag{2.127}$$

The result is

$$P_{av} = \frac{1}{2}\text{Re}\ [V(l)I^*(l)]$$

$$= \frac{|V_0^+|^2}{2Z_0}\text{Re}\ (1 - \Gamma_l^* + \Gamma_l - |\Gamma_l|^2)$$

$$= \frac{|V_0^+|^2}{2Z_0}(1 - |\Gamma_l|^2) \,. \tag{2.128}$$

To achieve a maximum power delivery to the load, the reflection coefficient Γ_l must be zero, leading to the relation $Z_l = Z_0$, which is called a matched condition.

2.6.2 Voltage Standing-Wave Ratio

Investigating a voltage wave pattern on a transmission line is instructive. The total voltage on a transmission line is

$$V(z) = V_0^+ e^{ikz} + V_0^- e^{-ikz}$$

$$= V_0^+ e^{ikz} \left[1 + \Gamma_l e^{i2k(l-z)}\right] . \qquad (2.129)$$

Assume that the reflection coefficient at the load is

$$\Gamma_l = |\Gamma_l| e^{-i\theta} . \qquad (2.130)$$

The magnitude of $V(z)$ then becomes

$$|V(z)| = \left|V_0^+\right| \left|1 + |\Gamma_l| e^{i2k(l-z)-i\theta}\right| . \qquad (2.131)$$

The maximum of $|V(z)|$ occurs when $2k(l - z_1) - \theta = 2n\pi$ (n: integer) as

$$|V(z)|_{max} = \left|V_0^+\right| |1 + |\Gamma_l|| \qquad (2.132)$$

while the minimum occurs when $2k(l - z_2) - \theta = (2n - 1)\pi$ (n: integer) as

$$|V(z)|_{min} = \left|V_0^+\right| |1 - |\Gamma_l|| . \qquad (2.133)$$

The distance $d = z_2 - z_1$ between the minimum and maximum points on a transmission line is

$$kd = \frac{\pi}{2} \quad \left(d = \frac{\lambda}{4}\right) . \qquad (2.134)$$

The ratio of the maximum to minimum voltage magnitudes is called the voltage standing-wave ratio (VSWR) and is expressed as

$$\mathrm{VSWR} = \frac{|V(z)|_{max}}{|V(z)|_{min}}$$

$$= \frac{1 + |\Gamma_l|}{1 - |\Gamma_l|} . \qquad (2.135)$$

2.6.3 Cascaded Lines

Consider a two-stage transmission line, as shown in Fig. 2.13 (a). Note that Z_1 and Z_2 are the characteristic impedances of cascaded transmission lines, respectively. At $z = z_a$ on the second transmission line (Z_2), the voltage reflection coefficient Γ_a and impedance Z_a are

$$\Gamma_a = \frac{Z_l - Z_2}{Z_l + Z_2} \qquad (2.136)$$

$$Z_a = Z_2 \frac{1 + \Gamma_a}{1 - \Gamma_a} . \qquad (2.137)$$

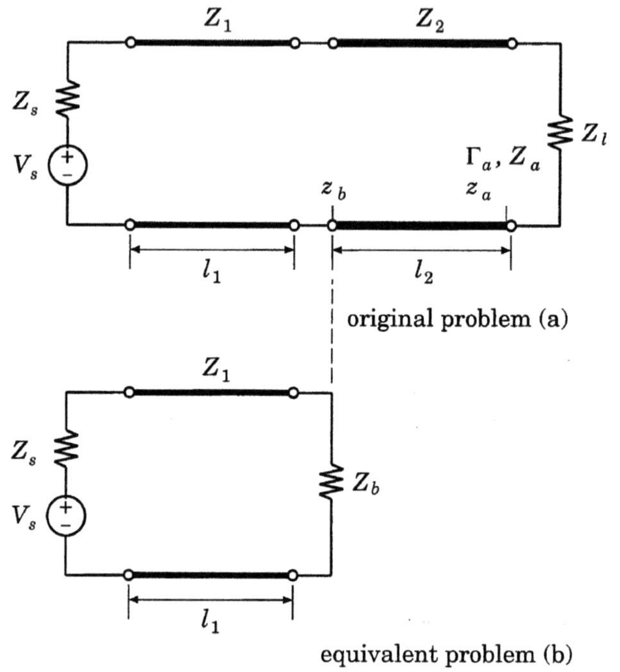

Fig. 2.13. Two-stage transmission line.

At $z = z_b$ on the second transmission line (Z_2), the voltage reflection coefficient Γ_b and impedance Z_b are

$$\Gamma_b = \Gamma_a e^{2ikl_2} \tag{2.138}$$

$$Z_b = Z_2 \frac{1+\Gamma_b}{1-\Gamma_b} \; . \tag{2.139}$$

Therefore, the original problem of a two-stage transmission line is simplified to a single-stage transmission line with the equivalent impedance Z_b, as shown in Fig. 2.13 (b). The geometry in Fig. 2.13 (b) is identical with that of the single-stage transmission line in Fig. 2.12. The solution to the problem in Fig. 2.12 is thus available from (2.122) and (2.123).

It is often desirable to eliminate the reflected wave on Z_1 by appropriately choosing Z_2. No reflection on Z_1 requires the matched condition

$$Z_b = Z_1 \; . \tag{2.140}$$

To realize the matched condition, the length $l_2 = \dfrac{\lambda}{4}$ is chosen, thereby resulting in

$$\Gamma_b = -\Gamma_a$$

$$= -\frac{Z_l - Z_2}{Z_l + Z_2} \tag{2.141}$$

$$Z_b = \frac{Z_2^2}{Z_l} . \tag{2.142}$$

Assume that the load impedance Z_l is real. Equating (2.140) with (2.142) gives

$$Z_2 = \sqrt{Z_1 Z_l} . \tag{2.143}$$

The inserted $\frac{\lambda}{4}$-long transmission line with $Z_2 = \sqrt{Z_1 Z_l}$ is called the quarter-wave transformer.

Problems for Chapter 2

1. Derive (2.40) and (2.44).
2. Derive (2.61) and (2.69).

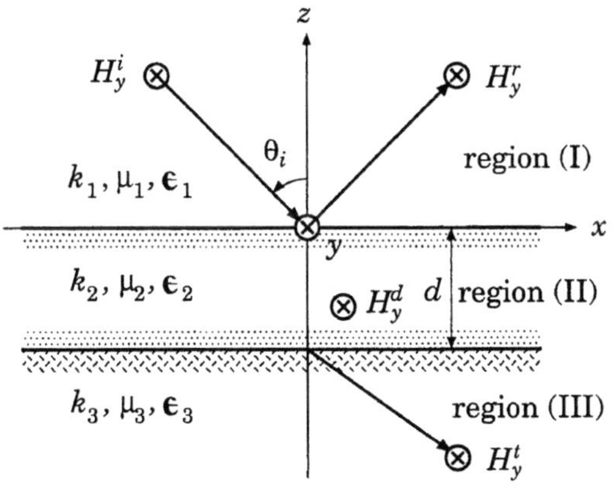

Fig. 2.14. TM wave impinging on dielectric slab of thickness d.

3. Consider a TM wave that impinges on a dielectric slab, as shown in Fig. 2.14. In regions (I), (II), and (III), the total magnetic field representations are given by

$$H_y^i(x,z) = \exp(ik_{xi}x - ik_{zi}z) \tag{2.144}$$

$$H_y^r(x,z) = R\exp(ik_{xr}x + ik_{zr}z) \tag{2.145}$$

$$H_y^d(x,z) = \exp(ik_{xd}x)\left[A\exp(ik_{zd}z) + B\exp(-ik_{zd}z)\right] \tag{2.146}$$

$$H_y^t(x,z) = T\exp(ik_{xt}x - ik_{zt}z)\,. \tag{2.147}$$

To determine the unknown coefficients, A, B, R, and T, the four boundary conditions for the field continuities, $H_y(x,0)$, $H_y(x,-d)$, $E_x(x,0)$, and $E_x(x,-d)$, must be enforced. Derive A, B, R, and T using the boundary conditions.

4. In circuit analysis, a lossless coaxial line is modeled as a transmission line with the shunt capacitances C and series inductances L. Given a coaxial line of inner and outer radii a and b, filled with dielectric of μ and ϵ, derive its shunt capacitance and series inductance per unit length.
5. Consider the terminated transmission line, as shown in Fig. 2.12. Derive the current reflection coefficient $\dfrac{I^-(z)}{I^+(z)}$.

3

Waveguides

3.1 Cylindrical Waveguides

The understanding of wave propagation along cylindrical waveguides is of practical interest in microwave and optical communications. Typical cylindrical waveguides include rectangular waveguides, optical fibers, and microstrip lines. This section investigates electromagnetic waves that can propagate along an infinitely long waveguide, as shown in Fig. 3.1. Wave propagation is conveniently analyzed in terms of the magnetic and electric vector potentials $\overline{A}$ and $\overline{F}$. The vector potentials in a source-free cylindrical waveguide satisfy the Helmholtz equations

$$\left(\nabla^2 + k^2\right) \overline{A} = 0 \tag{3.1}$$

$$\left(\nabla^2 + k^2\right) \overline{F} = 0 \tag{3.2}$$

where $k\,(= \omega\sqrt{\mu\epsilon})$ is the wavenumber. To understand the wave propagation behavior, the Helmholtz equations must be solved for $\overline{A}$ and $\overline{F}$ subject to the boundary conditions imposed at the waveguide boundary.

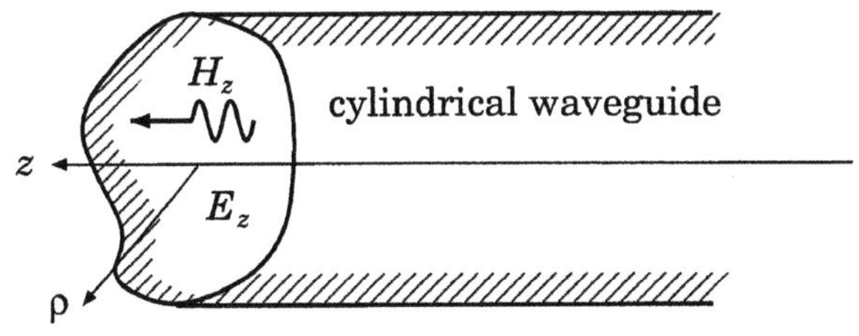

Fig. 3.1. Infinitely long waveguide that lies along z-direction.

Let us discuss the possible types of waves that can propagate along a waveguide. The propagating waves are conveniently represented in terms of the TE and TM waves. The TE wave refers to the wave whose electric field components are transverse to the wave propagation direction z ($E_z = 0$). Similarly, the TM wave refers to the wave whose magnetic field components are transverse to the wave propagation direction z ($H_z = 0$). The description of TM and TE waves is possible using the longitudinal components of $\overline{A}$ and $\overline{F}$, $A_z(x, y, z)$ and $F_z(x, y, z)$.

- When $\overline{F} = \hat{z}F_z(x, y, z)$ and $\overline{A} = 0$, then $E_z(x, y, z) = 0$; this type of wave is called a TE wave (transverse electric to the wave propagation direction z). A TE wave satisfies the Helmholtz equation

$$\left(\nabla^2 + k^2\right) F_z(x, y, z) = 0 . \tag{3.3}$$

  The explicit field expressions for a TE wave using rectangular and cylindrical coordinates are shown in Tables 3.1 and 3.2, respectively.

- When $\overline{A} = \hat{z}A_z(x, y, z)$ and $\overline{F} = 0$, $H_z(x, y, z) = 0$ and a TM wave (transverse magnetic to the z-direction) is obtained. A TM wave satisfies the Helmholtz equation

$$\left(\nabla^2 + k^2\right) A_z(x, y, z) = 0 . \tag{3.4}$$

  A hollow waveguide surrounded with metallic walls can support either TM or TE wave propagation.

- When the wavenumber in the z-direction is k, the propagating term takes the form of $e^{\pm ikz}$ and the longitudinal field components become zero $E_z(x, y, z) = H_z(x, y, z) = 0$, as seen in Tables 3.1 and 3.2. This type of wave is referred to as a TEM wave. Transmission lines (coaxial line, paired wire, etc.) consisting of two conducting terminals can support TEM wave propagation.

- A dielectric waveguide or a partially filled waveguide with metallic walls can support the propagation of a hybrid wave, which is a combination of TM and TE waves. Both $A_z(x, y, z)$ and $F_z(x, y, z)$ must be used for a hybrid wave analysis.

3.2 Rectangular Waveguide

A rectangular waveguide with conducting walls is commonly used to transmit power in microwave communication and antenna applications. The field analysis for a rectangular waveguide is a canonical boundary-value problem. This section investigates the behavior of TM and TE wave propagation along a rectangular waveguide filled with homogeneous dielectric material (μ, ϵ).

Table 3.1. Field representations using rectangular coordinates (x, y, z)

Fields	TM wave	TE wave
E_x	$\frac{i}{\omega\mu\epsilon}\frac{\partial^2 A_z}{\partial x\,\partial z}$	$-\frac{1}{\epsilon}\frac{\partial F_z}{\partial y}$
E_y	$\frac{i}{\omega\mu\epsilon}\frac{\partial^2 A_z}{\partial y\,\partial z}$	$\frac{1}{\epsilon}\frac{\partial F_z}{\partial x}$
E_z	$\frac{i}{\omega\mu\epsilon}\left(\frac{\partial^2}{\partial z^2}+k^2\right)A_z$	0
H_x	$\frac{1}{\mu}\frac{\partial A_z}{\partial y}$	$\frac{i}{\omega\mu\epsilon}\frac{\partial^2 F_z}{\partial x\,\partial z}$
H_y	$-\frac{1}{\mu}\frac{\partial A_z}{\partial x}$	$\frac{i}{\omega\mu\epsilon}\frac{\partial^2 F_z}{\partial y\,\partial z}$
H_z	0	$\frac{i}{\omega\mu\epsilon}\left(\frac{\partial^2}{\partial z^2}+k^2\right)F_z$

3.2.1 TM Wave

Consider TM wave propagation along a rectangular waveguide with perfect conducting walls, as shown in Fig. 3.2. A TM wave is assumed to propagate along the z-direction with the propagation constant k_z. The vector potentials associated with the TM wave are given by $F_z(x, y, z) = 0$ and

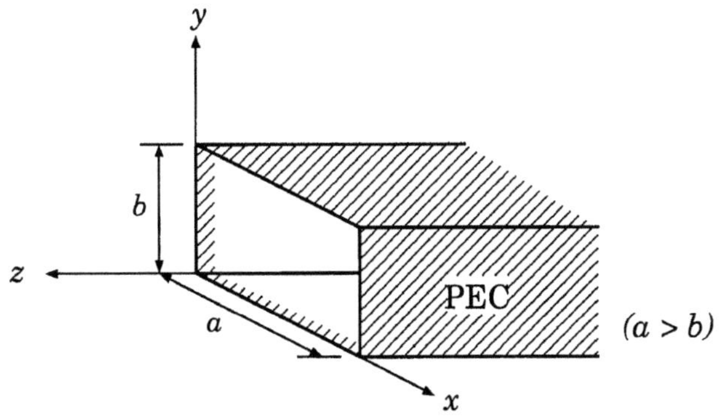

Fig. 3.2. Rectangular waveguide with cross section $a \times b$ $(a > b)$.

Table 3.2. Field representations using cylindrical coordinates (ρ, ϕ, z)

Fields	TM wave	TE wave
E_ρ	$\dfrac{i}{\omega\mu\epsilon}\dfrac{\partial^2 A_z}{\partial\rho\,\partial z}$	$-\dfrac{1}{\epsilon\rho}\dfrac{\partial F_z}{\partial\phi}$
E_ϕ	$\dfrac{i}{\omega\mu\epsilon\rho}\dfrac{\partial^2 A_z}{\partial\phi\,\partial z}$	$\dfrac{1}{\epsilon}\dfrac{\partial F_z}{\partial\rho}$
E_z	$\dfrac{i}{\omega\mu\epsilon}\left(\dfrac{\partial^2}{\partial z^2}+k^2\right)A_z$	0
H_ρ	$\dfrac{1}{\mu\rho}\dfrac{\partial A_z}{\partial\phi}$	$\dfrac{i}{\omega\mu\epsilon}\dfrac{\partial^2 F_z}{\partial\rho\,\partial z}$
H_ϕ	$-\dfrac{1}{\mu}\dfrac{\partial A_z}{\partial\rho}$	$\dfrac{i}{\omega\mu\epsilon\rho}\dfrac{\partial^2 F_z}{\partial\phi\,\partial z}$
H_z	0	$\dfrac{i}{\omega\mu\epsilon}\left(\dfrac{\partial^2}{\partial z^2}+k^2\right)F_z$

$A_z(x,y,z) = A_z(x,y)\exp(ik_z z)$. To determine the propagation constant k_z, it is necessary to solve the Helmholtz equation $\left(\nabla^2 + k^2\right)A_z(x,y,z) = 0$ subject to the boundary conditions. The Helmholtz equation can be rewritten as

$$\left(\frac{\partial^2}{\partial x^2} + \frac{\partial^2}{\partial y^2} + k^2 - k_z^2\right) A_z(x,y) = 0\,. \tag{3.5}$$

Based on the separation of variables technique, $A_z(x,y)$ takes the form of

$$A_z(x,y) = X(x)Y(y)\,. \tag{3.6}$$

Substituting $A_z(x,y)$ into the Helmholtz equation (3.5) and dividing by $X(x)Y(y)$ yields

$$\frac{1}{X(x)}\frac{d^2X(x)}{dx^2} + \frac{1}{Y(y)}\frac{d^2Y(y)}{dy^2} + k^2 - k_z^2 = 0. \tag{3.7}$$

To satisfy (3.7), the terms $\dfrac{1}{X(x)}\dfrac{d^2X(x)}{dx^2}$ and $\dfrac{1}{Y(y)}\dfrac{d^2Y(y)}{dy^2}$ must be equal to constants independent of x and y, respectively. Let the separation constants be $-k_x^2$ and $-k_y^2$ as

$$\frac{1}{X(x)}\frac{d^2X(x)}{dx^2} = -k_x^2 \tag{3.8}$$

$$\frac{1}{Y(y)}\frac{d^2Y(y)}{dy^2} = -k_y^2 \,. \tag{3.9}$$

Then, the dispersion relation is given by

$$k_x^2 + k_y^2 + k_z^2 = k^2 \,. \tag{3.10}$$

Since E_z (hence A_z) must vanish at the boundaries $x = 0$, a and $y = 0$, b, the vector potential $A_z(x, y, z)$ is chosen as

$$A_z(x, y, z) = \sin k_x x \, \sin k_y y \, \exp(ik_z z) \tag{3.11}$$

where $k_x = \dfrac{m\pi}{a}$, $k_y = \dfrac{n\pi}{b}$, and m and n are integers $(m, n \geq 1)$. The remaining field components are available by substituting $A_z(x, y, z)$ into Table 3.1 in Section 3.1.

The dispersion relation yields the propagation constant for the TM_{mn} mode

$$k_z = \sqrt{k^2 - \left(\frac{m\pi}{a}\right)^2 - \left(\frac{n\pi}{b}\right)^2} \tag{3.12}$$

while its cutoff wavenumber is given by $k_c = \sqrt{\left(\dfrac{m\pi}{a}\right)^2 + \left(\dfrac{n\pi}{b}\right)^2}$. When the operating frequency is high enough to satisfy the condition $k > k_c$, the TM_{mn} mode can propagate. When $k < k_c$, the TM_{mn} mode cannot propagate and becomes attenuated. Note that the lowest-order TM wave is the TM_{11} mode.

3.2.2 TE Wave

The analysis of TE wave propagation within a rectangular waveguide is somewhat similar to the analysis of TM waves. Let us start with the assumption $A_z(x, y, z) = 0$ and $F_z(x, y, z) = F_z(x, y) \exp(ik_z z)$, where the vector potential $F_z(x, y)$ satisfies the Helmholtz equation

$$\left(\frac{\partial^2}{\partial x^2} + \frac{\partial^2}{\partial y^2} + k^2 - k_z^2\right) F_z(x, y) = 0 \,. \tag{3.13}$$

Solving the Helmholtz equation is possible using the separation of variables technique. The tangential electric fields must vanish on the metallic surfaces, implying that the normal derivatives of $F_z(x, y)$ must be zero, $\left.\dfrac{\partial F_z(x, y)}{\partial y}\right|_{y=0,b} = 0$ and $\left.\dfrac{\partial F_z(x, y)}{\partial x}\right|_{x=0,a} = 0$. Hence, $F_z(x, y, z)$ takes the form of

$$F_z(x,y,z) = \cos k_x x \cos k_y y \exp(ik_z z) \tag{3.14}$$

where $k_x = \frac{m\pi}{a}$, $k_y = \frac{n\pi}{b}$, and m and n are integers ($m \geq 0$, $n \geq 0$, $m + n \neq 0$). The propagation constant k_z and cutoff wavenumber k_c for the TE_{mn} mode are

$$k_z = \sqrt{k^2 - \left(\frac{m\pi}{a}\right)^2 - \left(\frac{n\pi}{b}\right)^2} \tag{3.15}$$

$$k_c = \sqrt{\left(\frac{m\pi}{a}\right)^2 + \left(\frac{n\pi}{b}\right)^2} . \tag{3.16}$$

When $a > b$, the lowest-order TE wave is the TE_{10} mode, where its nonzero fields are

$$H_z = H_0 \cos\left(\frac{\pi x}{a}\right) \exp(ik_z z) \tag{3.17}$$

$$H_x = \frac{-ik_z a}{\pi} H_0 \sin\left(\frac{\pi x}{a}\right) \exp(ik_z z) \tag{3.18}$$

$$E_y = \frac{i\omega\mu a}{\pi} H_0 \sin\left(\frac{\pi x}{a}\right) \exp(ik_z z) \tag{3.19}$$

where $H_0 = \frac{i}{\omega\mu\epsilon}\left(\frac{\pi}{a}\right)^2$ and $k_z = \sqrt{k^2 - \left(\frac{\pi}{a}\right)^2}$.

Let us consider a rectangular waveguide of dimension $a = 2b$, where the lowest propagating wave is TE_{10} and the next higher wave is TE_{20} or TE_{01}. If the operating frequency is chosen as $\left(\frac{\pi}{a}\right) < k < \left(\frac{2\pi}{a}\right)$, a rectangular waveguide will only support single-mode TE_{10} propagation. The power density that is delivered by the TE_{10} mode is

$$\begin{aligned} \overline{S}_{av} &= \frac{1}{2}\text{Re}\left(\overline{E} \times \overline{H}^*\right) \\ &= \frac{1}{2}\text{Re}\left(-\hat{z}E_y H_x^* + \hat{x}E_y H_z^*\right) . \end{aligned} \tag{3.20}$$

The time-average power delivery along the z-direction is

$$\begin{aligned} P_{av} &= \int_0^b \int_0^a \overline{S}_{av} \cdot \hat{z}\, dx dy \\ &= -\int_0^b \int_0^a \frac{1}{2}\text{Re}\left(E_y H_x^*\right) dx dy \\ &= \frac{b k_z \pi^2}{4\omega\mu\epsilon^2 a} . \end{aligned} \tag{3.21}$$

3.3 Dielectric Slab Waveguide

A dielectric slab waveguide is one of the major waveguide components used in optical and microwave communications. For simplicity, the dielectric slab waveguide analyzed in this section is two-dimensional, as illustrated in Fig. 3.3. A dielectric slab waveguide consists of a denser dielectric slab [region (II)] that lies between two less dense dielectric media [regions (I) and (III)]. While a dielectric slab waveguide can generally support either TM or TE wave propagation, the TM case is considered first.

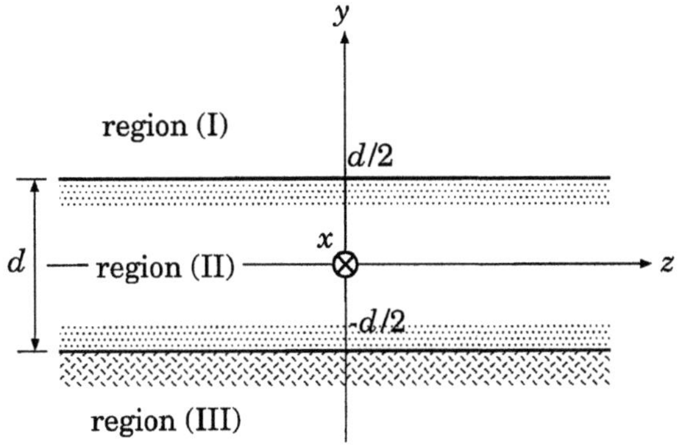

Fig. 3.3. Dielectric slab waveguide.

3.3.1 TM Wave

Assume that the TM wave ($H_z = 0$ and $E_z \neq 0$) is guided by a dielectric slab along the z-direction. The wavenumbers in regions (I), (II), and (III) are $k_1\,(=\omega\sqrt{\mu_1\epsilon_1})$, $k_2\,(=\omega\sqrt{\mu_2\epsilon_2})$, and $k_3\,(=\omega\sqrt{\mu_3\epsilon_3})$, respectively. There is no field variation in the x-direction $\left(\frac{\partial}{\partial x}=0\right)$. The vector potentials in each region also satisfy the respective Helmholtz equations

$$\left(\frac{\partial^2}{\partial y^2}+\frac{\partial^2}{\partial z^2}+k_1^2\right)A_z^I(y,z)=0 \tag{3.22}$$

$$\left(\frac{\partial^2}{\partial y^2}+\frac{\partial^2}{\partial z^2}+k_2^2\right)A_z^{II}(y,z)=0 \tag{3.23}$$

$$\left(\frac{\partial^2}{\partial y^2}+\frac{\partial^2}{\partial z^2}+k_3^2\right)A_z^{III}(y,z)=0\,. \tag{3.24}$$

Region (II) is assumed to be denser than regions (I) and (III) ($\epsilon_2 > \epsilon_1, \epsilon_3$) to guarantee wave guidance within region (II). The field in region (I) must be evanescent in the $+y$-direction for $y > 0$ and propagate along the z-direction. Hence, $A_z^I(y,z)$ is written as

$$A_z^I(y,z) = A_1 \exp(-\alpha_1 y) \ \exp(ik_z z) \tag{3.25}$$

where $\alpha_1 = \sqrt{k_z^2 - \omega^2 \mu_1 \epsilon_1}$.

The wave in region (II) experiences reflection at $y = \pm\frac{d}{2}$, thereby exhibiting standing wave characteristics along the y-direction. The vector potential in region (II) is

$$A_z^{II}(y,z) = (A_2' \cos k_y y + A_2'' \sin k_y y) \exp(ik_z z)$$

$$\equiv A_2 \cos(k_y y + \psi) \exp(ik_z z) \tag{3.26}$$

where $k_y^2 + k_z^2 = k_2^2 \equiv \omega^2 \mu_2 \epsilon_2$.

Similarly, since the field in region (III) must be evanescent along the $-y$-direction for $y < 0$,

$$A_z^{III}(y,z) = A_3 \exp(\alpha_3 y) \exp(ik_z z) \tag{3.27}$$

where $\alpha_3 = \sqrt{k_z^2 - \omega^2 \mu_3 \epsilon_3}$.

To represent the vector potentials, $A_z^I(y,z)$, $A_z^{II}(y,z)$, and $A_z^{III}(y,z)$, four unknown coefficients, A_2, ψ, A_3, and k_z, are introduced, respectively, whereas A_1 is considered a known coefficient. The aim is to determine the four unknown coefficients utilizing the four boundary conditions for the E_z and H_x field continuities at $y = \pm\frac{d}{2}$. The E_z and H_x field components are obtained by substituting $A_z^I(y,z)$, $A_z^{II}(y,z)$, and $A_z^{III}(y,z)$ into Table 3.1 in Section 3.1. The E_z continuities at $y = \pm\frac{d}{2}$ are

$$E_z^I\left(\frac{d}{2}, z\right) = E_z^{II}\left(\frac{d}{2}, z\right) \tag{3.28}$$

$$E_z^{II}\left(-\frac{d}{2}, z\right) = E_z^{III}\left(-\frac{d}{2}, z\right) . \tag{3.29}$$

Since $E_z = \frac{i}{\omega\mu\epsilon}\left(\frac{\partial^2}{\partial z^2} + k^2\right) A_z$, (3.28) and (3.29) can be rewritten as

$$-A_1 \frac{\alpha_1^2}{\mu_1 \epsilon_1} \exp(-\alpha_1 d/2) = A_2 \frac{k_y^2}{\mu_2 \epsilon_2} \cos\left(k_y \frac{d}{2} + \psi\right) \tag{3.30}$$

$$A_2 \frac{k_y^2}{\mu_2 \epsilon_2} \cos\left(-k_y \frac{d}{2} + \psi\right) = -A_3 \frac{\alpha_3^2}{\mu_3 \epsilon_3} \exp(-\alpha_3 d/2) . \tag{3.31}$$

Similarly, the $H_x \left(= -\frac{1}{\mu}\frac{\partial A_z}{\partial y}\right)$ continuities at $y = \pm\frac{d}{2}$ yield

$$\frac{\alpha_1}{\mu_1} A_1 \exp(-\alpha_1 d/2) = \frac{k_y}{\mu_2} A_2 \sin\left(k_y \frac{d}{2} + \psi\right) \tag{3.32}$$

$$-\frac{k_y}{\mu_2} A_2 \sin\left(-k_y \frac{d}{2} + \psi\right) = \frac{\alpha_3}{\mu_3} A_3 \exp(-\alpha_3 d/2) . \tag{3.33}$$

Taking the ratio of (3.30) and (3.32) results in

$$-\frac{\alpha_1}{\epsilon_1} \tan\left(k_y \frac{d}{2} + \psi\right) = \frac{k_y}{\epsilon_2} . \tag{3.34}$$

Similarly, (3.31) and (3.33) gives

$$\frac{\alpha_3}{\epsilon_3} \tan\left(-k_y \frac{d}{2} + \psi\right) = \frac{k_y}{\epsilon_2} . \tag{3.35}$$

Using the formula

$$\tan(\alpha \pm \beta) = \frac{\tan\alpha \pm \tan\beta}{1 \mp \tan\alpha \tan\beta} \tag{3.36}$$

it is possible to eliminate ψ in (3.34) and (3.35). The result yields the dispersion relation

$$\tan(k_y d) - \frac{\epsilon_2 \alpha_1}{\epsilon_1 k_y} = \frac{\epsilon_2 \alpha_3}{\epsilon_3 k_y}\left[1 + \tan(k_y d)\,\frac{\epsilon_2 \alpha_1}{\epsilon_1 k_y}\right] . \tag{3.37}$$

When region (III) is perfectly conducting (PEC, $\epsilon_3 \to \infty$), the dispersion relation reduces to

$$\tan(k_y d) - \frac{\epsilon_2 \alpha_1}{\epsilon_1 k_y} = 0 . \tag{3.38}$$

The transcendental relation (3.37) can be solved for k_z numerically. In general more than one solution exist.

3.3.2 TE Wave

The analysis of TE wave propagation is quite analogous to that of TM wave propagation. The TE wave ($E_z = 0$ and $H_z \neq 0$) is assumed to propagate along a slab waveguide with the propagation constant k_z. The associated vector potential $F_z(y, z)$ takes the form of

$$F_z^I(y, z) = B_1 \exp(-\alpha_1 y) \exp(ik_z z) \tag{3.39}$$

$$F_z^{II}(y, z) = B_2 \cos(k_y y + \psi) \exp(ik_z z) \tag{3.40}$$

$$F_z^{III}(y, z) = B_3 \exp(\alpha_3 y) \exp(ik_z z) . \tag{3.41}$$

Enforcing the boundary conditions for the E_x and H_z continuities at $y = \pm\frac{d}{2}$ results in the dispersion relation for the TE wave

$$\tan(k_y d) - \frac{\mu_2 \alpha_1}{\mu_1 k_y} = \frac{\mu_2 \alpha_3}{\mu_3 k_y}\left[1 + \tan(k_y d)\frac{\mu_2 \alpha_1}{\mu_1 k_y}\right] . \tag{3.42}$$

Note that the dispersion relation (3.42) can also be directly obtained from (3.37) based on the duality theorem (Section 6.5), which permits interchanging $\epsilon_{1,2,3} \rightarrow \mu_{1,2,3}$.

3.4 Circular Waveguide

In addition to rectangular waveguides, circular conducting tubes, known as circular waveguides, are also popular guiding structures for microwave applications. This section investigates wave propagation within a circular waveguide that is surrounded with a conducting wall. Both TM and TE waves are considered in the following subsections.

3.4.1 TM Wave

Consider a TM wave ($H_z = 0$ and $E_z \neq 0$) that propagates inside a circular waveguide, as shown in Fig. 3.4. Assume that the circular waveguide is filled with dielectric material of (μ, ϵ). When the wave propagates along the z-direction with the propagation constant k_z, the vector potential $\overline{A}(\rho, \phi, z)$ takes the form of

$$\overline{A}(\rho, \phi, z) = \hat{z} A_z(\rho, \phi, z) = \hat{z} A_z(\rho, \phi) \exp(ik_z z) . \tag{3.43}$$

The corresponding Helmholtz equation $\left(\nabla^2 + k^2\right) A_z(\rho, \phi, z) = 0$ using cylindrical coordinates is rewritten as

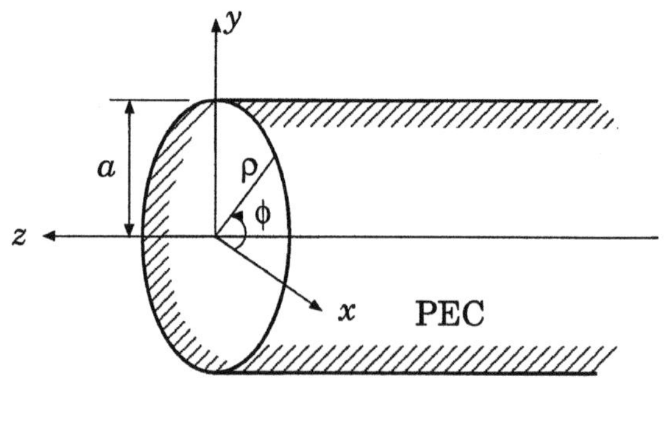

Fig. 3.4. Circular waveguide with radius a.

$$\left[\frac{1}{\rho}\frac{\partial}{\partial\rho}\left(\rho\frac{\partial}{\partial\rho}\right)+\frac{1}{\rho^2}\frac{\partial^2}{\partial\phi^2}+\frac{\partial^2}{\partial z^2}+k^2\right]A_z(\rho,\phi,z)=0 \tag{3.44}$$

or

$$\left[\frac{1}{\rho}\frac{\partial}{\partial\rho}\left(\rho\frac{\partial}{\partial\rho}\right)+\frac{1}{\rho^2}\frac{\partial^2}{\partial\phi^2}+k_\rho^2\right]A_z(\rho,\phi)=0 \tag{3.45}$$

where $k_\rho^2 \equiv k^2 - k_z^2$.

Based on the separation of variables technique, let

$$A_z(\rho,\phi)=R(\rho)\Phi(\phi)\,. \tag{3.46}$$

Substituting $A_z(\rho,\phi)$ into the Helmholtz equation and dividing by $R(\rho)\Phi(\phi)$ yields

$$\frac{1}{\rho R(\rho)}\frac{d}{d\rho}\left[\rho\frac{dR(\rho)}{d\rho}\right]+\frac{1}{\rho^2}\frac{1}{\Phi(\phi)}\frac{d^2\Phi(\phi)}{d\phi^2}+k_\rho^2=0\,. \tag{3.47}$$

Substituting

$$\frac{1}{\Phi(\phi)}\frac{d^2\Phi(\phi)}{d\phi^2}=-m^2 \tag{3.48}$$

into (3.47) results in the following two ordinary differential equations

$$\left(\frac{d^2}{d\phi^2}+m^2\right)\Phi(\phi)=0 \tag{3.49}$$

$$\left[\frac{1}{\rho}\frac{d}{d\rho}\left(\rho\frac{d}{d\rho}\right)+k_\rho^2-\frac{m^2}{\rho^2}\right]R(\rho)=0\,. \tag{3.50}$$

The solution to (3.49) is

$$\Phi(\phi)=\left\{\begin{array}{c}\sin m\phi\\ \\ \cos m\phi\end{array}\right\}. \tag{3.51}$$

Since the field is periodic in the azimuthal direction with a 2π-periodicity, the parameter m should be an integer.

Equation (3.50) is known as Bessel's equation (see Appendix B) and its solution is represented as the linear combination of the Bessel functions – $J_m(k_\rho\rho)$, $N_m(k_\rho\rho)$, $H_m^{(1),(2)}(k_\rho\rho)$, $\cdots$. Among these Bessel functions, the appropriate choice for $R(\rho)$ is $J_m(k_\rho\rho)$, since $N_m(k_\rho\rho)$ and $H_m^{(1),(2)}(k_\rho\rho)$ become infinite at $\rho=0$. Hence, $A_z(\rho,\phi,z)$ is given by

$$A_z(\rho,\phi,z)=J_m(k_\rho\rho)\left\{\begin{array}{c}\sin m\phi\\ \\ \cos m\phi\end{array}\right\}\exp(ik_z z)\,. \tag{3.52}$$

The explicit field components are obtained by substituting $A_z(\rho, \phi, z)$ into Table 3.2 in Section 3.1.

Since E_z and E_ϕ must vanish at $\rho = a$, $J_m(k_{\rho mn}a) = 0$, where $k_{\rho mn}a$ is the nth root of the mth order Bessel function. The TM wave with the eigenvalue $k_{\rho mn}$ is called the TM_{mn} mode, where the roots of $J_m(k_{\rho mn}a) = 0$ are tabulated in Table 3.3. The propagation constant for the TM_{mn} mode is given as

$$k_z = \sqrt{k^2 - k_{\rho mn}^2} \, . \tag{3.53}$$

In order for the TM_{mn} mode to propagate, the wavenumber must be higher than $k_{\rho mn}$

$$k \;(= \omega\sqrt{\mu\epsilon}) > k_{\rho mn} \tag{3.54}$$

where $k_{\rho mn}$ is the cutoff wavenumber for the TM_{mn} mode. The lowest cutoff wavenumber is $k_{\rho 01} \approx \dfrac{2.405}{a}$ (TM_{01} mode).

Table 3.3. Values of $k_{\rho mn}a$

	m=0	m=1	m=2	m=3
n=1	2.405	3.832	5.136	6.380
n=2	5.520	7.016	8.417	9.761
n=3	8.654	10.173	11.620	13.015

3.4.2 Power Delivery by TM_{mn} Mode

Let us evaluate the power that is delivered by the TM_{mn} mode when $k > k_{\rho mn}$. Since

$$A_z(\rho, \phi, z) = J_m(k_\rho \rho)(A \sin m\phi + B \cos m\phi) \exp(ik_z z) \tag{3.55}$$

the transverse field components are

$$E_\rho = \frac{i}{\omega\mu\epsilon}\frac{\partial^2 A_z(\rho,\phi,z)}{\partial\rho\,\partial z}$$

$$= -\frac{k_\rho k_z}{\omega\mu\epsilon} J'_m(k_\rho\rho)(A\sin m\phi + B\cos m\phi)\exp(ik_z z) \tag{3.56}$$

$$E_\phi = \frac{i}{\omega\mu\epsilon\rho}\frac{\partial^2 A_z(\rho,\phi,z)}{\partial\phi\,\partial z}$$

$$= -\frac{mk_z}{\omega\mu\epsilon\rho} J_m(k_\rho\rho)(A\cos m\phi - B\sin m\phi)\exp(ik_z z) \tag{3.57}$$

$$H_\rho = \frac{1}{\mu\rho}\frac{\partial A_z(\rho,\phi,z)}{\partial\phi}$$

$$= \frac{m}{\mu\rho} J_m(k_\rho\rho)(A\cos m\phi - B\sin m\phi)\exp(ik_z z) \tag{3.58}$$

$$H_\phi = -\frac{1}{\mu}\frac{\partial A_z(\rho,\phi,z)}{\partial\rho}$$

$$= -\frac{k_\rho}{\mu} J'_m(k_\rho\rho)(A\sin m\phi + B\cos m\phi)\exp(ik_z z) \tag{3.59}$$

where $J'_m(k_\rho\rho)$ denotes differentiation with respect to the argument $\dfrac{dJ_m(k_\rho\rho)}{d(k_\rho\rho)}$.

The total time-average power P_{av} is given by

$$P_{av} = \frac{1}{2}\mathrm{Re}\int_0^a\int_0^{2\pi}\left(\overline{E}\times\overline{H}^*\right)\cdot\hat{z}\rho\,d\phi\,d\rho$$

$$= \frac{1}{2}\mathrm{Re}\int_0^a\int_0^{2\pi}\left(E_\rho H_\phi^* - E_\phi H_\rho^*\right)\rho\,d\phi\,d\rho\,. \tag{3.60}$$

Let us first consider a case in which $m \neq 0$. The ϕ integrations for $m \neq 0$ yield

$$\int_0^{2\pi} E_\rho H_\phi^*\,d\phi = \frac{\pi k_\rho^2 k_z}{\omega\mu^2\epsilon} {J'_m}^2(k_\rho\rho)\left(|A|^2 + |B|^2\right) \tag{3.61}$$

$$\int_0^{2\pi} E_\phi H_\rho^*\,d\phi = -\frac{\pi m^2 k_z}{\omega\mu^2\epsilon\rho^2} J_m^2(k_\rho\rho)\left(|A|^2 + |B|^2\right)\,. \tag{3.62}$$

Therefore, P_{av} becomes

$$P_{av} = \left(|A|^2 + |B|^2\right)\frac{\pi k_z}{2\omega\mu^2\epsilon}$$

$$\cdot\int_0^a\left[k_\rho^2 {J'_m}^2(k_\rho\rho) + \frac{m^2}{\rho^2}J_m^2(k_\rho\rho)\right]\rho\,d\rho\,. \tag{3.63}$$

The evaluation of P_{av} is tedious but straightforward. Let $k_\rho \rho = u$, then

$$\int_0^a \left[k_\rho^2 J_m'^2(k_\rho \rho) + \frac{m^2}{\rho^2} J_m^2(k_\rho \rho)\right] \rho \, d\rho$$

$$= \int_0^{k_\rho a} \left[J_m'^2(u) + \frac{m^2}{u^2} J_m^2(u)\right] u \, du$$

$$\equiv I \,. \tag{3.64}$$

Using recurrence formulas for the Bessel function

$$J_m'(u) = \frac{1}{2}\left[-J_{m+1}(u) + J_{m-1}(u)\right] \tag{3.65}$$

$$\frac{m}{u} J_m(u) = \frac{1}{2}\left[J_{m+1}(u) + J_{m-1}(u)\right] \tag{3.66}$$

gives

$$I = \int_0^x \frac{1}{2}\left[J_{m+1}^2(u) + J_{m-1}^2(u)\right] u \, du \tag{3.67}$$

where $k_\rho a = x$. Based on the formula

$$\int J_m^2(u) u \, du = \frac{u^2}{2}\left[J_m^2(u) - J_{m-1}(u) J_{m+1}(u)\right] \tag{3.68}$$

I is integrated to

$$I = \frac{x^2}{4}\left[J_{m+1}^2(x) - J_m(x) J_{m+2}(x) + J_{m-1}^2(x) - J_{m-2}(x) J_m(x)\right] . \tag{3.69}$$

Since $J_m(x) = 0$,

$$I = \frac{x^2}{4}\left[J_{m+1}^2(x) + J_{m-1}^2(x)\right] . \tag{3.70}$$

Further simplification is possible using the relations

$$J_{m+1}(x) = -J_{m-1}(x) = -J_m'(x) . \tag{3.71}$$

The final result for $m = 1, 2, 3, \cdots$ is

$$I = \frac{x^2}{2} J_m'^2(x) \tag{3.72}$$

$$P_{av} = \left(|A|^2 + |B|^2\right) \frac{\pi k_z}{4\omega\mu^2\epsilon} (k_\rho a)^2 J_m'^2(k_\rho a) . \tag{3.73}$$

Similarly, it can be shown that for $m = 0$

$$P_{av} = |B|^2 \frac{\pi k_z}{2\omega\mu^2\epsilon} (k_\rho a)^2 J_0'^2(k_\rho a) . \tag{3.74}$$

3.4.3 TE Wave

The analysis of TE wave ($H_z \neq 0$ and $E_z = 0$) propagation inside a metallic circular waveguide is similar to the analysis of TM wave propagation. The solution to the Helmholtz equation

$$\left[\frac{1}{\rho}\frac{\partial}{\partial\rho}\left(\rho\frac{\partial}{\partial\rho}\right) + \frac{1}{\rho^2}\frac{\partial^2}{\partial\phi^2} + k_\rho^2\right] F_z(\rho, \phi) = 0 \qquad (3.75)$$

is

$$F_z(\rho, \phi, z) = J_m(k_\rho \rho) \begin{Bmatrix} \sin m\phi \\ \\ \cos m\phi \end{Bmatrix} \exp(ik_z z) \,. \qquad (3.76)$$

The explicit field components are obtained by substituting $F_z(\rho, \phi, z)$ into Table 3.2 in Section 3.1. Since E_ϕ vanishes at $\rho = a$, $J'_m(k'_{\rho mn}a) = 0$, where $k'_{\rho mn}a$ is the nth root of the derivative of the mth order Bessel function. The roots of $J'_m(k'_{\rho mn}a) = 0$ are tabulated in Table 3.4. Note that the propagation constant is given by $k_z = \sqrt{k^2 - (k'_{\rho mn})^2}$. The lowest cutoff wavenumber of the TE$_{mn}$ mode is $k'_{\rho 11} \approx \dfrac{1.841}{a}$, indicating that the lowest propagating mode among TE and TM waves is the TE$_{11}$ mode.

Table 3.4. Values of $k'_{\rho mn}a$

	m=0	m=1	m=2	m=3
n=1	3.832	1.841	3.054	4.201
n=2	7.016	5.331	6.706	8.015
n=3	10.173	8.536	9.969	11.346

3.5 Circular Dielectric Waveguide

Consider electromagnetic wave guidance by an infinitely long circular dielectric waveguide, as shown in Fig. 3.5. A circular waveguide (μ_1, ϵ_1) is embedded in a background medium (μ, ϵ), thus constituting a model for an optical fiber with a core and cladding. It is assumed that the circular dielectric waveguide is denser than the background medium ($\epsilon_1 > \epsilon$). Let us represent the propagating waves in regions (μ_1, ϵ_1) and (μ, ϵ) based on the boundary conditions. The

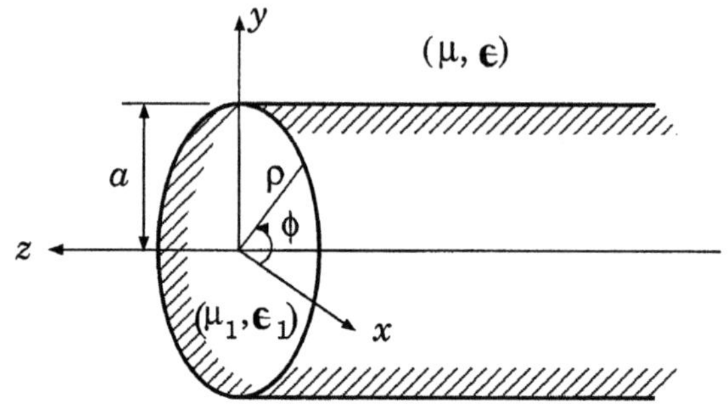

Fig. 3.5. Circular dielectric waveguide with radius a.

boundary conditions require that the four tangential electric and magnetic field (E_ϕ, E_z, H_ϕ, and H_z) continuities must be simultaneously satisfied at $\rho = a$. The four boundary conditions are available by assuming that both the E_z and H_z components are nonzero. Therefore, the guided wave is a hybrid mode that is a combination of TE ($H_z \neq 0$) and TM ($E_z \neq 0$) waves.

Inside the waveguide ($\rho \leq a$), both E_z (equivalently A_z) and H_z (equivalently F_z) satisfy the Helmholtz equations

$$(\nabla^2 + k_1^2) \begin{Bmatrix} A_z^I \\ F_z^I \end{Bmatrix} = 0 . \tag{3.77}$$

Hence, A_z^I and F_z^I take the form of

$$A_z^I = A J_m(k_\rho \rho) \cos m\phi \exp(ik_z z) \tag{3.78}$$

$$F_z^I = B J_m(k_\rho \rho) \sin m\phi \exp(ik_z z) \tag{3.79}$$

where the dispersion relation is

$$k_z^2 + k_\rho^2 = k_1^2 = \omega^2 \mu_1 \epsilon_1 . \tag{3.80}$$

The choice of the azimuthal field variation, $\cos m\phi$ and $\sin m\phi$, in A_z and F_z facilitates the enforcement of the boundary conditions later on.

Outside the waveguide ($\rho \geq a$), the fields are chosen as evanescent types that vanish at infinity $\rho \to \infty$. The vector potentials satisfy the Helmholtz equations

$$(\nabla^2 + k^2) \begin{Bmatrix} A_z^{II} \\ F_z^{II} \end{Bmatrix} = 0 . \tag{3.81}$$

The corresponding A_z^{II} and F_z^{II} are assumed to be

$$A_z^{II} = CK_m(\alpha_\rho \rho) \cos m\phi \, \exp(ik_z z) \tag{3.82}$$

$$F_z^{II} = DK_m(\alpha_\rho \rho) \sin m\phi \, \exp(ik_z z) \tag{3.83}$$

where K_m is the modified Bessel function of the second kind and the dispersion relation is

$$k_z^2 - \alpha_\rho^2 = k^2 = \omega^2 \mu\epsilon \, . \tag{3.84}$$

In order for the wave to be properly guided by the dielectric waveguide, the parameters k_z, k_ρ, and α_ρ must all be real numbers. The boundary conditions and field components must be used to find the propagation constant k_z. The field components in the respective media for $\rho \leq a$ and $\rho \geq a$ are straightforwardly available from Table 3.2 in Section 3.1.

The condition of E_ϕ continuity at $\rho = a$ then gives

$$\frac{mk_z}{\omega\mu_1\epsilon_1 a} J_m(k_\rho a) A + \frac{k_\rho}{\epsilon_1} J'_m(k_\rho a) B$$
$$= \frac{mk_z}{\omega\mu\epsilon a} K_m(\alpha_\rho a) C + \frac{\alpha_\rho}{\epsilon} K'_m(\alpha_\rho a) D \tag{3.85}$$

where the prime denotes differentiation with respect to the argument as $J'_m(k_\rho a) = \left.\frac{dJ_m(k_\rho \rho)}{d(k_\rho \rho)}\right|_{\rho=a}$.

Similarly, matching the remaining boundary conditions for the E_z, H_ϕ, and H_z continuities at $\rho = a$ produces the other three equations for A, B, C, and D. A set of the four simultaneous equations can be written in a matrix form

$$\begin{pmatrix} \frac{mk_z}{\omega\mu_1\epsilon_1 a} J_m(k_\rho a) & \frac{k_\rho}{\epsilon_1} J'_m(k_\rho a) & -\frac{mk_z}{\omega\mu\epsilon a} K_m(\alpha_\rho a) & -\frac{\alpha_\rho}{\epsilon} K'_m(\alpha_\rho a) \\ \frac{k_\rho^2}{\mu_1\epsilon_1} J_m(k_\rho a) & 0 & \frac{\alpha_\rho^2}{\mu\epsilon} K_m(\alpha_\rho a) & 0 \\ -\frac{k_\rho}{\mu_1} J'_m(k_\rho a) & -\frac{mk_z}{\omega\mu_1\epsilon_1 a} J_m(k_\rho a) & \frac{\alpha_\rho}{\mu} K'_m(\alpha_\rho a) & \frac{mk_z}{\omega\mu\epsilon a} K_m(\alpha_\rho a) \\ 0 & \frac{k_\rho^2}{\mu_1\epsilon_1} J_m(k_\rho a) & 0 & \frac{\alpha_\rho^2}{\mu\epsilon} K_m(\alpha_\rho a) \end{pmatrix}$$

$$\begin{pmatrix} A \\ B \\ C \\ D \end{pmatrix} = 0\,. \tag{3.86}$$

The dispersion relation is obtained by setting the determinant to zero as

$$\begin{vmatrix} \frac{p}{\epsilon_1} & J & -\frac{p}{\epsilon} & -K \\ \frac{u_1^2}{\epsilon_1} & 0 & \frac{u_2^2}{\epsilon} & 0 \\ -J & -\frac{p}{\mu_1} & K & \frac{p}{\mu} \\ 0 & \frac{u_1^2}{\mu_1} & 0 & \frac{u_2^2}{\mu} \end{vmatrix} = 0 \tag{3.87}$$

where $k_\rho a \equiv u_1$, $\alpha_\rho a \equiv u_2$, $\frac{mk_z}{\omega} \equiv p$, $J \equiv u_1 \frac{J'_m(u_1)}{J_m(u_1)}$, and $K \equiv u_2 \frac{K'_m(u_2)}{K_m(u_2)}$. The dispersion relation can be rewritten as

$$\frac{u_1^2}{\mu_1} \begin{vmatrix} \frac{p}{\epsilon_1} & -\frac{p}{\epsilon} & -K \\ \frac{u_1^2}{\epsilon_1} & \frac{u_2^2}{\epsilon} & 0 \\ -J & K & \frac{p}{\mu} \end{vmatrix} + \frac{u_2^2}{\mu} \begin{vmatrix} \frac{p}{\epsilon_1} & J & -\frac{p}{\epsilon} \\ \frac{u_1^2}{\epsilon_1} & 0 & \frac{u_2^2}{\epsilon} \\ -J & -\frac{p}{\mu_1} & K \end{vmatrix} = 0 \tag{3.88}$$

or

$$\left(\frac{u_1^2}{\mu_1} K + \frac{u_2^2}{\mu} J\right)\left(\frac{u_1^2}{\epsilon_1} K + \frac{u_2^2}{\epsilon} J\right) = \frac{p^2}{\mu_1 \epsilon_1 \mu \epsilon}\left(u_1^2 + u_2^2\right)^2 \tag{3.89}$$

or

$$\left[\frac{\mu}{u_2}\frac{K'_m(u_2)}{K_m(u_2)} + \frac{\mu_1}{u_1}\frac{J'_m(u_1)}{J_m(u_1)}\right]\left[\frac{\epsilon}{u_2}\frac{K'_m(u_2)}{K_m(u_2)} + \frac{\epsilon_1}{u_1}\frac{J'_m(u_1)}{J_m(u_1)}\right] = p^2\left(\frac{1}{u_1^2} + \frac{1}{u_2^2}\right)^2 . \tag{3.90}$$

In principle, the propagation constant k_z can be determined by solving the dispersion relation (3.90). Let us introduce the normalized frequency V as

$$V^2 = u_1^2 + u_2^2 = \left(k_1^2 - k^2\right) a^2 \,. \tag{3.91}$$

An alternative way of determining k_z is to solve the simultaneous equations (3.90) and (3.91) graphically. Plotting (3.90) and (3.91) on the u_1-u_2 plane and identifying intersections of two curves yields the solution k_z. Note that (3.91) represents a circle of radius V on the u_1-u_2 plane. For each index m, there are a number of solutions $k_z = k_{zmn}$. Once k_{zmn} is determined, the ratios of the field amplitudes A, B, C, and D can subsequently be obtained from (3.86). The hybrid waves are designated as HE_{mn} or EH_{mn} modes. The HE_{mn} mode refers to when the H_z component is dominant, while the EH_{mn} mode refers to when the E_z component is dominant.

Table 3.5. Mode cutoff frequencies

Modes	$J_m(u_1) = 0$	u_1
HE_{11}	$J_1(u_1) = 0$	$u_1 = 0$
$\mathrm{TE}_{01}, \mathrm{TM}_{01}, \mathrm{HE}_{21}$	$J_0(u_1) = 0$	$u_1 = 2.4048$

Table 3.5 shows that the HE_{11} mode is the lowest order with zero cutoff frequency. Single-mode HE_{11} propagation along a dielectric circular waveguide is possible if $0 < u_1 < 2.4048$. Considering the constraint (3.91), the radius of circle V on the u_1-u_2 plane must be $V < 2.4048$ to guarantee single-mode HE_{11} propagation. Hence, the condition for single-mode propagation is written as

$$0 < \sqrt{k_1^2 - k^2 a} < 2.4048. \tag{3.92}$$

Consider a practical single-mode silica fiber that operates at 1.55 (μm), where the refractive indices for the core and cladding are $n_1 = 1.46$ and $n = 1.46 - 0.002$, respectively. The condition (3.92) shows that the core radius a must be $a < 7.77$ (μm) to support single-mode propagation.

3.6 Shielded Stripline

Due to recent advances in integrated-circuit technologies, planar transmission lines such as striplines are now commonly used in microwave circuits. This section investigates a shielded stripline, as shown in Fig. 3.6. Regions (I), (II),

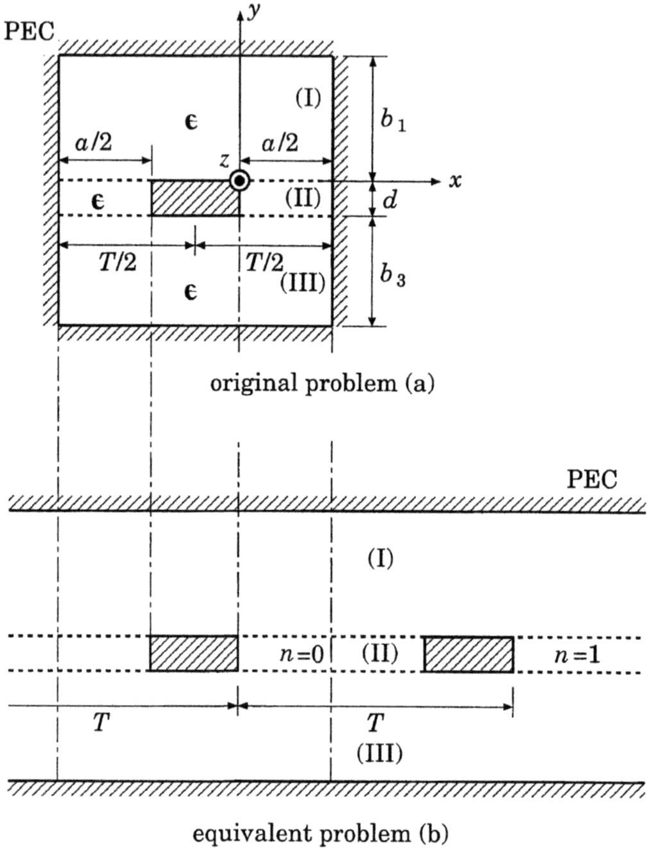

Fig. 3.6. Cross section of shielded stripline.

and (III) are filled with a dielectric material and a conducting strip is embedded in the center of region (II). Regions (I), (II), and (III) are covered with PEC shielding walls. When a voltage source is applied between the conducting strip and the shielding plane, an electromagnetic wave propagates along the z-direction through regions (I), (II), and (III). For the analysis of the shielded transmission lines, it is possible to use the image method and Fourier transform to obtain rigorous series solutions [3]. The image method and Fourier transform are utilized in this section to obtain an analytic series solution for a shielded stripline. Although the derivation procedure is somewhat involved, the final series solution for the dispersion is fast convergent, and is thus useful for practical application.

3.6.1 Field Representations

Consider the original problem of Fig. 3.6 (a), where regions (I), (II), and (III) are filled with a dielectric material of permittivity ϵ. Due to the electric walls placed at $x = \frac{a}{2}$ and $\frac{a}{2} - T$, it is possible to apply the image method to transform the problem in Fig. 3.6 (a) into the equivalent problem in Fig. 3.6 (b). Figure 3.6 (b) illustrates the problem geometry of an infinite number of striplines that are placed between two PEC parallel plates. Since regions (I) through (III) are homogeneous, either a TM or TE wave, transverse to the z-direction, propagates in regions (I) through (III) in Fig. 3.6 (b). This section investigates TE wave propagation ($E_z = 0$ and $H_z \neq 0$).

The electric vector potential is necessary for field representation as $F_z(x, y, z) = F_z(x, y) \exp(ik_z z)$. The vector potentials should satisfy the Helmholtz equations

$$\left(\nabla^2 + k^2\right) F_z^{I,II,III}(x, y, z) = 0 \tag{3.93}$$

where $k = \omega\sqrt{\mu\epsilon}$. In each region, the guided waves are written as

$$F_z^I(x, y) = \frac{1}{2\pi} \int_{-\infty}^{\infty} \left(\tilde{F}_I^+ e^{i\eta y} + \tilde{F}_I^- e^{-i\eta y}\right) e^{-i\zeta x}\, d\zeta \tag{3.94}$$

$$\begin{aligned} F_z^{II}(x, y) = & \sum_{m=0,2,4\cdots}^{\infty} \left[S_m^n \cos(\xi_m y) + Q_m^n \sin(\xi_m y)\right] \\ & \cdot \cos a_m (x - nT) \end{aligned} \tag{3.95}$$

$$F_z^{III}(x, y) = \frac{1}{2\pi} \int_{-\infty}^{\infty} \left[\tilde{F}_{III}^+ e^{i\eta(y+d)} + \tilde{F}_{III}^- e^{-i\eta(y+d)}\right] e^{-i\zeta x}\, d\zeta \tag{3.96}$$

where $a_m = \frac{m\pi}{a}$, $\xi_m = \sqrt{k^2 - a_m^2 - k_z^2}$, $\eta = \sqrt{k^2 - \zeta^2 - k_z^2}$, and $n =$ integers. Note that the constraint ($m = 0, 2, 4, \cdots$) is required to make the tangential electric field in region (II) vanish at the vertical PEC walls as

$$E_y^{II}\left(\frac{a}{2} + nT, y\right) = 0\,. \tag{3.97}$$

In view of Table 3.1 in Section 3.1, the fields in the respective regions can be obtained in terms of the vector potentials. In region (I)

$$E_x^I(x, y) = -\frac{1}{\epsilon}\frac{1}{2\pi} \int_{-\infty}^{\infty} i\eta \left(\tilde{F}_I^+ e^{i\eta y} - \tilde{F}_I^- e^{-i\eta y}\right) e^{-i\zeta x}\, d\zeta \tag{3.98}$$

$$H_x^I(x, y) = \frac{k_z}{\omega\mu\epsilon}\frac{1}{2\pi} \int_{-\infty}^{\infty} i\zeta \left(\tilde{F}_I^+ e^{i\eta y} + \tilde{F}_I^- e^{-i\eta y}\right) e^{-i\zeta x}\, d\zeta \tag{3.99}$$

$$H_z^I(x, y) = \frac{i}{\omega\mu\epsilon} \left(-k_z^2 + k^2\right) F_z^I(x, y)\,. \tag{3.100}$$

In region (II)

$$E_x^{II}(x,y) = -\frac{1}{\epsilon} \sum_{m=0,2,4\cdots}^{\infty} \xi_m \left[-S_m^n \sin(\xi_m y) + Q_m^n \cos(\xi_m y)\right] \cdot \cos a_m(x - nT) \tag{3.101}$$

$$H_x^{II}(x,y) = \frac{k_z}{\omega\mu\epsilon} \sum_{m=0,2,4\cdots}^{\infty} a_m \left[S_m^n \cos(\xi_m y) + Q_m^n \sin(\xi_m y)\right] \cdot \sin a_m(x - nT) \tag{3.102}$$

$$H_z^{II}(x,y) = \frac{i}{\omega\mu\epsilon}\left(-k_z^2 + k^2\right) F_z^{II}(x,y)\,. \tag{3.103}$$

In region (III)

$$E_x^{III}(x,y) = -\frac{1}{\epsilon}\frac{1}{2\pi}\int_{-\infty}^{\infty} i\eta \cdot \left[\tilde{F}_{III}^{+} e^{i\eta(y+d)} - \tilde{F}_{III}^{-} e^{-i\eta(y+d)}\right] e^{-i\zeta x}\, d\zeta \tag{3.104}$$

$$H_x^{III}(x,y) = \frac{k_z}{\omega\mu\epsilon}\frac{1}{2\pi}\int_{-\infty}^{\infty} i\zeta \cdot \left[\tilde{F}_{III}^{+} e^{i\eta(y+d)} + \tilde{F}_{III}^{-} e^{-i\eta(y+d)}\right] e^{-i\zeta x}\, d\zeta \tag{3.105}$$

$$H_z^{III}(x,y) = \frac{i}{\omega\mu\epsilon}\left(-k_z^2 + k^2\right) F_z^{III}(x,y)\,. \tag{3.106}$$

Since the number of striplines is infinite, the modal coefficients S_m^n and Q_m^n should be independent of the stripline location n. The shielded stripline in the original problem can have either an even or odd excitation. When even (or odd), the field configuration becomes even (or odd) with respect to the center $\left(x = \frac{a}{2} - \frac{T}{2}\right)$ of the shielded stripline. The even and odd fields are illustrated in Fig. 3.7. Therefore, it is possible to let $S_m^n \to (\pm 1)^n s_m$ and $Q_m^n \to (\pm 1)^n q_m$ for the even $[(+1)^n]$ and odd $[(-1)^n]$ excitations.

3.6.2 Boundary Conditions

To determine the propagation constant k_z, matching the boundary conditions is necessary. The vector potential expressions (3.94) through (3.96) employ six unknown modal coefficients–$\tilde{F}_I^{\pm}$, q_m, s_m, and $\tilde{F}_{III}^{\pm}$. This means that six simultaneous equations must be otained by matching the boundary conditions. The boundary conditions require that the tangential electric and magnetic

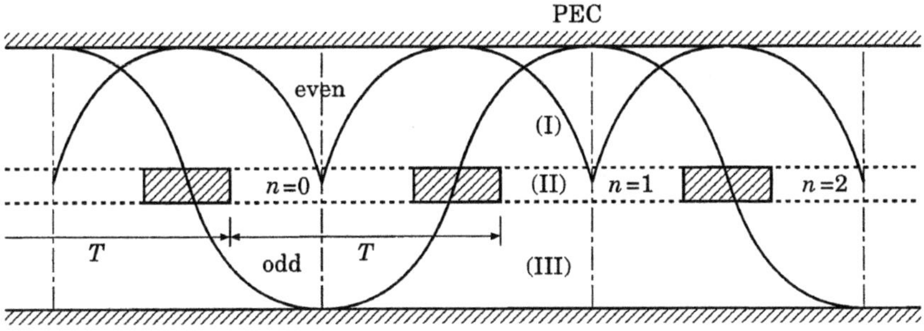

Fig. 3.7. Even and odd field configurations.

fields must be continuous at the boundaries $y = b_1$, 0, $-d$, and $(-d - b_3)$. The boundary conditions to be enforced at each boundary are given as

$$E_x^I(x, b_1) = 0 \tag{3.107}$$

$$E_x^I(x,0) = \begin{cases} E_x^{II}(x,0) & \text{for } nT < x < nT + a \\ 0 & \text{otherwise} \end{cases} \tag{3.108}$$

$$H_z^I(x,0) = H_z^{II}(x,0) \tag{3.109}$$

$$E_x^{III}(x, -d - b_3) = 0 \tag{3.110}$$

$$E_x^{III}(x,-d) = \begin{cases} E_x^{II}(x,-d) & \text{for } nT < x < nT + a \\ 0 & \text{otherwise} \end{cases} \tag{3.111}$$

$$H_z^{II}(x,-d) = H_z^{III}(x,-d) \ . \tag{3.112}$$

Note that the boundary conditions, $H_x^I(x,0) = H_x^{II}(x,0)$ and $H_x^{II}(x,-d) = H_x^{III}(x,-d)$, are automatically satisfied due to (3.109) and (3.112). It is necessary to obtain a set of simultaneous equations for the modal coefficients q_m and s_m.

- The condition $E_x^I(x, b_1) = 0$ gives

$$\tilde{F}_I^- = \tilde{F}_I^+ e^{i2\eta b_1} \ . \tag{3.113}$$

- Next consider the condition

$$E_x^I(x,0) = \begin{cases} E_x^{II}(x,0) & \text{for } nT < x < nT + a \\ 0 & \text{otherwise} \end{cases} \tag{3.114}$$

which can be written as

$$\frac{1}{2\pi}\int_{-\infty}^{\infty} i\eta\left(\tilde{F}_I^+ - \tilde{F}_I^-\right) e^{-i\zeta x}\, d\zeta$$

$$= \begin{cases} \sum_{m=0,2,4\cdots}^{\infty} \xi_m (\pm 1)^n q_m & \\ \cdot \cos a_m (x - nT) & \text{for } nT < x < nT + a \\ 0 & \text{otherwise}\,. \end{cases} \tag{3.115}$$

Applying the inverse Fourier transform $\left[\frac{1}{2\pi}\int_{-\infty}^{\infty} (\cdot) e^{i\zeta x}\, dx\right]$ to (3.115) yields

$$\begin{aligned} & i\eta\left(\tilde{F}_I^+ - \tilde{F}_I^-\right) \\ &= \sum_{n=-\infty}^{\infty} \sum_{m=0,2,4\cdots}^{\infty} \xi_m (\pm 1)^n q_m \int_{nT}^{nT+a} \cos a_m (x - nT) e^{i\zeta x}\, dx \\ &= \sum_{n=-\infty}^{\infty} \sum_{m=0,2,4\cdots}^{\infty} \xi_m (\pm 1)^n q_m G_m^n(\zeta) \end{aligned} \tag{3.116}$$

where

$$G_m^n(\zeta) = \frac{i\zeta\left(1 - e^{i\zeta a}\right)}{\zeta^2 - a_m^2} e^{i\zeta nT}\,. \tag{3.117}$$

Solving (3.113) and (3.116) for $\tilde{F}_I^\pm$ gives

$$\tilde{F}_I^+ = \sum_{n=-\infty}^{\infty} \sum_{m=0,2,4,\cdots}^{\infty} \frac{1}{i\eta\left(1 - e^{i2\eta b_1}\right)} (\pm 1)^n q_m \xi_m G_m^n(\zeta) \tag{3.118}$$

$$\tilde{F}_I^- = \sum_{n=-\infty}^{\infty} \sum_{m=0,2,4,\cdots}^{\infty} \frac{e^{i2\eta b_1}}{i\eta\left(1 - e^{i2\eta b_1}\right)} (\pm 1)^n q_m \xi_m G_m^n(\zeta)\,. \tag{3.119}$$

It is possible to show that the tangential electric field in region (I) vanishes at the vertical PEC walls $\left[E_y^I\left(\frac{a}{2} + pT, y\right) = 0\right]$ using $\tilde{F}_I^+$ and $\tilde{F}_I^-$ as follows:

$$E_y^I\left(\frac{a}{2}+pT,y\right) = \frac{1}{\epsilon}\frac{\partial F_z^I(x,y)}{\partial x}\bigg|_{x=\frac{a}{2}+pT}$$

$$= -\frac{i}{2\pi\epsilon}\int_{-\infty}^{\infty}\zeta\left(\tilde{F}_I^+ e^{i\eta y}+\tilde{F}_I^- e^{-i\eta y}\right)e^{-i\zeta(\frac{a}{2}+pT)}\,d\zeta$$

$$= -\frac{i}{2\pi\epsilon}\sum_{m=0,2,4,\cdots}^{\infty} q_m\xi_m$$

$$\cdot\int_{-\infty}^{\infty}\left[\sum_{n=-\infty}^{\infty}\frac{e^{i\eta y}+e^{i2\eta b_1}e^{-i\eta y}}{i\eta\left(1-e^{i2\eta b_1}\right)}\right.$$

$$\left.\cdot(\pm 1)^n\zeta G_m^n(\zeta)e^{-i\zeta(\frac{a}{2}+pT)}\right]d\zeta\,. \qquad (3.120)$$

Since the integrand $[\,\cdot\,]$ is an odd function of ζ, the condition $E_y^I\left(\frac{a}{2}+pT,y\right)$ $=0$ is satisfied irrespective of y.

- Consider the condition for $pT<x<pT+a$

$$H_z^I(x,0)=H_z^{II}(x,0) \qquad (3.121)$$

which can be written as

$$\frac{1}{2\pi}\int_{-\infty}^{\infty}\left(\tilde{F}_I^+ + \tilde{F}_I^-\right)e^{-i\zeta x}\,d\zeta$$

$$=\sum_{m=0,2,4\cdots}^{\infty}(\pm 1)^p s_m\cos a_m(x-pT)\,. \qquad (3.122)$$

Substituting

$$\tilde{F}_I^+ + \tilde{F}_I^- = \sum_{n=-\infty}^{\infty}\sum_{m=0,2,4,\cdots}^{\infty}\frac{1}{\eta\tan\eta b_1}(\pm 1)^n q_m\xi_m G_m^n(\zeta) \qquad (3.123)$$

into (3.122), multiplying (3.122) by $\cos a_l(x-pT)$, and integrating over $pT<x<pT+a$ yields

$$\sum_{n=-\infty}^{\infty}\sum_{m=0,2,4,\cdots}^{\infty}(\pm 1)^n q_m\xi_m$$

$$\cdot\frac{1}{2\pi}\int_{-\infty}^{\infty}\frac{1}{\eta\tan\eta b_1}G_m^n(\zeta)\left[\int_{pT}^{pT+a}\cos a_l(x-pT)e^{-i\zeta x}\,dx\right]d\zeta$$

$$=\sum_{m=0,2,4\cdots}^{\infty}(\pm 1)^p s_m$$

$$\cdot\left[\int_{pT}^{pT+a}\cos a_l(x-pT)\cos a_m(x-pT)\,dx\right]\,. \qquad (3.124)$$

Note

$$\int_{pT}^{pT+a} \cos a_l(x-pT)e^{-i\zeta x}\,dx = G_l^p(-\zeta) \tag{3.125}$$

$$\int_{pT}^{pT+a} \cos a_l(x-pT)\cos a_m(x-pT)\,dx$$

$$= \delta_{ml}\varepsilon_m\frac{a}{2} \tag{3.126}$$

where δ_{ml} is the Kronecker delta, $\varepsilon_0 = 2$, and $\varepsilon_m = 1$ $(m = 2,4,6\cdots)$. Therefore, (3.124) becomes

$$\sum_{n=-\infty}^{\infty}\sum_{m=0,2,4,\cdots}^{\infty} (\pm 1)^n q_m\xi_m\frac{1}{2\pi}\int_{-\infty}^{\infty}\frac{1}{\eta\tan\eta b_1}G_m^n(\zeta)G_l^p(-\zeta)\,d\zeta$$

$$= (\pm 1)^p s_l\varepsilon_l\frac{a}{2}\,. \tag{3.127}$$

where $l = 0,2,4,\cdots$ and $p =$ integers. Therefore, (3.127) is rewritten as

$$(\pm 1)^p s_l\frac{a}{2}\varepsilon_l - \sum_{m=0,2,4\cdots}^{\infty} q_m\xi_m J_1^{\pm} = 0 \tag{3.128}$$

where

$$J_1^{\pm} = \sum_{n=-\infty}^{\infty}\frac{(\pm 1)^n}{2\pi}\int_{-\infty}^{\infty}\frac{1}{\eta\tan(\eta b_1)}G_m^n(\zeta)G_l^p(-\zeta)\,d\zeta\,. \tag{3.129}$$

$J_1^{\pm}$ can be transformed into a fast convergent series based on residue calculus. The detailed derivation of $J_1^{\pm}$ is given in the next subsection.

- The condition $E_x^{III}(x,-d-b_3) = 0$ yields

$$\tilde{F}_{III}^{-} = \tilde{F}_{III}^{+}e^{-i2\eta b_3}\,. \tag{3.130}$$

- Consider the boundary condition (3.111). Taking the inverse Fourier transform of (3.111) results in

$$\tilde{F}_{III}^{+} - \tilde{F}_{III}^{-}$$

$$= \sum_{n=-\infty}^{\infty}\sum_{m=0,2,4\cdots}^{\infty}\frac{\xi_m}{i\eta}$$

$$\cdot\left[(\pm 1)^n s_m\sin(\xi_m d) + (\pm 1)^n q_m\cos(\xi_m d)\right]G_m^n(\zeta)\,. \tag{3.131}$$

- Consider the boundary condition $H_z^{II}(x,-d) = H_z^{III}(x,-d)$, which is rewritten as

$$\sum_{m=0,2,4\cdots}^{\infty} \left[(\pm 1)^p s_m \cos(\xi_m d) - (\pm 1)^p q_m \sin(\xi_m d)\right]$$
$$\cdot \cos a_m(x - pT)$$

$$= \frac{1}{2\pi} \int_{-\infty}^{\infty} \left(\tilde{F}_{III}^{+} + \tilde{F}_{III}^{-}\right) e^{-i\zeta x}\, d\zeta \,. \tag{3.132}$$

Substituting

$$\tilde{F}_{III}^{+} + \tilde{F}_{III}^{-} = - \sum_{n=-\infty}^{\infty} \sum_{m=0,2,4,\cdots}^{\infty} \frac{1}{\eta \tan \eta b_3} \xi_m \left[(\pm 1)^n s_m \sin(\xi_m d) + (\pm 1)^n q_m \cos(\xi_m d)\right] \cdot G_m^n(\zeta) \tag{3.133}$$

into (3.132), multiplying (3.132) by $\cos a_l(x - pT)$, and integrating over $pT < x < pT + a$ yields

$$(\pm 1)^p \left[s_l \cos(\xi_l d) - q_l \sin(\xi_l d)\right] \frac{a}{2} \varepsilon_l + \sum_{m=0,2,4\cdots}^{\infty} \xi_m \left[s_m \sin(\xi_m d) + q_m \cos(\xi_m d)\right] J_3^{\pm} = 0 \tag{3.134}$$

where

$$J_3^{\pm} = \sum_{n=-\infty}^{\infty} \frac{(\pm 1)^n}{2\pi} \int_{-\infty}^{\infty} \frac{1}{\eta \tan(\eta b_3)} G_m^n(\zeta) G_l^p(-\zeta)\, d\zeta \,. \tag{3.135}$$

To determine the propagation constant k_z, the dispersion relation must be obtained by setting the determinant of the simultaneous equations (3.128) and (3.134) to zero. In particular, when the strip width $d \to 0$, the dispersion relation is given by

$$\left|\xi_m (J_1^{\pm} + J_3^{\pm})\right| = 0 \tag{3.136}$$

where the sign $\pm$ corresponds to the even and odd excitations, respectively. Although the preceding derivation of the dispersion relation is rather lengthy, its final expression (3.136) is remarkably simple and thus efficient for numerical evaluation.

3.6.3 Residue Calculus for $J_q^{\pm}$

Let us transform the integral $J_q^{\pm} (q = 1, 3)$ into a fast convergent series based on residue calculus. The integral J_q is given by

$$J_q^{\pm} = \sum_{n=-\infty}^{\infty} \frac{(\pm 1)^n}{2\pi} \int_{-\infty}^{\infty} \frac{1}{\eta \tan(\eta b_q)} G_m^n(\zeta) G_l^p(-\zeta) d\zeta \tag{3.137}$$

where $G_m^n(\zeta) = \frac{i\zeta\left(1 - e^{i\zeta a}\right)}{\zeta^2 - a_m^2} e^{i\zeta nT}$, $a_m = \frac{m\pi}{a}$, and $\eta = \sqrt{k^2 - \zeta^2 - k_z^2}$.

Let

$$J_q^{\pm} = \sum_{n=-\infty}^{\infty} (\pm 1)^n I_q \tag{3.138}$$

where

$$I_q = \frac{1}{2\pi} \int_{-\infty}^{\infty} \frac{1}{\eta \tan(\eta b_q)} \frac{\zeta^2 \left(2 - e^{i\zeta a} - e^{-i\zeta a}\right)}{(\zeta^2 - a_m^2)(\zeta^2 - a_l^2)} e^{i\zeta T(n-p)} d\zeta$$
$$\equiv \int_{-\infty}^{\infty} f(\zeta)\, d\zeta \,. \tag{3.139}$$

Let us evaluate I_q for three different cases, namely, $n - p = 0$, $n - p > 0$, and $n - p < 0$.

- Consider I_q when $n = p$. It is more convenient to transform I_q into

$$I_q = \frac{1}{2\pi} \int_{-\infty}^{\infty} \frac{1}{\eta \tan(\eta b_q)} \frac{2\zeta^2 \left(1 - e^{i\zeta a}\right)}{(\zeta^2 - a_m^2)(\zeta^2 - a_l^2)} d\zeta$$
$$\equiv \int_{-\infty}^{\infty} f(\zeta)\, d\zeta \,. \tag{3.140}$$

Note that $f(\zeta)$ contains simple poles at $\zeta = \pm a_m$ when $m = l$ and other simple poles at $\zeta = \pm\zeta_v = \pm\sqrt{k^2 - \left(\frac{v\pi}{b_q}\right)^2 - k_z^2}$, which are the solutions to the characteristic equation $\eta \tan(\eta b_q) = 0$.

Consider the contour integral along the closed path Γ, which consists of Γ_1, Γ_2, Γ_3, and Γ_4 in the complex ζ-plane with $R \to \infty$, as shown in Fig. 3.8. The paths Γ_2, Γ_3, and Γ_4 denote semicircles. Assume that an imaginary part of the wavenumber k is slightly positive. The imaginary part of $+\zeta_v$ is therefore always positive. The residue theorem gives

$$\oint_{\Gamma} f(\zeta)\, d\zeta = 2\pi i \sum_{v=0}^{\infty} \operatorname*{Res}_{\zeta=\zeta_v} f(\zeta) \tag{3.141}$$

where

$$2\pi i \sum_{v=0}^{\infty} \operatorname*{Res}_{\zeta=\zeta_v} f(\zeta)$$
$$= i \sum_{v=0}^{\infty} \frac{2\zeta^2 (1 - e^{i\zeta a})}{(\zeta^2 - a_m^2)(\zeta^2 - a_l^2)} \frac{(\zeta - \zeta_v)}{\eta \tan(\eta b_q)}\bigg|_{\zeta \to \zeta_v} \,. \tag{3.142}$$

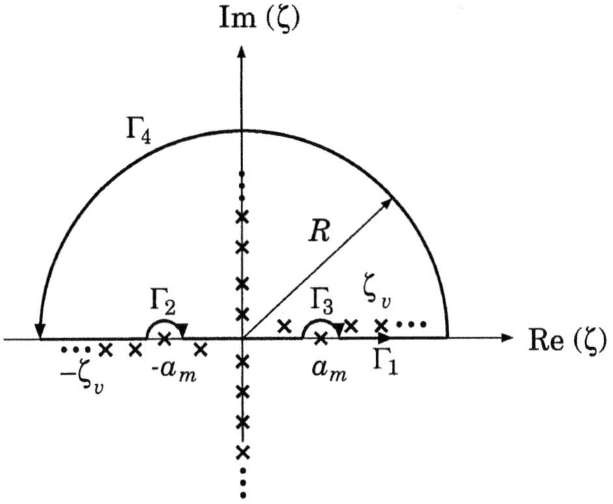

Fig. 3.8. Complex ζ-plane with simple poles.

L'Hôpital's rule gives

$$\left.\frac{(\zeta-\zeta_v)}{\eta\tan(\eta b_q)}\right|_{\zeta\to\zeta_v} = \left.\frac{1}{\dfrac{d}{d\zeta}\left[\eta\tan(\eta b_q)\right]}\right|_{\zeta\to\zeta_v} = -\frac{1}{\varepsilon_v\zeta_v b_q}\,. \tag{3.143}$$

Hence, the residue is written as

$$2\pi i\sum_{v=0}^{\infty}\operatorname*{Res}_{\zeta=\zeta_v} f(\zeta) = -\frac{i}{b_q}\sum_{v=0}^{\infty}\frac{2\zeta_v(1-e^{i\zeta_v a})}{\varepsilon_v(\zeta_v^2-a_m^2)(\zeta_v^2-a_l^2)}\,. \tag{3.144}$$

In addition, the contour integrations along the path Γ are necessary. Consider

$$\int_{\Gamma_1} f(\zeta)\,d\zeta = I_q \tag{3.145}$$

$$\int_{\Gamma_4} f(\zeta)\,d\zeta = 0\,. \tag{3.146}$$

For $m = 2,4,6,\cdots$, two different simple poles at $\zeta = \pm a_m$ exist. Their residue contributions are

$$\int_{\Gamma_2} f(\zeta)\,d\zeta = -\pi i \delta_{ml} \operatorname*{Res}_{\zeta=-a_m} f(\zeta)$$
$$= -\frac{i}{2}\delta_{ml}\frac{1}{\eta\tan(\eta b_q)}\frac{2\zeta^2}{(\zeta-a_m)^2}\frac{(1-e^{i\zeta a})}{\zeta+a_m}\bigg|_{\zeta\to -a_m}$$
$$= -\delta_{ml}\frac{a}{4\eta\tan(\eta b_q)}\bigg|_{\zeta\to -a_m} . \tag{3.147}$$

$$\int_{\Gamma_3} f(\zeta)\,d\zeta = -\pi i \delta_{ml} \operatorname*{Res}_{\zeta=a_m} f(\zeta)$$
$$= -\delta_{ml}\frac{a}{4\eta\tan(\eta b_q)}\bigg|_{\zeta\to a_m} . \tag{3.148}$$

For $m = 0$, a simple pole at $\zeta = 0$ exists. Its residue contribution is

$$\int_{\Gamma_2} f(\zeta)\,d\zeta = -\pi i \delta_{0l} \operatorname*{Res}_{\zeta=0} f(\zeta)$$
$$= -\frac{i}{2}\delta_{0l}\frac{1}{\eta\tan(\eta b_q)}\frac{2\zeta^2(1-e^{i\zeta a})}{\zeta^3}\bigg|_{\zeta\to 0}$$
$$= -\delta_{0l}\frac{a}{\eta\tan(\eta b_q)}\bigg|_{\zeta\to 0} . \tag{3.149}$$

Substituting (3.144) through (3.149) into (3.141) yields

$$I_q = \delta_{ml}\frac{\varepsilon_m a}{2\eta\tan(\eta b_q)}\bigg|_{\zeta\to a_m} - \frac{i}{b_q}\sum_{v=0}^{\infty}\frac{2\zeta_v(1-e^{i\zeta_v a})}{\varepsilon_v(\zeta_v^2-a_m^2)(\zeta_v^2-a_l^2)} . \tag{3.150}$$

- Consider I_q when $n > p$. The complex ζ-plane is shown in Fig. 3.9. Note that the function $f(\zeta)$ contains simple poles at $\pm\zeta_v = \pm\sqrt{k^2 - \left(\frac{v\pi}{b_q}\right)^2 - k_z^2}$, which are the solutions to the equation $\eta\tan(\eta b_q) = 0$. Thus

$$\int_{\Gamma_1} f(\zeta)\,d\zeta + \int_{\Gamma_2} f(\zeta)\,d\zeta = 2\pi i\sum_{v=0}^{\infty}\operatorname*{Res}_{\zeta=\zeta_v} f(\zeta) . \tag{3.151}$$

Since

$$\int_{\Gamma_1} f(\zeta)\,d\zeta = I_q \tag{3.152}$$

$$\int_{\Gamma_2} f(\zeta)\,d\zeta = 0 \tag{3.153}$$

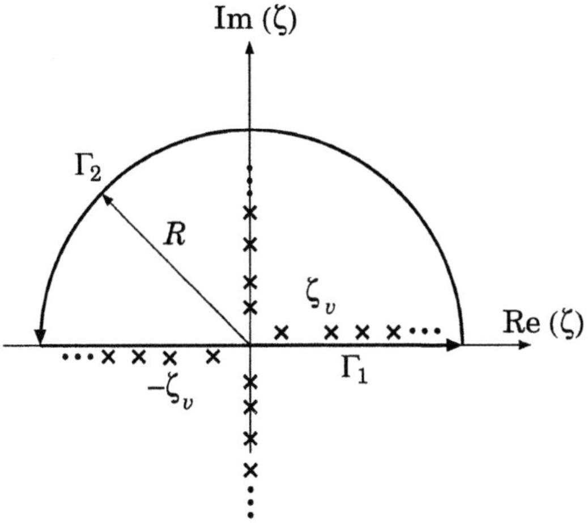

Fig. 3.9. Complex ζ-plane with simple poles.

I_q is given by

$$I_q = 2\pi i \sum_{v=0}^{\infty} \operatorname*{Res}_{\zeta=\zeta_v} f(\zeta)$$

$$= -\frac{i}{b_q} \sum_{v=0}^{\infty} \frac{\zeta_v \left(2 - e^{i\zeta_v a} - e^{-i\zeta_v a}\right)}{\varepsilon_v(\zeta_v^2 - a_m^2)(\zeta_v^2 - a_l^2)} e^{i\zeta_v T(n-p)} . \qquad (3.154)$$

- Consider I_q when $n < p$. Let us choose a semicircular path in the lower half-plane and perform residue calculus. The result is

$$I_q = -2\pi i \sum_{v=0}^{\infty} \operatorname*{Res}_{\zeta=-\zeta_v} f(\zeta)$$

$$= -\frac{i}{b_q} \sum_{v=0}^{\infty} \frac{\zeta_v \left(2 - e^{-i\zeta_v a} - e^{i\zeta_v a}\right)}{\varepsilon_v(\zeta_v^2 - a_m^2)(\zeta_v^2 - a_l^2)} e^{-i\zeta_v T(n-p)} . \qquad (3.155)$$

The expressions (3.150), (3.154), and (3.155) can be combined into

$$I_q = \delta_{ml}\delta_{np} \left. \frac{\varepsilon_m a}{2\eta \tan(\eta b_q)} \right|_{\zeta \to a_m} - \frac{i}{b_q} \sum_{v=0}^{\infty} \frac{\zeta_v A_n}{\varepsilon_v(\zeta_v^2 - a_m^2)(\zeta_v^2 - a_l^2)} \qquad (3.156)$$

where

$$A_n = 2e^{i\zeta_v T|n-p|} - e^{i\zeta_v |T(n-p)+a|} - e^{i\zeta_v |T(n-p)-a|} . \tag{3.157}$$

Substituting (3.156) into (3.138) gives

$$J_q^{\pm} = \sum_{n=-\infty}^{\infty} (\pm 1)^n \left[\delta_{ml}\delta_{np} \frac{\varepsilon_m a}{2\eta \tan(\eta b_q)} \bigg|_{\zeta \to a_m} - \frac{i}{b_q} \sum_{v=0}^{\infty} \frac{\zeta_v A_n}{\varepsilon_v (\zeta_v^2 - a_m^2)(\zeta_v^2 - a_l^2)} \right] . \tag{3.158}$$

Summing series over n in $J_q^{\pm}$ yields

$$\sum_{n=-\infty}^{\infty} (\pm 1)^n e^{i\zeta_v T|n-p|} = \sum_{n=-\infty}^{\infty} (\pm 1)^{(n+p)} e^{i\zeta_v T|n|} = (\pm 1)^p \frac{1 \pm e^{i\zeta_v T}}{1 \mp e^{i\zeta_v T}} \tag{3.159}$$

$$\sum_{n=-\infty}^{\infty} (\pm 1)^n e^{i\zeta_v |T(n-p)+a|} = (\pm 1)^p \frac{e^{i\zeta_v a} \pm e^{i\zeta_v (T-a)}}{1 \mp e^{i\zeta_v T}} \tag{3.160}$$

$$\sum_{n=-\infty}^{\infty} (\pm 1)^n e^{i\zeta_v |T(n-p)-a|} = (\pm 1)^p \frac{e^{i\zeta_v a} \pm e^{i\zeta_v (T-a)}}{1 \mp e^{i\zeta_v T}} . \tag{3.161}$$

Therefore

$$\sum_{n=-\infty}^{\infty} (\pm 1)^n A_n = (\pm 1)^p \frac{2\left[1 \pm e^{i\zeta_v T} - e^{i\zeta_v a} \mp e^{i\zeta_v (T-a)}\right]}{1 \mp e^{i\zeta_v T}} . \tag{3.162}$$

Substituting (3.162) into (3.158) gives the final series expression for $J_q^{\pm}$ as

$$J_q^{\pm} = (\pm 1)^p \left\{ \delta_{ml} \frac{\varepsilon_m a}{2\eta \tan(\eta b_q)} \bigg|_{\zeta \to a_m} - \frac{i}{b_q} \sum_{v=0}^{\infty} \frac{\zeta_v}{\varepsilon_v (\zeta_v^2 - a_m^2)(\zeta_v^2 - a_l^2)} \cdot \frac{2\left[1 \pm e^{i\zeta_v T} - e^{i\zeta_v a} \mp e^{i\zeta_v (T-a)}\right]}{1 \mp e^{i\zeta_v T}} \right\} . \tag{3.163}$$

Problems for Chapter 3

1. The TE_{mn} mode propagates along a rectangular waveguide with dimensions $(a \times b)$, as shown in Fig. 3.2. Compute the power delivered by the TE_{mn} mode.

2. Consider the dielectric slab waveguide shown in Fig. 3.3. The dielectric medium in region (III) is replaced with a perfect magnetic conductor (PMC). Determine the dispersion relation of a TE wave.
3. Construct (3.86) using the boundary conditions for the field continuities.
4. Derive (3.134) from (3.132) and (3.133).
5. Derive (3.155) using residue calculus.
6. Consider the shielded stripline shown in Fig. 3.6. Assume TM wave propagation in the z-direction ($H_z = 0$ and $E_z \neq 0$). Following the TE wave analysis in Section 3.6, derive the dispersion relation for a TM wave.

4

Cavity Resonators and Coupler

4.1 Rectangular Cavity Resonator

Resonant cavities are basic microwave components that store electromagnetic energy. Microwave resonant cavities are known to have large quality factors. A rectangular cavity resonator is relatively easy to analyze, yet it provides physical insight into the resonance mechanism. This section investigates wave resonance in a rectangular resonant cavity that is surrounded by electric walls.

4.1.1 TE Mode

Consider a rectangular cavity resonator that is surrounded by perfectly conducting walls, as shown in Fig. 4.1. A resonant cavity is filled with a dielectric medium with permeability μ and permittivity ϵ. A rectangular cavity resonator of dimensions $(a \times b \times d)$ is viewed as a rectangular waveguide of dimensions $(a \times b)$ with perfectly conducting walls placed at $z = 0$ and d.

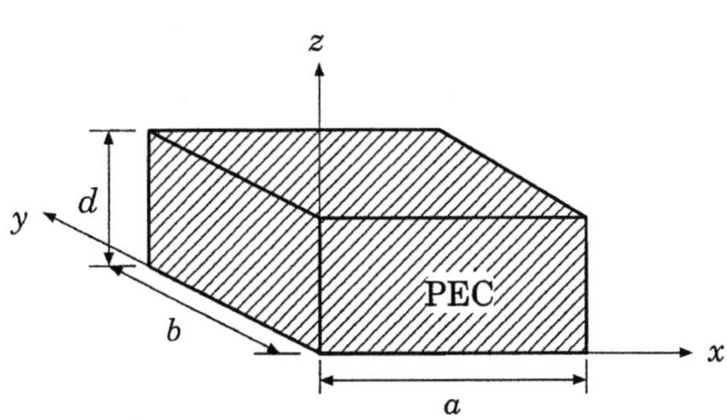

Fig. 4.1. Rectangular cavity resonator surrounded by perfectly conducting walls.

Assume that the TE_{mn} mode propagates along the $\pm z$-direction with $\overline{A}(x,y,z) = 0$ and $\overline{F}(x,y,z) = \hat{z}F_z(x,y,z)$. The Helmholtz equation for $F_z(x,y,z)$ is given by

$$\left(\nabla^2 + k^2\right) F_z(x,y,z) = 0 . \tag{4.1}$$

Let

$$F_z(x,y,z) = \cos\frac{m\pi}{a}x \,\cos\frac{n\pi}{b}y \left(Ae^{ik_z z} + Be^{-ik_z z}\right) \tag{4.2}$$

where

$$\left(\frac{m\pi}{a}\right)^2 + \left(\frac{n\pi}{b}\right)^2 + k_z^2 = k^2 \tag{4.3}$$

and A and B are unknown amplitudes.

Since the tangential electric fields at $z = 0$ and d must vanish,

$$E_x\big|_{z=0,d} = -\frac{1}{\epsilon}\frac{\partial F_z(x,y,z)}{\partial y}\bigg|_{z=0,d} = 0 \tag{4.4}$$

$$E_y\big|_{z=0,d} = \frac{1}{\epsilon}\frac{\partial F_z(x,y,z)}{\partial x}\bigg|_{z=0,d} = 0 . \tag{4.5}$$

Hence, the unknown coefficients A and B are chosen accordingly to yield

$$F_z(x,y,z) = C\cos\frac{m\pi}{a}x \,\cos\frac{n\pi}{b}y \,\sin k_z z \tag{4.6}$$

where $k_z = \frac{p\pi}{d}$, $(m,n) = 0,1,2,\cdots$, $(m \geq 0,\ n \geq 0,\ m+n \neq 0)$, and $p = 1,2,3,\cdots$. The wave associated with $F_z(x,y,z)$ is called the rectangular cavity TE_{mnp} mode, whereas its field components can be obtained by substituting $F_z(x,y,z)$ into Table 3.1 in Section 3.1.

4.1.2 TM Mode

Similarly, it is possible to obtain the vector potentials for the rectangular cavity TM_{mnp} mode.

Assume

$$A_z(x,y,z) = \sin\frac{m\pi}{a}x \,\sin\frac{n\pi}{b}y \left(A'e^{ik_z z} + B'e^{-ik_z z}\right) \tag{4.7}$$

$$F_z(x,y,z) = 0 . \tag{4.8}$$

The tangential electric fields at $z = 0$ and d must vanish. Therefore, the boundary conditions

$$E_x\Big|_{z=0,d} = \frac{i}{\omega\mu\epsilon}\frac{\partial^2 A_z(x,y,z)}{\partial x\,\partial z}\Bigg|_{z=0,d} = 0 \tag{4.9}$$

$$E_y\Big|_{z=0,d} = \frac{i}{\omega\mu\epsilon}\frac{\partial^2 A_z(x,y,z)}{\partial y\,\partial z}\Bigg|_{z=0,d} = 0 \tag{4.10}$$

yield

$$A_z(x,y,z) = C' \sin\frac{m\pi}{a}x \sin\frac{n\pi}{b}y \cos\frac{p\pi}{d}z \tag{4.11}$$

where $m, n = 1, 2, 3, \cdots$ and $p = 0, 1, 2, \cdots$.

The resonant wavenumber k_{mnp} for the TE_{mnp} and TM_{mnp} modes is given by

$$\begin{aligned} k_{mnp} &= \sqrt{\left(\frac{m\pi}{a}\right)^2 + \left(\frac{n\pi}{b}\right)^2 + \left(\frac{p\pi}{d}\right)^2} \\ &= \omega\sqrt{\mu\epsilon} \end{aligned} \tag{4.12}$$

where ω is the resonant angular frequency.

4.1.3 Quality Factor

When conducting walls that surround a cavity are imperfect with finite conductivity σ, ohmic energy loss due to the finite σ occurs. To measure the energy storage and loss for a resonant cavity, the quality factor Q of a resonant cavity is used as

$$Q = \frac{\omega U}{W_l} \tag{4.13}$$

where U is the energy storage and W_l is the time-average power loss. Let us evaluate Q for a rectangular cavity that is surrounded by imperfect but good conducting walls. The field configuration near the surface is shown in Fig. 4.2. Assume that current flow through good conductors exists almost near the conducting surface due to the small skin depth. For a good conductor like copper, the skin depth is $\delta = 6.6 \times 10^{-3}$ (mm) at 100 (MHz). The surface current density $\overline{J}_s$ on a good conductor is approximated by the surface current on a PEC surface as

$$\overline{J}_s \approx \hat{n} \times \overline{H} \tag{4.14}$$

where $\overline{H}$ is the magnetic field on an idealized PEC surface. The average power loss on the imperfect conductor surface S is then

$$W_l = \frac{1}{2}\mathrm{Re}\int_S \overline{E}\cdot\overline{J}_s^*\,ds\,. \tag{4.15}$$

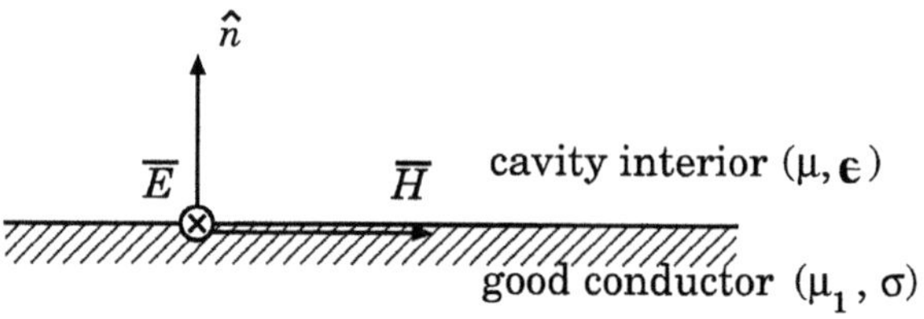

Fig. 4.2. Fields in good conducting surface.

According to (2.29), the relation between the electric and magnetic fields inside a good conductor is given in terms of the wave impedance Z as

$$\overline{E} \approx Z\,\hat{n} \times \overline{H}$$

$$= Z\,\overline{J}_s \tag{4.16}$$

where

$$Z = (1 - i)\sqrt{\frac{\omega\mu_1}{2\sigma}}\,. \tag{4.17}$$

Hence

$$W_l = \frac{1}{2}\int_S R\,\left|\overline{J}_s\right|^2\,ds \tag{4.18}$$

where $R = \sqrt{\dfrac{\omega\mu_1}{2\sigma}}$ is the surface resistivity of a good conductor.

For illustration, let us consider a case when the TE_{101} mode is excited within a rectangular resonant cavity. Substituting the electric vector potential for the TE_{101} mode

$$F_z(x, y, z) = C\cos\left(\frac{\pi}{a}x\right)\sin\left(\frac{\pi}{d}z\right) \tag{4.19}$$

into Table 3.1 of Section 3.1 results in the field components. They are given by

$$E_x = E_z = 0 \tag{4.20}$$

$$E_y = -\frac{C}{\epsilon}\left(\frac{\pi}{a}\right)\sin\left(\frac{\pi}{a}x\right)\sin\left(\frac{\pi}{d}z\right) \tag{4.21}$$

$$H_x = -\frac{i}{\omega\mu\epsilon} C \frac{\pi^2}{ad} \sin\left(\frac{\pi}{a}x\right) \cos\left(\frac{\pi}{d}z\right) \tag{4.22}$$

$$H_y = 0 \tag{4.23}$$

$$H_z = \frac{i}{\omega\mu\epsilon} C \left(\frac{\pi}{a}\right)^2 \cos\left(\frac{\pi}{a}x\right) \sin\left(\frac{\pi}{d}z\right) . \tag{4.24}$$

First, consider the current flow on the top surface ($z = d$) as

$$\begin{aligned} \overline{J}_s\Big|_{z=d} &= \hat{n} \times \overline{H}\Big|_{z=d} \\ &= -\hat{z} \times (\hat{x}H_x + \hat{z}H_z)\Big|_{z=d} \\ &= -\hat{y}H_x\Big|_{z=d} \\ &= -\hat{y}\frac{i}{\omega\mu\epsilon} C \frac{\pi^2}{ad} \sin\left(\frac{\pi}{a}x\right) . \end{aligned} \tag{4.25}$$

The power loss from the top surface is

$$\begin{aligned} W_l(top) &= \frac{1}{2}\int_S R \left|\overline{J}_s\right|^2_{z=d} ds \\ &= \frac{1}{2}\int_0^b \int_0^a R \left(\frac{1}{\omega\mu\epsilon}\right)^2 |C|^2 \left(\frac{\pi^2}{ad}\right)^2 \sin^2\left(\frac{\pi x}{a}\right) dx\, dy \\ &= \frac{R}{4}\left(\frac{|C|\pi^2}{\omega\mu\epsilon}\right)^2 \frac{b}{ad^2} \end{aligned} \tag{4.26}$$

where

$$\omega = \frac{\pi}{\sqrt{\mu\epsilon}} \frac{\sqrt{a^2+d^2}}{ad} . \tag{4.27}$$

Similarly, it is possible to evaluate the power loss associated with the bottom and side surfaces. The total power loss is shown to be

$$W_l = \frac{R}{4}\left(\frac{|C|\pi^2}{\omega\mu\epsilon}\right)^2 \frac{2b(a^3+d^3)+ad(a^2+d^2)}{a^4d^2} . \tag{4.28}$$

The energy storage is the sum of the electric and magnetic energy as

$$U = \frac{1}{4}\int_V \epsilon|\overline{E}|^2 dv + \frac{1}{4}\int_V \mu|\overline{H}|^2 dv \tag{4.29}$$

where

$$\frac{1}{4}\int_V \epsilon|\overline{E}|^2\,dv = \frac{1}{4}\int_0^d\int_0^b\int_0^a \epsilon|E_y|^2\,dx\,dy\,dz$$
$$= \frac{bd}{16a\epsilon}\left(|C|\pi\right)^2 \tag{4.30}$$

$$\frac{1}{4}\int_V \mu|\overline{H}|^2\,dv = \frac{1}{4}\int_0^d\int_0^b\int_0^a \mu\left(|H_x|^2+|H_z|^2\right)dx\,dy\,dz$$
$$= \frac{bd}{16a\epsilon}\left(|C|\pi\right)^2 . \tag{4.31}$$

Note that the stored time-average electric energy is equal to the stored time-average magnetic energy. Substituting U and W_l into the quality factor Q finally yields

$$Q = \frac{\omega U}{W_l}$$
$$= \frac{\pi}{2R}\sqrt{\frac{\mu}{\epsilon}}\frac{b(a^2+d^2)^{1.5}}{2b(a^3+d^3)+ad(a^2+d^2)} . \tag{4.32}$$

For a cube cavity ($a=b=d$),

$$Q = \frac{\sqrt{2}\pi}{6R}\sqrt{\frac{\mu}{\epsilon}} . \tag{4.33}$$

4.2 Circular Cavity Resonator

A hollow circular cylinder is another simple cavity structure often used for confining electromagnetic wave energy. This section analyzes electromagnetic resonance within a circular cavity resonator (μ and ϵ) covered with perfect conducting walls, as shown in Fig. 4.3.

4.2.1 TM Mode

Consider a TM_{mn} mode that propagates along the $\pm z$-directions in a circular cavity resonator. The vector potentials associated with a TM mode are $\overline{F}(\rho,\phi,z)=0$ and $\overline{A}(\rho,\phi,z)=\hat{z}A_z(\rho,\phi,z)$. The magnetic vector potential $A_z(\rho,\phi,z)$ should satisfy the Helmholtz equation

$$\left(\nabla^2+k^2\right)A_z(\rho,\phi,z)=0 \tag{4.34}$$

where $k\ (=\omega\sqrt{\mu\epsilon})$ is the wavenumber. In cylindrical coordinates, $A_z(\rho,\phi,z)$ takes the form of

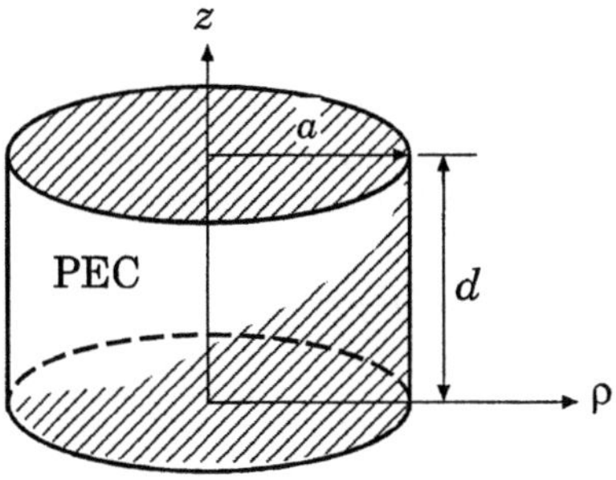

Fig. 4.3. Circular cavity resonator surrounded by perfect conducting walls.

$$A_z(\rho, \phi, z) = J_m(k_\rho \rho) \begin{Bmatrix} \cos m\phi \\ \\ \sin m\phi \end{Bmatrix} \left(Ae^{ik_z z} + Be^{-ik_z z}\right) \tag{4.35}$$

where $m = 0, 1, 2, \cdots$, A and B are unknown coefficients, and $k_\rho^2 + k_z^2 = k^2$.

Since the tangential electric fields, E_ϕ and E_ρ, must vanish at $z = 0$ and d, the boundary conditions

$$E_\rho\Big|_{z=0,d} = \frac{i}{\omega\mu\epsilon} \frac{\partial^2 A_z(\rho, \phi, z)}{\partial\rho\, \partial z}\Bigg|_{z=0,d} = 0 \tag{4.36}$$

$$E_\phi\Big|_{z=0,d} = \frac{i}{\omega\mu\epsilon\rho} \frac{\partial^2 A_z(\rho, \phi, z)}{\partial\phi\, \partial z}\Bigg|_{z=0,d} = 0 \tag{4.37}$$

yield

$$A_z(\rho, \phi, z) = H_0 J_m(k_\rho \rho) \begin{Bmatrix} \cos m\phi \\ \\ \sin m\phi \end{Bmatrix} \cos k_z z \tag{4.38}$$

where $k_z = \dfrac{p\pi}{d}$ and $p = 0, 1, 2, \cdots$.

Since the tangential electric fields, E_ϕ and E_z, must vanish at $\rho = a$, the boundary conditions require

$$E_z\Big|_{\rho=a} = \frac{i}{\omega\mu\epsilon} \left(\frac{\partial^2}{\partial z^2} + k^2\right) A_z(\rho, \phi, z)\Bigg|_{\rho=a} = 0 \tag{4.39}$$

$$E_\phi\Big|_{\rho=a} = \frac{i}{\omega\mu\epsilon\rho} \frac{\partial^2 A_z(\rho, \phi, z)}{\partial\phi\, \partial z}\Bigg|_{\rho=a} = 0\,. \tag{4.40}$$

The resonance condition is thus determined by

$$J_m(k_\rho a) = 0 \,. \tag{4.41}$$

There exist a number of roots that satisfy the resonance condition. Let the nth root be denoted by $k_\rho = k_{\rho mn}$, where $J_m(k_{\rho mn}a) = 0$, $m = 0, 1, 2, \cdots$, and $n = 1, 2, 3, \cdots$. The resonant wavenumber k_{mnp} for the TM$_{mnp}$ mode is

$$\begin{aligned} k_{mnp} \, (= \omega\sqrt{\mu\epsilon}) &= \sqrt{k_{\rho mn}^2 + k_z^2} \\ &= \sqrt{k_{\rho mn}^2 + \left(\frac{p\pi}{d}\right)^2} \,. \end{aligned} \tag{4.42}$$

The field expressions for the TM$_{mnp}$ mode can subsequently be obtained by substituting $A_z(\rho, \phi, z)$ into Table 3.2 in Section 3.1.

4.2.2 Quality Factor for TM$_{010}$ Mode

When a conducting surface is imperfect, power loss due to the conduction current occurs. Let us evaluate the quality factor Q that is associated with the simple TM$_{010}$ mode ($m = 0$, $n = 1$, and $p = 0$), where $k_\rho = k$ and $ka = 2.405$. The magnetic vector potential for the TM$_{010}$ mode is

$$A_z(\rho, \phi, z) = H_0 J_0(k\rho) \tag{4.43}$$

and its respective field components are

$$E_z = \frac{ik^2}{\omega\mu\epsilon} H_0 J_0(k\rho) \tag{4.44}$$

$$\begin{aligned} H_\phi &= -\frac{k}{\mu} H_0 J_0'(k\rho) \\ &= \frac{k}{\mu} H_0 J_1(k\rho) \,. \end{aligned} \tag{4.45}$$

The time-average electric and magnetic energy storage U is given by

$$\begin{aligned} U &= \frac{1}{4}\int_V \epsilon |E_z|^2 \, dv + \frac{1}{4}\int_V \mu |H_\phi|^2 \, dv \\ &= \frac{1}{4}\int_0^a 2\pi d\epsilon |E_z|^2 \rho \, d\rho + \frac{1}{4}\int_0^a 2\pi d\mu |H_\phi|^2 \rho \, d\rho \\ &= \frac{k^2 |H_0|^2 \pi d}{2\mu}\int_0^a \left[J_0^2(k\rho) + J_1^2(k\rho)\right] \rho \, d\rho \,. \end{aligned} \tag{4.46}$$

Since

$$\int J_m^2(u) u \, du = \frac{u^2}{2}\left[J_m^2(u) - J_{m-1}(u) J_{m+1}(u)\right] \tag{4.47}$$

U can be rewritten as

$$U = \frac{|H_0|^2 \pi d}{2\mu}(ka)^2 J_1^2(ka) \ . \tag{4.48}$$

The power loss from the top surface with surface resistivity R is

$$\begin{aligned} W_l(top) &= \frac{1}{2}\int_S R \ \left|\overline{J}_s\right|^2_{z=d} \ ds \\ &= \frac{1}{2}\int_0^a R \left[\frac{|H_0|k}{\mu}J_1(k\rho)\right]^2 2\pi\rho \, d\rho \\ &= \frac{\pi}{2}R\frac{|H_0|^2}{\mu^2}(ka)^2 J_1^2(ka) \ . \end{aligned} \tag{4.49}$$

The power loss from the side is thus

$$\begin{aligned} W_l(side) &= \frac{1}{2}\int_S R \ \left|\overline{J}_s\right|^2_{\rho=a} \ ds \\ &= \pi R\frac{|H_0|^2}{\mu^2}k^2 a d J_1^2(ka) \ . \end{aligned} \tag{4.50}$$

The total power loss is

$$\begin{aligned} W_l &= 2W_l(top) + W_l(side) \\ &= \pi R\frac{|H_0|^2}{\mu^2}k^2 a(a+d) J_1^2(ka) \ . \end{aligned} \tag{4.51}$$

Hence, the quality factor Q is shown to be

$$\begin{aligned} Q &= \frac{\omega U}{W_l} \\ &= \sqrt{\frac{\mu}{\epsilon}}\frac{1.202d}{R(a+d)} \ . \end{aligned} \tag{4.52}$$

4.2.3 TE Mode

The wave analysis for the circular cavity TE_{mnp} mode is similar to the previous TM_{mnp} case. It is assumed that the TE_{mn} mode propagates along the $\pm z$-directions with the vector potentials $\overline{A}(\rho,\phi,z) = 0$ and $\overline{F}(\rho,\phi,z) = \hat{z}F_z(\rho,\phi,z)$. The electric vector potential $F_z(\rho,\phi,z)$ is given by

$$F_z(\rho,\phi,z) = J_m(k_\rho \rho)\begin{Bmatrix} \cos m\phi \\ \\ \sin m\phi \end{Bmatrix}\left(Ae^{ik_z z} + Be^{-ik_z z}\right) \tag{4.53}$$

where $k_\rho^2 + k_z^2 = k^2$.

The boundary conditions

$$E_\rho\Big|_{z=0,d} = -\frac{1}{\epsilon\rho}\frac{\partial F_z(\rho,\phi,z)}{\partial\phi}\Bigg|_{z=0,d} = 0 \tag{4.54}$$

$$E_\phi\Big|_{z=0,d} = \frac{1}{\epsilon}\frac{\partial F_z(\rho,\phi,z)}{\partial\rho}\Bigg|_{z=0,d} = 0 \tag{4.55}$$

yield

$$F_z(\rho,\phi,z) = E_0 J_m(k_\rho\rho)\begin{Bmatrix}\cos m\phi\\ \\ \sin m\phi\end{Bmatrix}\sin k_z z \tag{4.56}$$

where $k_z = \dfrac{p\pi}{d}$ and $p = 1, 2, 3, \cdots$.

Another boundary condition requires

$$E_\phi\Big|_{\rho=a} = \frac{1}{\epsilon}\frac{\partial F_z(\rho,\phi,z)}{\partial\rho}\Bigg|_{\rho=a} = 0\,. \tag{4.57}$$

The parameter $k_\rho = k'_{\rho mn}$ is chosen such that

$$J'_m(k'_{\rho mn}a) = \frac{dJ_m(k'_{\rho mn}\rho)}{d(k'_{\rho mn}\rho)}\Bigg|_{\rho=a} = 0 \tag{4.58}$$

where $m = 0, 1, 2, \cdots$ and $n = 1, 2, 3, \cdots$.

Hence, the field expressions for the TE_{mnp} mode are made available by substitution of the electric vector potential

$$F_z(\rho,\phi,z) = E_0 J_m\left(k'_{\rho mn}\rho\right)\begin{Bmatrix}\cos m\phi\\ \\ \sin m\phi\end{Bmatrix}\sin k_z z \tag{4.59}$$

into Table 3.2 in Section 3.1.

The resonant wavenumber k'_{mnp} for the TE_{mnp} mode is

$$\begin{aligned} k'_{mnp} &= \sqrt{\left(k'_{\rho mn}\right)^2 + k_z^2} \\ &= \sqrt{\left(k'_{\rho mn}\right)^2 + \left(\frac{p\pi}{d}\right)^2}\,. \end{aligned} \tag{4.60}$$

4.3 Spherical Cavity Resonator

Let us consider electromagnetic resonance in a perfectly conducting spherical cavity resonator, as shown in Fig. 4.4. A cavity is filled with a dielectric material of permeability μ and permittivity ϵ. A boundary-value problem of a spherical cavity resonator can be solved using the separation of variables technique. It is convenient to decompose a wave into TE and TM components to the r-direction where the TE (or TM) components refer to the waves whose electric (or magnetic) field is transverse to the r-direction, respectively.

4.3.1 TM Mode

First, consider the TM case by choosing $\overline{A}(r,\theta,\phi) = \hat{r}A_r(r,\theta,\phi)$ and $\overline{F}(r,\theta,\phi) = 0$. Consider source-free Maxwell's equations

$$\nabla \times \overline{E} = i\omega\mu\overline{H} \tag{4.61}$$

$$\nabla \times \overline{H} = -i\omega\epsilon\overline{E}\,. \tag{4.62}$$

Substituting

$$\overline{H} = \frac{1}{\mu}\nabla \times \overline{A} \tag{4.63}$$

into (4.61) yields

$$\nabla \times \left(\overline{E} - i\omega\overline{A}\right) = 0\,. \tag{4.64}$$

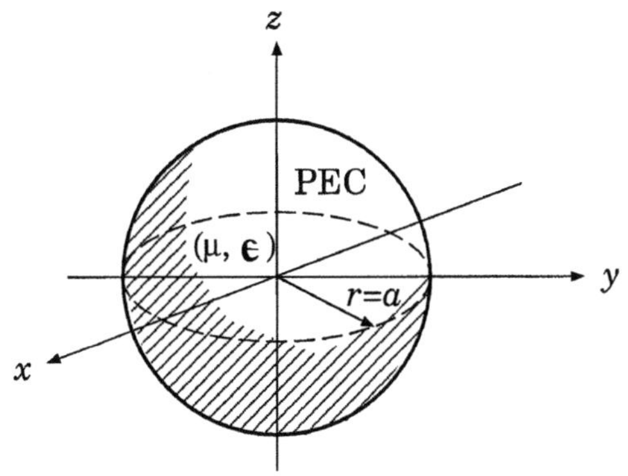

Fig. 4.4. Spherical cavity resonator surrounded by perfect conducting wall.

Let

$$\overline{E} - i\omega\overline{A} = -\nabla\phi_e\ . \tag{4.65}$$

Eliminating $\overline{E}$ and $\overline{H}$ from (4.62), (4.63), and (4.65) gives

$$-\nabla\times\nabla\times\overline{A} + \omega^2\mu\epsilon\overline{A} + i\omega\mu\epsilon\nabla\phi_e = 0\ . \tag{4.66}$$

The $\hat{r}$ component of $\nabla^2\overline{A}$ using spherical coordinates is not equal to the Laplacian of A_r, namely, $\left(\nabla^2\overline{A}\right)_r \neq \nabla^2 A_r$. Therefore, the Lorentz condition $\nabla\cdot\overline{A} = i\omega\mu\epsilon\phi_e$ will not reduce (4.66) into the simple scalar equation that we wish to obtain. Another gauge condition is, therefore, needed to make (4.66) a single scalar equation. To that end, let us rewrite (4.66) explicitly using spherical coordinates

$$\frac{\hat{r}}{r^2\sin\theta}\left[\frac{\partial}{\partial\theta}\left(\sin\theta\frac{\partial A_r}{\partial\theta}\right) + \frac{1}{\sin\theta}\frac{\partial^2 A_r}{\partial\phi^2}\right] - \frac{\hat{\theta}}{r}\frac{\partial^2 A_r}{\partial r\,\partial\theta} - \frac{\hat{\phi}}{r\sin\theta}\frac{\partial^2 A_r}{\partial r\,\partial\phi}$$

$$+\hat{r}\omega^2\mu\epsilon A_r + i\omega\mu\epsilon\left(\hat{r}\frac{\partial\phi_e}{\partial r} + \hat{\theta}\frac{1}{r}\frac{\partial\phi_e}{\partial\theta} + \hat{\phi}\frac{1}{r\sin\theta}\frac{\partial\phi_e}{\partial\phi}\right) = 0\ . \tag{4.67}$$

Consider the condition that makes the $\hat{\theta}$ and $\hat{\phi}$ components of (4.67) zero as

$$\frac{1}{r}\frac{\partial}{\partial\theta}\left(-\frac{\partial A_r}{\partial r} + i\omega\mu\epsilon\phi_e\right) = 0 \tag{4.68}$$

$$\frac{1}{r\sin\theta}\frac{\partial}{\partial\phi}\left(-\frac{\partial A_r}{\partial r} + i\omega\mu\epsilon\phi_e\right) = 0\ . \tag{4.69}$$

If the gauge condition

$$-\frac{\partial A_r}{\partial r} + i\omega\mu\epsilon\phi_e = 0 \tag{4.70}$$

is chosen, (4.68) and (4.69) can be simultaneously satisfied and (4.67) becomes the scalar equation

$$\frac{\partial^2 A_r}{\partial r^2} + \frac{1}{r^2\sin\theta}\frac{\partial}{\partial\theta}\left(\sin\theta\frac{\partial A_r}{\partial\theta}\right) + \frac{1}{r^2\sin^2\theta}\frac{\partial^2 A_r}{\partial\phi^2} + k^2 A_r = 0 \tag{4.71}$$

or equivalently

$$\left(\nabla^2 + k^2\right)\frac{A_r}{r} = 0 \tag{4.72}$$

where $k\ (=\omega\sqrt{\mu\epsilon})$ is the wavenumber.

Based on the separation of variables technique, it is possible to assume

$$A_r(r,\theta,\phi) = \psi_1(r)\psi_2(\theta)\psi_3(\phi)\ . \tag{4.73}$$

Substituting (4.73) into (4.71), dividing (4.71) by $\psi_1(r)\psi_2(\theta)\psi_3(\phi)$, and multiplying by r^2 results in

$$\frac{r^2}{\psi_1}\frac{d^2\psi_1}{d^2r} + \frac{1}{\psi_2 \sin\theta}\frac{d}{d\theta}\left(\sin\theta\frac{d\psi_2}{d\theta}\right) + \frac{1}{\sin^2\theta\psi_3}\frac{d^2\psi_3}{d\phi^2} + k^2r^2 = 0 \,. \quad (4.74)$$

Let us introduce the separation constants m and n as

$$\frac{1}{\psi_3}\frac{d^2\psi_3}{d\phi^2} = -m^2 \quad (4.75)$$

$$\frac{1}{\psi_2 \sin\theta}\frac{d}{d\theta}\left(\sin\theta\frac{d\psi_2}{d\theta}\right) - \frac{m^2}{\sin^2\theta} = -n(n+1) \,. \quad (4.76)$$

Then, (4.74) results in the following three ordinary differential equations

$$\frac{d^2\psi_3}{d\phi^2} + m^2\psi_3 = 0 \quad (4.77)$$

$$\frac{1}{\sin\theta}\frac{d}{d\theta}\left(\sin\theta\frac{d\psi_2}{d\theta}\right) + \left[n(n+1) - \frac{m^2}{\sin^2\theta}\right]\psi_2 = 0 \quad (4.78)$$

$$r^2\frac{d^2\psi_1}{dr^2} + \left[k^2r^2 - n(n+1)\right]\psi_1 = 0 \,. \quad (4.79)$$

- The solution to (4.77) is the circular harmonic functions, $\cos m\phi$ and $\sin m\phi$. Since A_r is periodic with a 2π-periodicity, the parameter m must be an integer ($m = 0, 1, 2, \cdots$).
- Two independent solutions to (4.78) are the associated Legendre functions of the first and second kinds, $P_n^m(\cos\theta)$ and $Q_n^m(\cos\theta)$, respectively. A discussion on the associated Legendre functions is available in Appendix D. When $n \neq$ an integer, $P_n^m(\cos\theta)$ becomes singular (unbounded) at $\theta = \pi$ and 0. For $P_n^m(\cos\theta)$ to be a solution to the spherical cavity problem considered in this section, n should be an integer. Since $P_n^m(\cos\theta) = 0$ when $m > n$, the integer m should be $0, 1, 2, \cdots, n$. The function $P_0^0(\cos\theta)$ ($= 1$) is an unacceptable solution since the resulting electric and magnetic fields all vanish. Furthermore, since $P_0^m(\cos\theta) = 0$, the integer n should be $n = 1, 2, 3, \cdots$. Since $Q_n^m(\cos\theta)$ becomes singular at $\theta = \pi$ and 0, $Q_n^m(\cos\theta)$ cannot be solutions. Hence, $\psi_2(\theta)$ is written as

$$\psi_2(\theta) = P_n^m(\cos\theta) \,. \quad (4.80)$$

- The solution to (4.79) is $\psi_1(r) = \hat{B}_n(kr)\ \left[\hat{J}_n(kr),\ \hat{N}_n(kr),\ \hat{H}_n^{(1)}(kr), \cdots\right]$, which are related to the spherical Bessel functions $b_n(kr)$ as

$$\hat{B}_n(kr) = krb_n(kr) \,. \quad (4.81)$$

The spherical Bessel functions $b_n(kr)$ satisfy

$$r^2 \frac{d^2 b_n(kr)}{dr^2} + 2r \frac{db_n(kr)}{dr} + \left[k^2 r^2 - n(n+1)\right] b_n(kr) = 0 \,. \quad (4.82)$$

A discussion on the spherical Bessel functions is provided in Appendix B. Since $\psi_1(r)$ must remain finite at $r = 0$, the spherical Bessel function of the first kind is chosen as the solution

$$\psi_1(r) = \hat{J}_n(kr) \,. \quad (4.83)$$

The solution A_r, therefore, is written as

$$A_r = \hat{J}_n(kr)\, P_n^m(\cos\theta) \begin{Bmatrix} \cos m\phi \\ \\ \sin m\phi \end{Bmatrix} . \quad (4.84)$$

The resulting TM wave components can be derived from

$$\overline{E} = \frac{i}{\omega\mu\epsilon} \nabla \times \nabla \times \overline{A} \quad (4.85)$$

$$\overline{H} = \frac{1}{\mu} \nabla \times \overline{A} \,. \quad (4.86)$$

The explicit relations between the TM wave components and A_r are shown in Table 4.1.

Since the tangential electric fields, E_θ and E_ϕ, must vanish at $r = a$, the resonant condition is given by

$$\left. \frac{d\hat{J}_n(kr)}{d(kr)} \right|_{r=a} = 0 \,. \quad (4.87)$$

A number of solutions to (4.87) exist. The pth root of (4.87) $k = k_{mnp}$ is the resonant wavenumber that is associated with the TM_{mnp} mode of a spherical cavity resonator.

4.3.2 Quality Factor for TM_{011} Mode

The lowest-order mode corresponds to the cases ($m = 0$, $n = 1$, $p = 1$) and ($m = 1$, $n = 1$, $p = 1$). Consider the lowest TM_{011} mode

$$A_r = \hat{J}_1(kr)\, P_1^0(\cos\theta) \quad (4.88)$$

where

$$\hat{J}_1(kr) = \sqrt{\frac{\pi kr}{2}} J_{1.5}(kr)$$
$$= \left(\frac{\sin kr}{kr} - \cos kr \right) \quad (4.89)$$

$$P_1^0(\cos\theta) = P_1(\cos\theta) = \cos\theta \,. \quad (4.90)$$

Table 4.1. Field representations using spherical coordinates

Fields	TM wave	TE wave
E_r	$\frac{i}{\omega\mu\epsilon}\left(\frac{\partial^2}{\partial r^2}+k^2\right)A_r$	0
E_θ	$\frac{i}{\omega\mu\epsilon r}\frac{\partial^2 A_r}{\partial r\,\partial\theta}$	$-\frac{1}{\epsilon r\sin\theta}\frac{\partial F_r}{\partial\phi}$
E_ϕ	$\frac{i}{\omega\mu\epsilon r\sin\theta}\frac{\partial^2 A_r}{\partial r\,\partial\phi}$	$\frac{1}{\epsilon r}\frac{\partial F_r}{\partial\theta}$
H_r	0	$\frac{i}{\omega\mu\epsilon}\left(\frac{\partial^2}{\partial r^2}+k^2\right)F_r$
H_θ	$\frac{1}{\mu r\sin\theta}\frac{\partial A_r}{\partial\phi}$	$\frac{i}{\omega\mu\epsilon r}\frac{\partial^2 F_r}{\partial r\,\partial\theta}$
H_ϕ	$-\frac{1}{\mu r}\frac{\partial A_r}{\partial\theta}$	$\frac{i}{\omega\mu\epsilon r\sin\theta}\frac{\partial^2 F_r}{\partial r\,\partial\phi}$

Substituting A_r into the TM wave components of Table 4.1 yields three nonzero components, E_r, E_θ, and H_ϕ. For instance, the H_ϕ component is given by

$$\begin{aligned} H_\phi &= -\frac{1}{\mu r}\frac{\partial A_r}{\partial\theta} \\ &= \frac{1}{\mu r}\sin\theta\left(\frac{\sin kr}{kr}-\cos kr\right) . \end{aligned} \tag{4.91}$$

The resonant frequency is determined from the boundary condition

$$E_\theta\Big|_{r=a}=0 . \tag{4.92}$$

The resonant condition

$$\left.\frac{d\hat{J}_1(kr)}{d(kr)}\right|_{r=a}=0 \tag{4.93}$$

gives the characteristic equation

$$\frac{ka}{1-(ka)^2}=\tan ka \tag{4.94}$$

where the solution $ka = 2.744$ determines the cavity resonance frequency.

The time-average electric and magnetic energy storage (or equivalently peak magnetic energy storage) U is given by

$$\begin{aligned}
U &= \frac{1}{2}\int_V \mu |H_\phi|^2 \, dv \\
&= \frac{1}{2}\int_0^{2\pi}\int_0^{\pi}\int_0^{a} \mu \left[\frac{1}{\mu r}\sin\theta \hat{J}_1(kr)\right]^2 r^2 \sin\theta \, dr \, d\theta \, d\phi \\
&= \frac{4\pi}{3\mu}\int_0^a \left[\hat{J}_1(kr)\right]^2 dr \\
&= \frac{2\pi^2}{3\mu}\int_0^a kr \left[J_{1.5}(kr)\right]^2 dr \, .
\end{aligned} \tag{4.95}$$

Since

$$\int \left[J_m(u)\right]^2 u \, du = \frac{u^2}{2}\left\{\left[J_m(u)\right]^2 - J_{m-1}(u)J_{m+1}(u)\right\} \tag{4.96}$$

U can be rewritten as

$$\begin{aligned}
U &= \frac{\pi^2 ka^2}{3\mu}\left\{\left[J_{1.5}(ka)\right]^2 - J_{0.5}(ka)J_{2.5}(ka)\right\} \\
&= \frac{\pi^2 ka^2}{3\mu}\left[0.512^2 - 0.187 \times 0.373\right] \\
&= \frac{\pi^2 ka^2}{3\mu} \times 0.192 \, .
\end{aligned} \tag{4.97}$$

The power loss from a cavity inner surface with surface resistivity R is

$$\begin{aligned}
W_l &= \frac{1}{2}\int_S R \left|\overline{J}_s\right|^2_{r=a} ds \\
&= \frac{1}{2}\int_S R \left|-\hat{r} \times \hat{\phi} H_\phi\right|^2_{r=a} ds \\
&= \frac{1}{2}\int_0^{2\pi}\int_0^{\pi} R \left[\frac{\sin\theta}{\mu a}\hat{J}_1(ka)\right]^2 a^2 \sin\theta \, d\theta \, d\phi \\
&= \frac{4\pi R}{3\mu^2}\left[\hat{J}_1(ka)\right]^2 \\
&= \frac{2\pi^2 Rka}{3\mu^2}\left[J_{1.5}(ka)\right]^2 \\
&= \frac{2\pi^2 Rka}{3\mu^2} \times 0.512^2 \, .
\end{aligned} \tag{4.98}$$

Hence, the quality factor Q is shown to be

$$Q = \frac{\omega U}{W_l} = \frac{1.005}{R}\sqrt{\frac{\mu}{\epsilon}} \, . \tag{4.99}$$

4.3.3 TE Mode

Similarly, a TE wave in a spherical cavity resonator can be obtained from the assumption $\overline{A}(r,\theta,\phi) = 0$ and $\overline{F}(r,\theta,\phi) = \hat{r}F_r(r,\theta,\phi)$. Its analysis is somewhat analogous to that of TM waves. The governing equation for the electric vector potential $\overline{F}$ is

$$-\nabla \times \nabla \times \overline{F} + \omega^2\mu\epsilon\overline{F} + i\omega\mu\epsilon\nabla\phi_m = 0 \, . \tag{4.100}$$

Using the gauge condition

$$\phi_m = -\frac{i}{\omega\mu\epsilon}\frac{\partial F_r}{\partial r} \tag{4.101}$$

gives

$$\frac{\partial^2 F_r}{\partial r^2} + \frac{1}{r^2\sin\theta}\frac{\partial}{\partial\theta}\left(\sin\theta\frac{\partial F_r}{\partial\theta}\right) + \frac{1}{r^2\sin^2\theta}\frac{\partial^2 F_r}{\partial\phi^2} + k^2F_r = 0 \tag{4.102}$$

where its solution is given by

$$F_r = \hat{J}_n(kr)\,P_n^m(\cos\theta)\begin{Bmatrix}\cos m\phi \\ \\ \sin m\phi\end{Bmatrix} \, . \tag{4.103}$$

A TE wave is represented in terms of $\overline{F}$ as

$$\overline{E} = -\frac{1}{\epsilon}\nabla \times \overline{F} \tag{4.104}$$

$$\overline{H} = \frac{i}{\omega\mu\epsilon}\nabla \times \nabla \times \overline{F} \, . \tag{4.105}$$

For convenience, the explicit field components for a TE wave are listed in Table 4.1.

Since the tangential electric fields, E_θ and E_ϕ, must vanish at $r = a$, the resonance condition is given by

$$\hat{J}_n(ka) = 0 \tag{4.106}$$

where the pth root $k = k_{mnp}$ gives the resonant wavenumber for the TE_{mnp} mode of a spherical cavity resonator.

4.4 Groove Guide Coupler

An open groove guide is known to have a low-loss capacity at millimeter wavelengths. An open groove guide consists of rectangular grooves that are covered with PEC walls, as shown in Fig. 4.5. The groove guide interior is filled with a homogeneous dielectric medium, thereby supporting either TE ($E_z = 0$ and $H_z \neq 0$) or TM ($H_z = 0$ and $E_z \neq 0$) wave propagation. TE waves are known to be dominant since TE waves have a lower cutoff frequency than TM waves. A groove guide coupler, which consists of a double groove guide, is potentially a high-power coupling structure. A cross-sectional view of a groove guide coupler is shown in Fig. 4.6, where a wave is assumed to propagate along the z-direction. This section investigates the guiding and coupling characteristics of a groove guide coupler using the Fourier transform and mode matching [4]. The Fourier transform and mode matching approach provides a rigorous way of analyzing a groove guide coupler.

4.4.1 Field Analysis

Consider a double rectangular groove guide that is filled with a homogeneous dielectric material with permittivity ϵ and permeability μ. A TE wave propagates along the z-direction with the electric vector potential

$$\overline{F}(x,y,z) = \hat{z}F_z(x,y,z)$$

$$= \hat{z}F_z(x,y)\exp(ik_z z) \,. \tag{4.107}$$

The electric vector potential $F_z(x,y,z)$ should satisfy the Helmholtz equation $\left(\nabla^2 + k^2\right) F_z(x,y,z) = 0$. To determine the propagation constant k_z, the boundary conditions for the field continuities must be utilized.

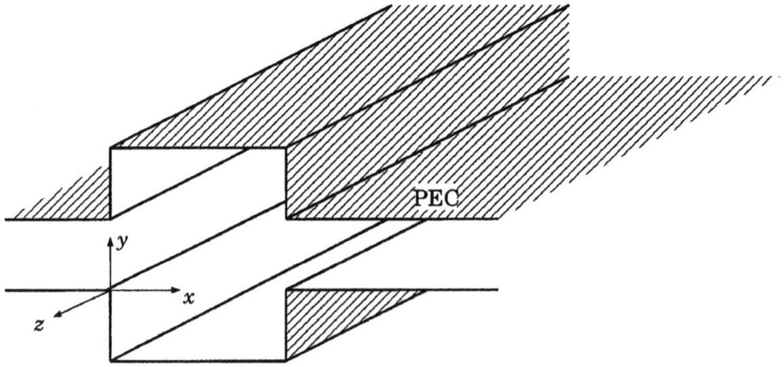

Fig. 4.5. Single groove guide.

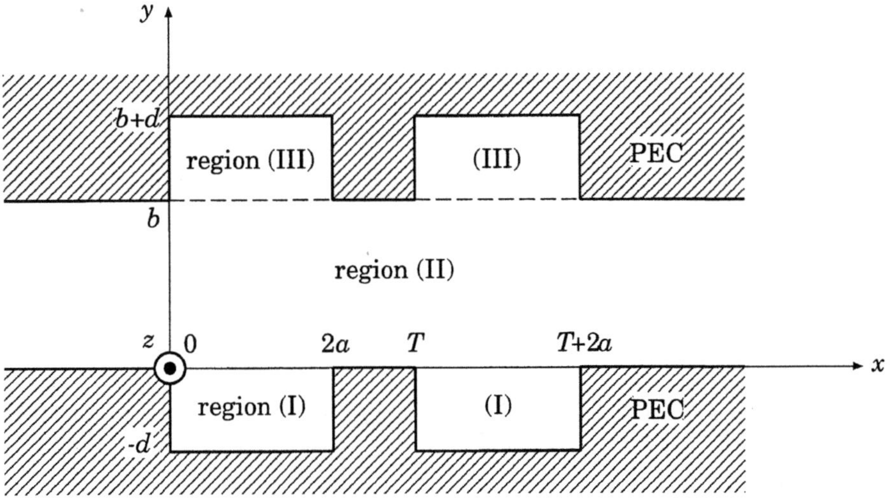

Fig. 4.6. Groove guide coupler: double groove guide.

For analysis, the guide interior is divided into regions (I) ($-d \le y \le 0$), (II) ($0 \le y \le b$), and (III) ($b \le y \le b+d$), where their respective F_z components are given as

$$F_z^I(x,y) = \sum_{m=0}^{\infty} Q_m^n \cos a_m(x-nT) \cos \xi_m(y+d) \tag{4.108}$$

$$F_z^{II}(x,y) = \int_{-\infty}^{\infty} \left(F^+ e^{i\kappa y} + F^- e^{-i\kappa y}\right) e^{-i\zeta x}\, d\zeta \tag{4.109}$$

$$F_z^{III}(x,y) = \sum_{m=0}^{\infty} S_m^n \cos a_m(x-nT) \cos \xi_m(y-b-d) \tag{4.110}$$

where

$$\xi_m = \sqrt{k^2 - a_m^2 - k_z^2} \tag{4.111}$$

$$\kappa = \sqrt{k^2 - \zeta^2 - k_z^2} \tag{4.112}$$

$a_m = \dfrac{m\pi}{2a}$, $k = \omega\sqrt{\mu\epsilon} = \dfrac{2\pi}{\lambda}$, and $n = 0, 1$. In (4.108) through (4.110), the modal coefficients–Q_m^n, S_m^n, F^+, and F^-–are introduced to represent the fields. The aim is to constitute a set of simultaneous equations for the discrete modal coefficients Q_m^n and S_m^n. Note that (4.108) and (4.110) are constructed so as to satisfy the boundary conditions at the PEC walls where the tangential electric field is required to vanish.

Since $E_x(x,y,z) = E_x(x,y)\exp(ik_z z)$, the continuity of E_x $\left(= -\dfrac{1}{\epsilon}\dfrac{\partial F_z}{\partial y}\right)$ at $y=0$ is written as

$$E_x^{II}(x,0) = \begin{cases} E_x^I(x,0) & \text{for } nT < x < nT+2a \quad (n=0,1) \\ 0 & \text{otherwise}\,. \end{cases} \tag{4.113}$$

Applying the inverse Fourier transform $\dfrac{1}{2\pi}\displaystyle\int_{-\infty}^{\infty}(\cdot)e^{i\zeta x}\,dx$ to (4.113) yields

$$F^+ - F^- = -\sum_{n=0}^{1}\sum_{m=0}^{\infty}\frac{1}{2\pi i\kappa}Q_m^n\xi_m\sin(\xi_m d)G_m^n(\zeta) \tag{4.114}$$

where

$$G_m^n(\zeta) = \frac{i\zeta[1-(-1)^m e^{i\zeta 2a}]}{\zeta^2 - a_m^2}e^{i\zeta nT}\,. \tag{4.115}$$

The E_x continuity at $y=b$ is written as

$$E_x^{II}(x,b) = \begin{cases} E_x^{III}(x,b) & \text{for } nT < x < nT+2a \\ 0 & \text{otherwise}\,. \end{cases} \tag{4.116}$$

Applying the inverse Fourier transform to (4.116) yields

$$F^+e^{i\kappa b} - F^-e^{-i\kappa b} = \sum_{n=0}^{1}\sum_{m=0}^{\infty}\frac{1}{2\pi i\kappa}S_m^n\xi_m\sin(\xi_m d)G_m^n(\zeta)\,. \tag{4.117}$$

Solving (4.114) and (4.117) for F^+ and F^- results in

$$F^+ = -\sum_{n=0}^{1}\sum_{m=0}^{\infty}\left(\frac{e^{-i\kappa b}Q_m^n + S_m^n}{4\pi\kappa\sin\kappa b}\right)\xi_m\sin(\xi_m d)G_m^n(\zeta) \tag{4.118}$$

$$F^- = -\sum_{n=0}^{1}\sum_{m=0}^{\infty}\left(\frac{e^{i\kappa b}Q_m^n + S_m^n}{4\pi\kappa\sin\kappa b}\right)\xi_m\sin(\xi_m d)G_m^n(\zeta)\,. \tag{4.119}$$

Note that H_z is given by

$$H_z(x,y,z) = \frac{i}{\omega\mu\epsilon}\left(\frac{\partial^2}{\partial z^2} + k^2\right)F_z(x,y,z) \tag{4.120}$$

where $H_z(x,y,z) = H_z(x,y)\exp(ik_z z)$. Hence, the H_z continuity at $y=0$ and $pT < x < pT+2a$ $(p=0,1)$

$$H_z^{II}(x,0) = H_z^I(x,0) \tag{4.121}$$

can be rewritten as

$$\int_{-\infty}^{\infty} \left(F^+ + F^-\right) e^{-i\zeta x}\, d\zeta$$

$$= \sum_{m=0}^{\infty} Q_m^p \cos a_m(x - pT)\cos(\xi_m d)\ . \tag{4.122}$$

Since $H_x(x,y,z) = \dfrac{i}{\omega\mu\epsilon}\dfrac{\partial^2 F_z(x,y,z)}{\partial x\,\partial z}$, the H_x continuity at $y = 0$ and $(pT < x < pT + 2a)$ is seen to be automatically satisfied from differentiating (4.122) with respect to x. Substituting F^+ and F^- into (4.122) gives

$$-\sum_{n=0}^{1}\sum_{m=0}^{\infty}\int_{-\infty}^{\infty} \frac{1}{2\pi\kappa}(Q_m^n \cot\kappa b + S_m^n \csc\kappa b)\xi_m \sin(\xi_m d)$$
$$\cdot G_m^n(\zeta) e^{-i\zeta x}\, d\zeta$$

$$= \sum_{m=0}^{\infty} Q_m^p \cos a_m(x - pT)\cos(\xi_m d)\ . \tag{4.123}$$

It is possible to simplify (4.123) by using the orthogonality of the sinusoidal function $\cos a_m(x - pT)$. Multiplying (4.123) by $\cos a_l(x - pT)$, $(p = 0, 1)$ and integrating over $(pT < x < pT + 2a)$ with respect to x yields

$$\sum_{n=0}^{1}\sum_{m=0}^{\infty}\Big\{Q_m^n\left[\xi_m \sin(\xi_m d) I_1 + a\cos(\xi_m d)\delta_{ml}\delta_{np}\varepsilon_m\right]$$
$$+ S_m^n \xi_m \sin(\xi_m d) I_2\Big\} = 0 \tag{4.124}$$

where $\varepsilon_0 = 2$, $\varepsilon_1 = \varepsilon_2 = \cdots = 1$, δ_{ml} is the Kronecker delta, and

$$I_1 = \int_{-\infty}^{\infty} \frac{\cot(\kappa b)}{2\pi\kappa} G_m^n(\zeta) G_l^p(-\zeta)\, d\zeta \tag{4.125}$$

$$I_2 = \int_{-\infty}^{\infty} \frac{\csc(\kappa b)}{2\pi\kappa} G_m^n(\zeta) G_l^p(-\zeta)\, d\zeta\ . \tag{4.126}$$

Based on residue calculus, the integrals I_1 and I_2 can be transformed into a rapidly convergent series. An evaluation of I_1 and I_2 is given in the next subsection.

It is necessary to derive another set of simultaneous equations based on the boundary condition at $y = b$. The analytic procedure for enforcing the boundary condition at $y = b$ is similar to that for enforcing the boundary condition at $y = 0$. Applying the orthogonality property to the H_z continuity at $y = b$ and $pT < x < pT + 2a$

$$H_z^{II}(x,b) = H_z^{III}(x,b) \tag{4.127}$$

yields

$$\sum_{n=0}^{1}\sum_{m=0}^{\infty}\Big\{Q_m^n \xi_m \sin(\xi_m d) I_2$$
$$+ S_m^n \left[\xi_m \sin(\xi_m d) I_1 + a\cos(\xi_m d)\delta_{ml}\delta_{np}\varepsilon_m\right]\Big\} = 0\,. \tag{4.128}$$

Equations (4.124) and (4.128) constitute a set of simultaneous equations for the discrete modal coefficients Q_m^n and S_m^n. The dispersion relation is formed by setting the determinant of the simultaneous equations to zero

$$\begin{vmatrix} \Psi_1 & \Psi_2 \\ \Psi_2 & \Psi_1 \end{vmatrix} = 0 \tag{4.129}$$

where the elements of Ψ_1 and Ψ_2 are

$$\psi_{1,ml}^{np} = \xi_m \sin(\xi_m d) I_1 + a\cos(\xi_m d)\delta_{ml}\delta_{np}\varepsilon_m \tag{4.130}$$

$$\psi_{2,ml}^{np} = \xi_m \sin(\xi_m d) I_2\,. \tag{4.131}$$

Solving the dispersion relation (4.129) finally yields the propagation constant k_z.

4.4.2 Residue Calculus for I_1 and I_2

Let us rewrite I_1 as

$$I_1 = \int_{-\infty}^{\infty} \frac{\cot(\kappa b)}{2\pi\kappa} G_m^n(\zeta) G_l^p(-\zeta)\, d\zeta$$
$$= \int_{-\infty}^{\infty} \frac{\zeta^2 \cot(\kappa b) A}{2\pi\kappa(\zeta^2 - a_m^2)(\zeta^2 - a_l^2)}\, d\zeta \tag{4.132}$$

where

$$A = \left[(-1)^{m+l} + 1\right] e^{i\zeta(n-p)T}$$
$$-(-1)^m e^{i\zeta[(n-p)T+2a]} - (-1)^l e^{i\zeta[(n-p)T-2a]}\,. \tag{4.133}$$

- First, consider the case $n = p$.
  When $m + l=$ odd, $I_1 = 0$.
  When $m + l=$ even, I_1 becomes

$$I_1 = \int_{-\infty}^{\infty} \frac{\zeta^2}{\pi\kappa\tan(\kappa b)} \frac{\left[1-(-1)^m e^{i2\zeta a}\right]}{(\zeta^2 - a_m^2)(\zeta^2 - a_l^2)}\, d\zeta$$

$$\equiv \int_{-\infty}^{\infty} f(\zeta)\, d\zeta \,. \tag{4.134}$$

In the complex ζ-plane, as shown in Fig. 4.7, the term $\kappa\sin(\kappa b)$ in I_1 yields an infinite number of poles at $\zeta = \pm\zeta_v = \pm\sqrt{k^2 - \left(\frac{v\pi}{b}\right)^2 - k_z^2}$ for $v = 0, 1, 2, \cdots$. When $m = l$, the condition $\zeta = \pm a_m$ yields two simple poles for $m = 1, 2, 3, \cdots$ and one simple pole for $m = 0$. Performing contour integration in the upper half-plane along Γ_1, Γ_2, Γ_3, and Γ_4 gives

$$\oint f(\zeta)\, d\zeta$$

$$= \int_{\Gamma_1} f(\zeta)\, d\zeta + \int_{\Gamma_2} f(\zeta)\, d\zeta + \int_{\Gamma_3} f(\zeta)\, d\zeta + \int_{\Gamma_4} f(\zeta)\, d\zeta$$

$$= 2\pi i \sum_{v=0}^{\infty} \operatorname*{Res}_{\zeta=\zeta_v} f(\zeta) \,. \tag{4.135}$$

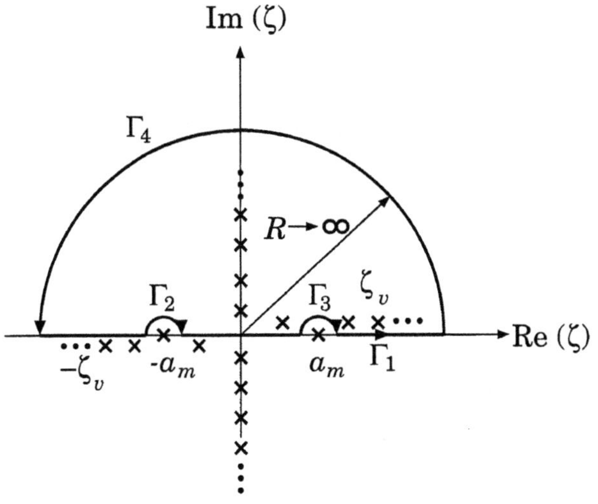

Fig. 4.7. Complex ζ-plane with simple poles.

Note

$$\int_{\Gamma_1} f(\zeta)\, d\zeta = I_1 \tag{4.136}$$

$$\int_{\Gamma_2} f(\zeta)\, d\zeta = -\pi i \operatorname*{Res}_{\zeta=-a_m} f(\zeta) \equiv I_3 \tag{4.137}$$

$$\int_{\Gamma_3} f(\zeta)\, d\zeta = -\pi i \operatorname*{Res}_{\zeta=a_m} f(\zeta) \equiv I_4 \tag{4.138}$$

$$\int_{\Gamma_4} f(\zeta)\, d\zeta = 0\,. \tag{4.139}$$

Hence

$$I_1 = 2\pi i \sum_{v=0}^{\infty} \operatorname*{Res}_{\zeta=\zeta_v} f(\zeta) - \begin{cases} I_3 + I_4 & \text{for } m = 1,2,3,\cdots \\ I_3 & \text{for } m = 0\,. \end{cases} \tag{4.140}$$

Consider

$$\begin{aligned} & I_3(m = 1,2,3,\cdots) \\ &= -\pi i \operatorname*{Res}_{\zeta=-a_m} \left\{ \frac{\zeta^2}{\pi\kappa \tan(\kappa b)} \frac{\left[1-(-1)^m e^{i2\zeta a}\right]}{(\zeta^2 - a_m^2)(\zeta^2 - a_l^2)} \delta_{ml}\delta_{np} \right\} \\ &= -\pi i \left\{ \frac{\zeta^2}{\pi\kappa \tan(\kappa b)} \frac{\left[1-(-1)^m e^{i2\zeta a}\right]}{(\zeta - a_m)^2(\zeta + a_m)} \delta_{ml}\delta_{np} \right\}_{\zeta=-a_m} . \end{aligned} \tag{4.141}$$

Since the term $\left[1-(-1)^m e^{i2\zeta a}\right]$ becomes zero at $\zeta = -a_m$, let us use l'Hôpital's rule

$$\left. \frac{\left[1-(-1)^m e^{i2\zeta a}\right]}{(\zeta + a_m)} \right|_{\zeta=-a_m} = -i2a\,. \tag{4.142}$$

Therefore

$$I_3(m = 1,2,3,\cdots) = -\frac{a}{2} \frac{\delta_{ml}\delta_{np}}{\xi_m \tan(\xi_m b)}\,. \tag{4.143}$$

Similarly, $I_4(m = 1,2,3,\cdots)$ and $I_3(m = 0)$ are

$$I_4(m = 1, 2, 3, \cdots)$$

$$= -\pi i \operatorname*{Res}_{\zeta=a_m} \left\{ \frac{\zeta^2}{\pi\kappa \tan(\kappa b)} \frac{\left[1 - (-1)^m e^{i2\zeta a}\right]}{(\zeta^2 - a_m^2)(\zeta^2 - a_l^2)} \delta_{ml}\delta_{np} \right\}$$

$$= -\pi i \left\{ \frac{\zeta^2}{\pi\kappa \tan(\kappa b)} \frac{\left[1 - (-1)^m e^{i2\zeta a}\right]}{(\zeta + a_m)^2(\zeta - a_m)} \delta_{ml}\delta_{np} \right\}_{\zeta=a_m}$$

$$= -\frac{a}{2} \frac{\delta_{ml}\delta_{np}}{\xi_m \tan(\xi_m b)} \tag{4.144}$$

$$I_3(m = 0) = -\pi i \operatorname*{Res}_{\zeta=0} \left[\frac{1}{\pi\kappa \tan(\kappa b)} \frac{(1 - e^{i2\zeta a})}{\zeta^2} \delta_{0l}\delta_{np} \right]$$

$$= -2a \frac{\delta_{0l}\delta_{np}}{\xi_0 \tan(\xi_0 b)} . \tag{4.145}$$

Therefore

$$I_1 = a \frac{\varepsilon_m \delta_{ml}\delta_{np}}{\xi_m \tan(\xi_m b)} + 2\pi i \sum_{v=0}^{\infty} \operatorname*{Res}_{\zeta=\zeta_v} f(\zeta) . \tag{4.146}$$

Consider

$$\operatorname*{Res}_{\zeta=\zeta_v} f(\zeta)$$

$$= \operatorname*{Res}_{\zeta=\zeta_v} \left\{ \frac{\zeta^2}{\pi\kappa \tan(\kappa b)} \frac{\left[1 - (-1)^m e^{i2\zeta a}\right]}{(\zeta^2 - a_m^2)(\zeta^2 - a_l^2)} \delta_{np} \right\}$$

$$= \left\{ \frac{\zeta_v^2}{\pi} \frac{\left[1 - (-1)^m e^{i2\zeta_v a}\right]}{(\zeta_v^2 - a_m^2)(\zeta_v^2 - a_l^2)} \delta_{np} \right\} \operatorname*{Res}_{\zeta=\zeta_v} \left[\frac{1}{\kappa \tan(\kappa b)} \right] . \tag{4.147}$$

When $\zeta = \zeta_0$ $(\kappa = 0)$,

$$\operatorname*{Res}_{\zeta=\zeta_0} \left[\frac{1}{\kappa \tan(\kappa b)} \right] = -\frac{1}{2\zeta_0 b} . \tag{4.148}$$

When $\zeta = \zeta_1 = \zeta_2, \cdots$,

$$\operatorname*{Res}_{\zeta=\zeta_v} \left[\frac{1}{\kappa \tan(\kappa b)} \right] = -\frac{1}{\zeta_v b} . \tag{4.149}$$

Therefore, I_1 can be rewritten as

$$I_1 = a \frac{\varepsilon_m \delta_{ml}\delta_{np}}{\xi_m \tan(\xi_m b)} - \frac{i}{b} \sum_{v=0}^{\infty} \frac{2\zeta_v \left[1 - (-1)^m e^{i2\zeta_v a}\right]}{\varepsilon_v (\zeta_v^2 - a_m^2)(\zeta_v^2 - a_l^2)} \delta_{np} . \tag{4.150}$$

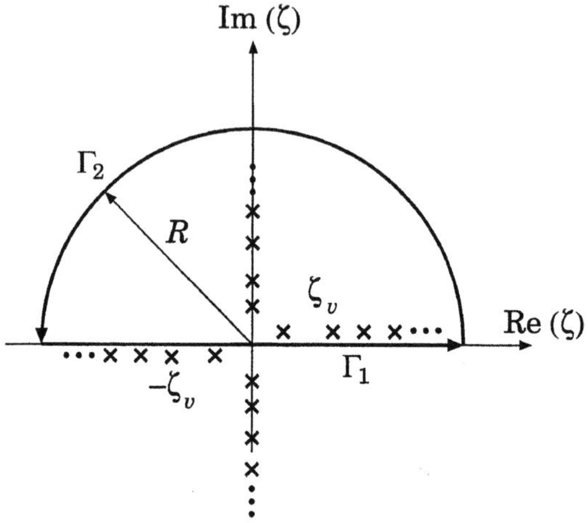

Fig. 4.8. Complex ζ-plane with simple poles.

- Second, consider the case $n \neq p$. Note that I_1 contains an infinite number of poles at $\zeta = \pm\zeta_v$, which are the solutions to $\kappa \sin(\kappa b) = 0$. When $n - p > 0$, let us use the contour shown in Fig. 4.8 to evaluate I_1.

$$\begin{aligned} I_1 &= 2\pi i \sum_{v=0}^{\infty} \operatorname*{Res}_{\zeta=\zeta_v} \left[\frac{\zeta^2 \cot(\kappa b) A}{2\pi\kappa(\zeta^2 - a_m^2)(\zeta^2 - a_l^2)} \right] \\ &= 2\pi i \sum_{v=0}^{\infty} \left. \frac{\zeta^2 A}{2\pi(\zeta^2 - a_m^2)(\zeta^2 - a_l^2)} \right|_{\zeta=\zeta_v} \operatorname*{Res}_{\zeta=\zeta_v} \left[\frac{1}{\kappa \tan(\kappa b)} \right] \\ &= -\frac{i}{b} \sum_{v=0}^{\infty} \left. \frac{\zeta A}{\varepsilon_v(\zeta^2 - a_m^2)(\zeta^2 - a_l^2)} \right|_{\zeta=\zeta_v} . \end{aligned} \tag{4.151}$$

Similarly, when $n - p < 0$,

$$I_1 = \frac{i}{b} \sum_{v=0}^{\infty} \left. \frac{\zeta A}{\varepsilon_v(\zeta^2 - a_m^2)(\zeta^2 - a_l^2)} \right|_{\zeta=-\zeta_v} . \tag{4.152}$$

It is possible to combine (4.151) and (4.152) into

$$I_1 = -\frac{i}{b} \sum_{v=0}^{\infty} \frac{\zeta_v \tilde{A}}{\varepsilon_v(\zeta_v^2 - a_m^2)(\zeta_v^2 - a_l^2)} \tag{4.153}$$

$$\begin{aligned} \tilde{A} &= \left[(-1)^{m+l} + 1\right] e^{i\zeta_v |n-p| T} \\ &\quad -(-1)^m e^{i\zeta_v |(n-p)T + 2a|} - (-1)^l e^{i\zeta_v |(n-p)T - 2a|} . \end{aligned} \tag{4.154}$$

Rewriting (4.150) and (4.153) in a compact form results in

$$I_1 = a\frac{\varepsilon_m \delta_{ml} \delta_{np}}{\xi_m \tan(\xi_m b)} - \frac{i}{b}\sum_{v=0}^{\infty} \frac{\zeta_v \tilde{A}}{\varepsilon_v(\zeta_v^2 - a_m^2)(\zeta_v^2 - a_l^2)} . \tag{4.155}$$

Similarly, it can be shown that

$$I_2 = a\frac{\varepsilon_m \delta_{ml} \delta_{np}}{\xi_m \sin(\xi_m b)} - \frac{i}{b}\sum_{v=0}^{\infty} (-1)^v \frac{\zeta_v \tilde{A}}{\varepsilon_v(\zeta_v^2 - a_m^2)(\zeta_v^2 - a_l^2)} . \tag{4.156}$$

Problems for Chapter 4

1. Consider the TE_{mnp} mode in the circular cavity resonator in Fig. 4.3. Determine the lowest resonant frequency and energy stored in the cavity resonator.
2. Consider the TE_{mnp} mode in the spherical cavity resonator in Fig. 4.4. Determine the lowest resonant frequency.
3. Consider the integral I_2 given by (4.126). Using residue calculus, show that (4.126) is represented by the series given by (4.156).
4. Consider the groove guide coupler shown in Fig. 4.6. Assume TM wave propagation in the z-direction ($H_z = 0$ and $E_z \neq 0$). Following the TE wave analysis in Section 4.4, derive the dispersion relation for a TM wave.

5

Propagation in Anisotropic Media

5.1 Propagation in Anisotropic Media

Materials such as crystals, magnetoplasma, and ferrites possess anisotropic properties at optical or microwave frequencies. In anisotropic materials, the direction of $\overline{E}$ (or $\overline{H}$) usually does not coincide with that of $\overline{D}$ (or $\overline{B}$). For instance, the relation between $\overline{D}$ and $\overline{E}$ using rectangular coordinates is

$$\begin{pmatrix} D_x \\ D_y \\ D_z \end{pmatrix} = \begin{pmatrix} \epsilon_{11} & \epsilon_{12} & \epsilon_{13} \\ \epsilon_{21} & \epsilon_{22} & \epsilon_{23} \\ \epsilon_{31} & \epsilon_{32} & \epsilon_{33} \end{pmatrix} \begin{pmatrix} E_x \\ E_y \\ E_z \end{pmatrix} . \tag{5.1}$$

Equation (5.1) can be rewritten as

$$\overline{D} = \overline{\overline{\epsilon}}\,\overline{E} \tag{5.2}$$

where $\overline{\overline{\epsilon}}$ is a tensor permittivity represented by a (3×3) matrix.

Similarly, it is possible to relate $\overline{B}$ to $\overline{H}$ in terms of the tensor permeability $\overline{\overline{\mu}}$ as

$$\overline{B} = \overline{\overline{\mu}}\,\overline{H} . \tag{5.3}$$

Wave propagation in anisotropic materials depends on the wave propagation direction with respect to the tensor permittivity or permeability.

5.1.1 Dispersion Relation

This section investigates the wave propagation characteristics in unbounded anisotropic materials and derives a dispersion relation using the tensor permittivity. Assume that a plane wave

$$\overline{E} = \overline{E}_0 \exp\left(i\overline{k} \cdot \overline{r}\right) = \overline{E}_0 \exp(ik_x x + ik_y y + ik_z z) \tag{5.4}$$

propagates within an unbounded, source-free, and anisotropic medium with tensor permittivity $\overline{\overline{\epsilon}}$. Note that for a constant vector $\overline{E}_0$

$$\nabla \times \overline{E} = \nabla \times \left[\overline{E}_0 \exp\left(i\overline{k} \cdot \overline{r}\right)\right] = i\overline{k} \times \overline{E} \,. \tag{5.5}$$

Therefore, Maxwell's equations

$$\nabla \times \overline{E} = i\omega\mu\overline{H} \tag{5.6}$$

$$\nabla \times \overline{H} = -i\omega\overline{\overline{\epsilon}}\,\overline{E} \tag{5.7}$$

become

$$i\overline{k} \times \overline{E} = i\omega\mu\overline{H} \tag{5.8}$$

$$i\overline{k} \times \overline{H} = -i\omega\overline{\overline{\epsilon}}\,\overline{E} \,. \tag{5.9}$$

Eliminating $\overline{H}$ from (5.8) and (5.9) gives

$$\overline{k} \times \left(\overline{k} \times \overline{E}\right) + \omega^2\mu\overline{\overline{\epsilon}}\,\overline{E} = 0 \tag{5.10}$$

where the vector identity

$$\overline{k} \times \left(\overline{k} \times \overline{E}\right) = -\left(\overline{k} \cdot \overline{k}\right)\overline{E} + \overline{k}\left(\overline{k} \cdot \overline{E}\right) \tag{5.11}$$

can be used for further simplification.

It is possible to express $\overline{\overline{\epsilon}}$ in a diagonal form using newly rotated coordinates. The axes of the newly rotated coordinates, which make the tensor permittivity $\overline{\overline{\epsilon}}$ diagonal, are called the principal axes. Assume that the tensor permittivity $\overline{\overline{\epsilon}}$ using the principal coordinates (x, y, z) is represented by

$$\overline{\overline{\epsilon}} = \begin{pmatrix} \epsilon_x & 0 & 0 \\ 0 & \epsilon_y & 0 \\ 0 & 0 & \epsilon_z \end{pmatrix} . \tag{5.12}$$

Substituting $\overline{\overline{\epsilon}}$ into (5.10) results in

$$\begin{pmatrix} \omega^2\mu\epsilon_x - k_y^2 - k_z^2 & k_x k_y & k_x k_z \\ k_x k_y & \omega^2\mu\epsilon_y - k_x^2 - k_z^2 & k_y k_z \\ k_x k_z & k_y k_z & \omega^2\mu\epsilon_z - k_x^2 - k_y^2 \end{pmatrix} \begin{pmatrix} E_x \\ E_y \\ E_z \end{pmatrix} = 0 \,. \tag{5.13}$$

Let

$$k^2 - \omega^2 \mu \epsilon_x = p^2 \tag{5.14}$$

$$k^2 - \omega^2 \mu \epsilon_y = q^2 \tag{5.15}$$

$$k^2 - \omega^2 \mu \epsilon_z = s^2 \tag{5.16}$$

where $k^2 = k_x^2 + k_y^2 + k_z^2$. Consequently, (5.13) can be rewritten as

$$\begin{pmatrix} k_x^2 - p^2 & k_x k_y & k_x k_z \\ k_x k_y & k_y^2 - q^2 & k_y k_z \\ k_x k_z & k_y k_z & k_z^2 - s^2 \end{pmatrix} \begin{pmatrix} E_x \\ E_y \\ E_z \end{pmatrix} = 0\,. \tag{5.17}$$

The dispersion relation is obtained by setting its determinant to zero

$$\begin{vmatrix} k_x^2 - p^2 & k_x k_y & k_x k_z \\ k_x k_y & k_y^2 - q^2 & k_y k_z \\ k_x k_z & k_y k_z & k_z^2 - s^2 \end{vmatrix} = 0 \tag{5.18}$$

which can be further simplified to

$$(pqk_z)^2 + (psk_y)^2 + (qsk_x)^2 = (pqs)^2\,. \tag{5.19}$$

The dispersion relation (5.19) is a quadratic equation of k^2. Therefore, there are two solutions for k^2. Once the wavenumbers k are determined from the dispersion relation, the field components E_x, E_y, and E_z can be obtained from (5.17).

5.1.2 Uniaxial Medium

For certain anisotropic crystals known as a uniaxial medium, $\bar{\bar{\epsilon}}$ takes the following special form using newly rotated coordinates (x, y, z)

$$\bar{\bar{\epsilon}} = \begin{pmatrix} \epsilon_1 & 0 & 0 \\ 0 & \epsilon_1 & 0 \\ 0 & 0 & \epsilon_2 \end{pmatrix} \tag{5.20}$$

where the z-direction is called the optical axis. Quartz and lithium niobate are uniaxial crystals. For lithium niobate, $\epsilon_1 = (2.300)^2\,\epsilon_0$ and $\epsilon_2 = (2.208)^2\,\epsilon_0$. Assume that a plane wave

$$\overline{E} = \overline{E}_0 \exp\left(i\overline{k}\cdot\overline{r}\right) \equiv \hat{x}E_x + \hat{y}E_y + \hat{z}E_z \tag{5.21}$$

propagates within an unbounded uniaxial medium, where the incident wave vector $\overline{k}$ is

$$\overline{k} = k\left(\hat{x}\sin\theta\cos\phi + \hat{y}\sin\theta\sin\phi + \hat{z}\cos\theta\right)$$

$$\equiv \hat{x}k_x + \hat{y}k_y + \hat{z}k_z\,. \tag{5.22}$$

The dispersion relation (5.19) becomes

$$p^2\left[-p^2k_z^2 - s^2(k_x^2+k_y^2) + s^2p^2\right] = 0. \tag{5.23}$$

Let us solve (5.23) to investigate the wave propagation characteristics.

- The first solution $p^2 = 0$ yields the wavenumber $k = \omega\sqrt{\mu\epsilon_1}$. The relation

$$\begin{pmatrix} k_x^2 & k_xk_y & k_xk_z \\ k_xk_y & k_y^2 & k_yk_z \\ k_xk_z & k_yk_z & k_z^2 - s^2 \end{pmatrix}\begin{pmatrix} E_x \\ E_y \\ E_z \end{pmatrix} = 0 \tag{5.24}$$

yields $E_z = 0$ and $\dfrac{E_x}{E_y} = -\dfrac{k_y}{k_x} = -\tan\phi$, which represents an ordinary wave that propagates isotropically through a uniaxial medium. The electric field vector of the ordinary wave is perpendicular to the wave vector $\overline{k}$ and optical axis z since

$$\overline{E}\cdot\overline{k} = E_xk_x + E_yk_y + E_zk_z = 0 \tag{5.25}$$

$$\overline{E}\cdot\hat{z} = 0\,. \tag{5.26}$$

The time-average Poynting vector $\overline{S}_{av}$ for the ordinary wave is

$$\begin{aligned}\overline{S}_{av} &= \frac{1}{2}\,\mathrm{Re}\left(\overline{E}\times\overline{H}^*\right)\\ &= \frac{1}{2\omega\mu}\,\mathrm{Re}\left[\overline{E}\times\left(\overline{k}\times\overline{E}^*\right)\right]\\ &= \frac{1}{2\omega\mu}\,\mathrm{Re}\left[\left(\overline{E}\cdot\overline{E}^*\right)\overline{k} - \overline{E}^*\underbrace{\left(\overline{E}\cdot\overline{k}\right)}_{0}\right]\\ &= \frac{|\overline{E}|^2}{2\omega\mu}\overline{k}\,.\end{aligned} \tag{5.27}$$

The expression (5.27) indicates that the power flows in the $\overline{k}$-direction. A graphical representation of the ordinary wave is shown in Fig. 5.1.

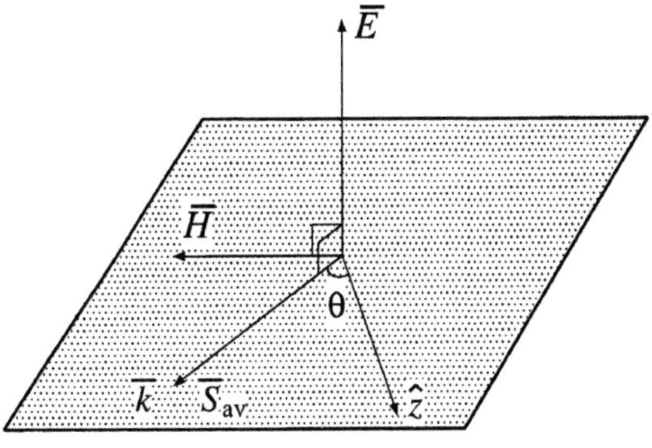

Fig. 5.1. Ordinary wave in uniaxial medium.

- The second solution

$$-p^2k_z^2 - s^2\left(k_x^2 + k_y^2\right) + s^2p^2 = 0 \tag{5.28}$$

yields the wavenumber

$$k = \omega\sqrt{\frac{\mu\epsilon_1\epsilon_2}{\epsilon_1\sin^2\theta + \epsilon_2\cos^2\theta}} \tag{5.29}$$

and field components

$$\frac{E_x}{E_z} = \frac{s^2k_x}{p^2k_z} \tag{5.30}$$

$$\frac{E_y}{E_z} = \frac{s^2k_y}{p^2k_z}\ . \tag{5.31}$$

The second solution represents an extraordinary wave that propagates anisotropically in a uniaxial medium. The time-average Poynting vector $\overline{S}_{av}$ for the extraordinary wave is

$$\begin{aligned}\overline{S}_{av} &= \frac{1}{2}\,\mathrm{Re}\left(\overline{E}\times\overline{H}^*\right)\\ &= \frac{1}{2\omega\mu}\,\mathrm{Re}\left[\overline{k}\left|\overline{E}\right|^2 - \overline{E}^*\left(\overline{E}\cdot\overline{k}\right)\right]\ .\end{aligned} \tag{5.32}$$

Since

$$\overline{E}\cdot\overline{k} = \frac{E_z}{p^2k_z}\left(s^2k_x^2 + s^2k_y^2 + p^2k_z^2\right) = E_z\frac{s^2}{k_z} \neq 0 \tag{5.33}$$

the electric field vector of the extraordinary wave is, in general, not perpendicular to the wave vector $\overline{k}$. The direction of the power flow does not coincide with the $\overline{k}$-direction. The electric field vectors of ordinary $(\overline{E}_{ord})$ and extraordinary $(\overline{E}_{ext})$ waves are perpendicular to each other since

$$\overline{E}_{ord} \cdot \overline{E}_{ext} = 0 . \tag{5.34}$$

A graphical representation of the extraordinary wave is shown in Fig. 5.2.

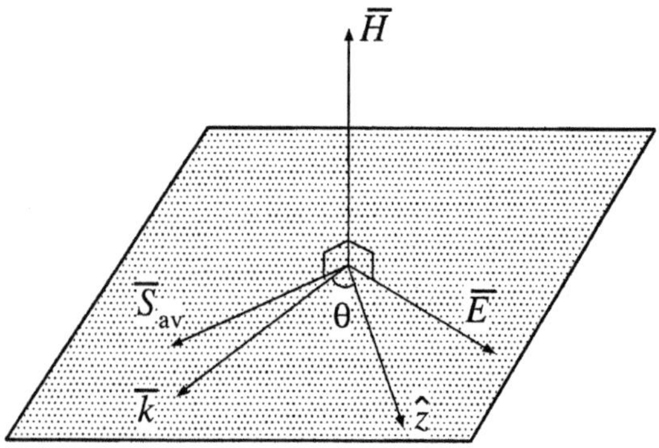

Fig. 5.2. Extraordinary wave in uniaxial medium.

A uniaxial medium permits two types of waves, namely, ordinary and extraordinary waves. A uniaxial medium is, therefore, said to possess a birefringence or double refraction property.

5.2 Propagation in Ferrites

Certain magnetic materials, called ferrites, find wide applications as phase-shifting elements in microwave circuits and antennas. Commonly used ferrite devices include the ferrite gyrators, ferrite isolators, and ferrite circulators. This section investigates the wave propagation characteristics in ferrite materials in terms of the dispersion relation.

5.2.1 Magnetized Ferrite

When ferrites are subject to a DC magnetic field, i.e. when ferrites are magnetized, they exhibit anisotropy and their permeability takes a tensor form $\overline{\overline{\mu}}$. The permeability of ferrite materials depends on the direction and state of the internal DC magnetic field. If the applied magnetic field is, for example, in the z-direction, the ferrite permeability $\overline{\overline{\mu}}$ is represented as

$$\overline{\overline{\mu}} = \begin{pmatrix} \mu & i\kappa & 0 \\ -i\kappa & \mu & 0 \\ 0 & 0 & \mu_0 \end{pmatrix} \tag{5.35}$$

where the explicit expressions for the ferrite parameters, μ and κ, are given in [2, page 235].

Assume that a plane wave $\overline{H} = \overline{H}_0 \exp\left(i\overline{k}\cdot\overline{r}\right)$ propagates in an unbounded ferrite medium, where the governing equation is written as

$$\overline{k} \times \left(\overline{k} \times \overline{H}\right) + \omega^2 \epsilon \overline{\overline{\mu}}\,\overline{H} = 0\,. \tag{5.36}$$

First, consider wave propagation in any arbitrary direction with the wavenumber $\overline{k} = \hat{x}k_x + \hat{y}k_y + \hat{z}k_z$. Substituting the vector identity

$$\overline{k} \times \left(\overline{k} \times \overline{H}\right) = \overline{k}\left(\overline{k}\cdot\overline{H}\right) - \left(\overline{k}\cdot\overline{k}\right)\overline{H} \tag{5.37}$$

into (5.36) yields

$$\begin{pmatrix} \omega^2\mu\epsilon - k_y^2 - k_z^2 & i\omega^2\kappa\epsilon + k_xk_y & k_xk_z \\ -i\omega^2\kappa\epsilon + k_xk_y & \omega^2\mu\epsilon - k_x^2 - k_z^2 & k_yk_z \\ k_xk_z & k_yk_z & \omega^2\mu_0\epsilon - k_x^2 - k_y^2 \end{pmatrix} \begin{pmatrix} H_x \\ H_y \\ H_z \end{pmatrix} = 0\,. \tag{5.38}$$

This can be rewritten as

$$\begin{pmatrix} k_x^2 - p^2 & i\omega^2\kappa\epsilon + k_xk_y & k_xk_z \\ -i\omega^2\kappa\epsilon + k_xk_y & k_y^2 - p^2 & k_yk_z \\ k_xk_z & k_yk_z & k_z^2 - q^2 \end{pmatrix} \begin{pmatrix} H_x \\ H_y \\ H_z \end{pmatrix} = 0 \tag{5.39}$$

where $p^2 = k^2 - \omega^2\mu\epsilon$, $q^2 = k^2 - \omega^2\mu_0\epsilon$, and $k^2 = k_x^2 + k_y^2 + k_z^2$.

Setting the determinant to zero

$$\begin{vmatrix} k_x^2 - p^2 & i\omega^2\kappa\epsilon + k_xk_y & k_xk_z \\ -i\omega^2\kappa\epsilon + k_xk_y & k_y^2 - p^2 & k_yk_z \\ k_xk_z & k_yk_z & k_z^2 - q^2 \end{vmatrix} = 0 \tag{5.40}$$

yields the dispersion relation

$$k_x^2p^2q^2 + k_y^2p^2q^2 + k_z^2\left(p^4 - \omega^4\kappa^2\epsilon^2\right) = \left(p^4 - \omega^4\kappa^2\epsilon^2\right)q^2\,. \tag{5.41}$$

When the wavenumber

$$\overline{k} = \hat{x}k_x + \hat{y}k_y + \hat{z}k_z$$

$$= k\left(\hat{x}\sin\theta\cos\phi + \hat{y}\sin\theta\sin\phi + \hat{z}\cos\theta\right) \tag{5.42}$$

is assumed, the dispersion relation becomes

$$Ak^4 + Bk^2 + C = 0 \tag{5.43}$$

where

$$A = \left(\mu_0\cos^2\theta + \mu\sin^2\theta\right)\omega^2\epsilon \tag{5.44}$$

$$B = -\omega^4\epsilon^2\left[\mu\mu_0\left(1+\cos^2\theta\right) + \left(\mu^2 - \kappa^2\right)\sin^2\theta\right] \tag{5.45}$$

$$C = \omega^6\left(\mu^2 - \kappa^2\right)\mu_0\epsilon^3 \,. \tag{5.46}$$

5.2.2 Transversely Magnetized Ferrite

As a special case, consider a wave that propagates in the direction transverse to the z-axis $\left(\overline{k} = \hat{x}k_x + \hat{y}k_y\right)$. Since $k_z = 0$ and $k_x^2 + k_y^2 = k^2$, (5.39) becomes

$$\begin{pmatrix} k_x^2 - p^2 & i\omega^2\kappa\epsilon + k_x k_y & 0 \\ -i\omega^2\kappa\epsilon + k_x k_y & k_y^2 - p^2 & 0 \\ 0 & 0 & -q^2 \end{pmatrix}\begin{pmatrix} H_x \\ H_y \\ H_z \end{pmatrix} = 0 \,. \tag{5.47}$$

The dispersion relation (5.41) reduces to

$$q^2\left(p^4 - \omega^4\kappa^2\epsilon^2 - k^2p^2\right) = 0 \tag{5.48}$$

which yields the two solutions $k = \omega\sqrt{\mu_0\epsilon}$ or $\omega\sqrt{\epsilon\left(\mu - \dfrac{\kappa^2}{\mu}\right)}$.

- When $k = \omega\sqrt{\mu_0\epsilon}$, a TEM wave $\left(H_z \neq 0 \text{ and } H_x = H_y = 0\right)$ propagates and the ferrite material behaves like an isotropic medium.

- When $k = \omega\sqrt{\epsilon\left(\mu - \dfrac{\kappa^2}{\mu}\right)}$, the field components are related by

$$\frac{H_x}{H_y} = -\frac{i\kappa\mu + \left(\mu^2 - \kappa^2\right)\cos\phi\sin\phi}{\mu^2\cos^2\phi + \kappa^2\sin^2\phi} \,. \tag{5.49}$$

5.2.3 Longitudinally Magnetized Ferrite

Let us consider another special case in which $k_x = k_y = 0$ and $\overline{k} = \hat{z}k_z$, where the direction of wave propagation coincides with the direction of the DC magnetic field. Then, the dispersion relation (5.41) gives

$$k_z = \omega\sqrt{\epsilon(\mu \pm \kappa)} \equiv k^{\pm} \, . \tag{5.50}$$

Substituting k^+ into (5.39) results in

$$\begin{pmatrix} -\kappa & i\kappa & 0 \\ -i\kappa & -\kappa & 0 \\ 0 & 0 & \mu_0 \end{pmatrix} \begin{pmatrix} H_x \\ H_y \\ H_z \end{pmatrix} = 0 \tag{5.51}$$

which gives

$$H_x = iH_y \, . \tag{5.52}$$

Substituting (5.52) into Ampère's law

$$\nabla \times \overline{H} = -i\omega\epsilon\overline{E} \tag{5.53}$$

gives

$$E_x = iE_y \, . \tag{5.54}$$

Similarly, the condition $k_z = k^-$ gives

$$H_x = -iH_y \tag{5.55}$$

$$E_x = -iE_y \, . \tag{5.56}$$

Therefore, the electric field is represented as

$$\overline{E}^{\pm} = \exp\left(ik^{\pm}z\right)(\hat{x} \mp \hat{y}i) \tag{5.57}$$

which indicates left-hand $\left(\overline{E}^{+}\right)$ and right-hand $\left(\overline{E}^{-}\right)$ circular polarizations.

Let us consider a linear combination of two circularly polarized waves

$$\begin{aligned} \overline{E} &= \overline{E}^{+} + \overline{E}^{-} \\ &= (\hat{x} - \hat{y}i)e^{ik^{+}z} + (\hat{x} + \hat{y}i)e^{ik^{-}z} \\ &= \hat{x}\left(e^{ik^{+}z} + e^{ik^{-}z}\right) + \hat{y}i\left(e^{ik^{-}z} - e^{ik^{+}z}\right) \end{aligned} \tag{5.58}$$

where their time-varying form is

$$E_x(z,t) = \cos(k^+ z - \omega t) + \cos(k^- z - \omega t)$$

$$= 2\cos\left(\frac{k^+ - k^-}{2} z\right)\cos\left(\frac{k^+ + k^-}{2} z - \omega t\right) \quad (5.59)$$

$$E_y(z,t) = \sin(k^+ z - \omega t) - \sin(k^- z - \omega t)$$

$$= 2\sin\left(\frac{k^+ - k^-}{2} z\right)\cos\left(\frac{k^+ + k^-}{2} z - \omega t\right) . \quad (5.60)$$

Taking the ratio of

$$\frac{E_y(z,t)}{E_x(z,t)} = \tan\left(\frac{k^+ - k^-}{2} z\right) \quad (5.61)$$

shows that the polarization state of (5.58) is not only linear but also rotates as the wave propagates along the z-direction. This behavior is known as a Faraday rotation. The principle of a ferrite gyrator, ferrite isolator, and ferrite circulator is based on the Faraday rotation of a wave that propagates in ferrites at microwave frequencies. Further discussion on the microwave gyrator, microwave isolator, and microwave circulator can be found, for instance, in [5].

5.3 Propagation Along Ferrite-Filled Parallel-Plate Waveguide

Waveguides containing magnetized ferrites find practical applications in microwave circuits and systems. For instance, a ferrite-loaded rectangular waveguide is used as a nonreciprocal phase shifter in microwave frequencies. This section considers a conducting parallel-plate waveguide filled with transversely magnetized ferrite. The problem geometry is shown in Fig. 5.3. Note that some relevant discussion is available in [6]. Assume that a DC magnetic field is applied along the z-direction and the ferrite tensor permeability $\overline{\overline{\mu}}$ is given by

$$\overline{\overline{\mu}} = \begin{pmatrix} \mu & i\kappa & 0 \\ -i\kappa & \mu & 0 \\ 0 & 0 & \mu_0 \end{pmatrix} . \quad (5.62)$$

A wave is assumed to propagate in the y-direction. Assume no field variation in the z-direction $\left(\frac{\partial}{\partial z} = 0\right)$.

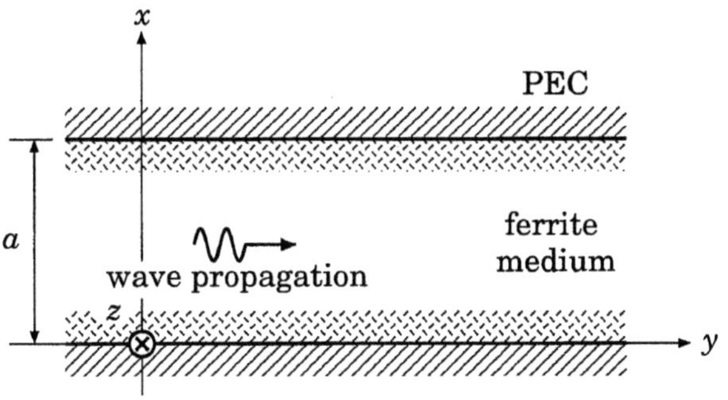

Fig. 5.3. Conducting parallel-plate waveguide filled with ferrite.

Faraday's law

$$\nabla \times \overline{E} = i\omega \overline{\overline{\mu}} \, \overline{H} \tag{5.63}$$

can be rewritten as

$$\frac{\partial E_z}{\partial y} = i\omega \left(\mu H_x + i\kappa H_y \right) \tag{5.64}$$

$$-\frac{\partial E_z}{\partial x} = i\omega \left(-i\kappa H_x + \mu H_y \right) \tag{5.65}$$

$$\frac{\partial E_y}{\partial x} - \frac{\partial E_x}{\partial y} = i\omega\mu_0 H_z \, . \tag{5.66}$$

Similarly, Ampère's law

$$\nabla \times \overline{H} = -i\omega\epsilon\overline{E} \tag{5.67}$$

can be rewritten as

$$\frac{\partial H_z}{\partial y} = -i\omega\epsilon E_x \tag{5.68}$$

$$-\frac{\partial H_z}{\partial x} = -i\omega\epsilon E_y \tag{5.69}$$

$$\frac{\partial H_y}{\partial x} - \frac{\partial H_x}{\partial y} = -i\omega\epsilon E_z \, . \tag{5.70}$$

Equations (5.64), (5.65), and (5.70) govern the field components (E_z, H_x, and H_y), whereas (5.66), (5.68), and (5.69) govern the field components (H_z, E_x, and E_y), respectively. The field components (E_z, H_x, and H_y) and (H_z,

E_x, and E_y) constitute TE ($E_y = 0$) and TM ($H_y = 0$) waves, respectively. Thus, a parallel-plate waveguide filled with transversely magnetized ferrite admits TE and TM waves.

- Let us first consider the TE case. Solving (5.64) and (5.65) for H_x and H_y yields

$$H_x = \frac{\left(-i\mu\frac{\partial E_z}{\partial y} + \kappa\frac{\partial E_z}{\partial x}\right)}{\omega\left(\mu^2 - \kappa^2\right)} \tag{5.71}$$

$$H_y = \frac{\left(i\mu\frac{\partial E_z}{\partial x} + \kappa\frac{\partial E_z}{\partial y}\right)}{\omega\left(\mu^2 - \kappa^2\right)} . \tag{5.72}$$

Substituting H_x and H_y into (5.70) gives the governing equation

$$\frac{\partial^2 E_z}{\partial x^2} + \frac{\partial^2 E_z}{\partial y^2} + \omega^2\left(\mu - \frac{\kappa^2}{\mu}\right)\epsilon E_z = 0 . \tag{5.73}$$

Based on the separation of variables technique, E_z takes the form of

$$E_z = A_n \sin\left(\frac{n\pi}{a}x\right)\exp(ik_y y) \tag{5.74}$$

where $n = 1, 2, 3\cdots$. The expression E_z is chosen to satisfy the boundary condition $E_z\Big|_{x=0,a} = 0$.

Substituting (5.74) into (5.73) produces the propagation constant k_y

$$k_y = \sqrt{\omega^2\left(\mu - \frac{\kappa^2}{\mu}\right)\epsilon - \left(\frac{n\pi}{a}\right)^2} . \tag{5.75}$$

Subsequently, substituting (5.74) into (5.71) and (5.72) results in the magnetic field components

$$H_x = \frac{A_n \exp(ik_y y)}{\omega(\mu^2 - \kappa^2)}\left[\mu k_y \sin\left(\frac{n\pi}{a}x\right) + \kappa\frac{n\pi}{a}\cos\left(\frac{n\pi}{a}x\right)\right] \tag{5.76}$$

$$H_y = \frac{A_n \exp(ik_y y)}{\omega(\mu^2 - \kappa^2)}\left[i\mu\frac{n\pi}{a}\cos\left(\frac{n\pi}{a}x\right) + i\kappa k_y \sin\left(\frac{n\pi}{a}x\right)\right] . \tag{5.77}$$

- Next, consider the TM wave case. Substituting E_x and E_y from (5.68) and (5.69) into (5.66) gives

$$\left(\frac{\partial^2}{\partial x^2} + \frac{\partial^2}{\partial y^2} + \omega^2\mu_0\epsilon\right)H_z = 0 . \tag{5.78}$$

Since E_y must vanish at $x = 0$ and a, H_z takes the form of

$$H_z = B_n \cos\left(\frac{n\pi}{a}x\right) \exp(ik_y y) \tag{5.79}$$

where $n = 0, 1, 2, \cdots$. Substituting (5.79) into (5.78) finally gives the solution

$$k_y = \sqrt{\omega^2 \mu_0 \epsilon - \left(\frac{n\pi}{a}\right)^2}\,. \tag{5.80}$$

The TM wave propagates as if the waveguide were filled with an isotropic medium with permittivity ϵ and permeability μ_0.

Problems for Chapter 5

1. Prove the vector identity $\overline{k} \times \left(\overline{k} \times \overline{E}\right) = -\left(\overline{k} \cdot \overline{k}\right) \overline{E} + \overline{k} \left(\overline{k} \cdot \overline{E}\right)$.
2. Show that (5.41) reduces to (5.43).
3. Consider a TE wave propagating in a ferrite-filled parallel-plate waveguide, as shown in Fig. 5.3. Compute the time-average power delivered by the TE wave.

6

Electromagnetic Theorems

When an electromagnetic wave impinges on scattering objects, the incident wave not only scatters off but also penetrates the scattering objects. If the scattering objects are complex-shaped, the scattering analysis becomes complicated; hence, electromagnetic theorems can be used to facilitate the problem formulation. This chapter introduces some useful theorems and principles that are often used in electromagnetic scattering analyses.

6.1 Uniqueness Theorem

When solving electromagnetic boundary-value problems, it is important to ensure the uniqueness of the solutions. The uniqueness theorem provides the boundary conditions that guarantee the uniqueness of solutions to boundary-value problems. Consider the electromagnetic fields $\overline{E}$ and $\overline{H}$ within volume V that contains the electric and magnetic sources $\overline{J}_i$ and $\overline{M}_i$. The problem geometry is shown in Fig. 6.1. The uniqueness theorem can be stated as follows: The fields $\overline{E}$ and $\overline{H}$ can be uniquely determined if one of the following

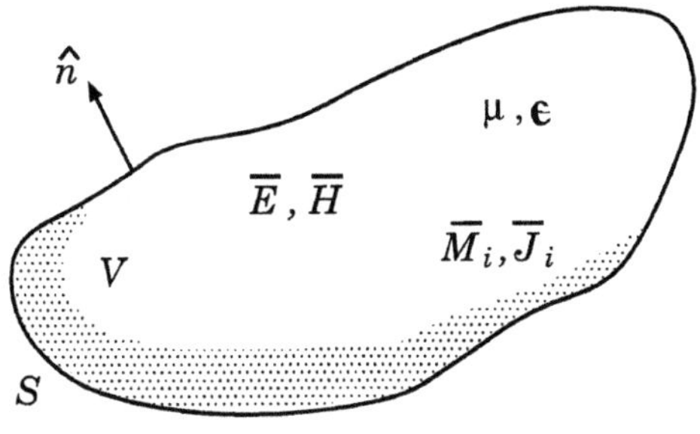

Fig. 6.1. Bounded volume V containing sources $\overline{J}_i$ and $\overline{M}_i$.

boundary conditions is specified on the surface S.

1. The tangential component $\hat{n} \times \overline{E}$ is given on S.
2. The tangential component $\hat{n} \times \overline{H}$ is given on S.
3. The tangential component $\hat{n} \times \overline{E}$ is given on a portion of S and the tangential component $\hat{n} \times \overline{H}$ is given on the rest of S.

Proof

Let us assume that two different fields $\left(\overline{E}_a, \overline{H}_a\right)$ and $\left(\overline{E}_b, \overline{H}_b\right)$ exist within V. Each field must obey Maxwell's equations as

$$\nabla \times \overline{E}_a = i\omega\mu\overline{H}_a - \overline{M}_i \tag{6.1}$$

$$\nabla \times \overline{H}_a = -i\omega\epsilon\overline{E}_a + \overline{J}_i \tag{6.2}$$

$$\nabla \times \overline{E}_b = i\omega\mu\overline{H}_b - \overline{M}_i \tag{6.3}$$

$$\nabla \times \overline{H}_b = -i\omega\epsilon\overline{E}_b + \overline{J}_i \,. \tag{6.4}$$

Subtracting (6.3) and (6.4) from (6.1) and (6.2), respectively, yields

$$\nabla \times \overline{E}_d = i\omega\mu\overline{H}_d \tag{6.5}$$

$$\nabla \times \overline{H}_d = -i\omega\epsilon\overline{E}_d \tag{6.6}$$

where $\overline{E}_d = \overline{E}_a - \overline{E}_b$ and $\overline{H}_d = \overline{H}_a - \overline{H}_b$.

Let us consider

$$\begin{aligned}\nabla \cdot \left(\overline{E}_d \times \overline{H}_d^*\right) &= \overline{H}_d^* \cdot \left(\nabla \times \overline{E}_d\right) - \overline{E}_d \cdot \left(\nabla \times \overline{H}_d^*\right) \\ &= \overline{H}_d^* \cdot \left(i\omega\mu\overline{H}_d\right) - \overline{E}_d \cdot \left(i\omega\epsilon^*\overline{E}_d^*\right) \\ &= i\omega\mu\left|\overline{H}_d\right|^2 - i\omega\epsilon^*\left|\overline{E}_d^*\right|^2\end{aligned} \tag{6.7}$$

where the symbol $(\cdot)^*$ denotes the complex conjugate of $(\cdot)$.

Integrating (6.7) over V and utilizing the divergence theorem gives

$$\begin{aligned}\int_V \nabla \cdot \left(\overline{E}_d \times \overline{H}_d^*\right) dv &= \oint_S \left(\overline{E}_d \times \overline{H}_d^*\right) \cdot d\overline{s} \\ &= \int_V i\omega\mu\left|\overline{H}_d\right|^2 dv - \int_V i\omega\epsilon^*\left|\overline{E}_d\right|^2 dv \,.\end{aligned} \tag{6.8}$$

For analytic convenience, the medium is considered to be lossy with a complex permeability $\mu = \mu_r + i\mu_i$ and complex permittivity $\epsilon = \epsilon_r + i\epsilon_i$, where μ_i and ϵ_i account for medium loss. Substituting μ and ϵ into (6.8) finally produces

$$\oint_S \left(\overline{E}_d \times \overline{H}_d^*\right) \cdot \overline{ds} = -\int_V \omega \left(\mu_i \left|\overline{H}_d\right|^2 + \epsilon_i \left|\overline{E}_d\right|^2\right) dv + i \int_V \omega \left(\mu_r \left|\overline{H}_d\right|^2 - \epsilon_r \left|\overline{E}_d\right|^2\right) dv \,. \tag{6.9}$$

Consider the vector identities

$$\left(\overline{E}_d \times \overline{H}_d^*\right) \cdot \hat{n} = \left(\hat{n} \times \overline{E}_d\right) \cdot \overline{H}_d^* = \left(\overline{H}_d^* \times \hat{n}\right) \cdot \overline{E}_d \,. \tag{6.10}$$

If one of the conditions 1, 2, or 3 is satisfied, then, either $\hat{n} \times \overline{E}_d = 0$ or $\hat{n} \times \overline{H}_d = 0$ on the surface S. Therefore

$$\left(\overline{E}_d \times \overline{H}_d^*\right) \cdot \hat{n} = 0 \tag{6.11}$$

indicating that the left-hand side of (6.9) becomes zero. The integrands of the right-hand side of (6.9) are all real numbers. The only way to make the right-hand side of (6.9) zero is to let

$$\overline{E}_d \;\left(= \overline{E}_a - \overline{E}_b\right) = 0 \tag{6.12}$$

$$\overline{H}_d \;\left(= \overline{H}_a - \overline{H}_b\right) = 0 \,. \tag{6.13}$$

Therefore, as long as either condition 1, 2, or 3 is specified on the boundary S, the field must be unique, such as $\left(\overline{E}_a = \overline{E}_b \text{ and } \overline{H}_a = \overline{H}_b\right)$. This completes the proof. A lossless medium is considered to be a limiting case of a lossy medium with an infinitesimally small ϵ_i and μ_i ($\epsilon_i \to 0$ and $\mu_i \to 0$), making the uniqueness theorem also applicable to a lossless medium.

6.2 Image Method

The image method is a useful concept for solving problems associated with infinitely extended PEC or PMC plane boundaries. This section introduces the image method and its application to radiation from a small current above a perfect electric conducting plane.

6.2.1 Image Method Using Infinite Planes

Let us consider the original problem in Fig. 6.2, illustrating the real sources $\overline{J}$ and $\overline{M}$ above a PEC plane boundary of infinite extent. The real source $\overline{J}$ represents a small electric current density, which is a small electric dipole

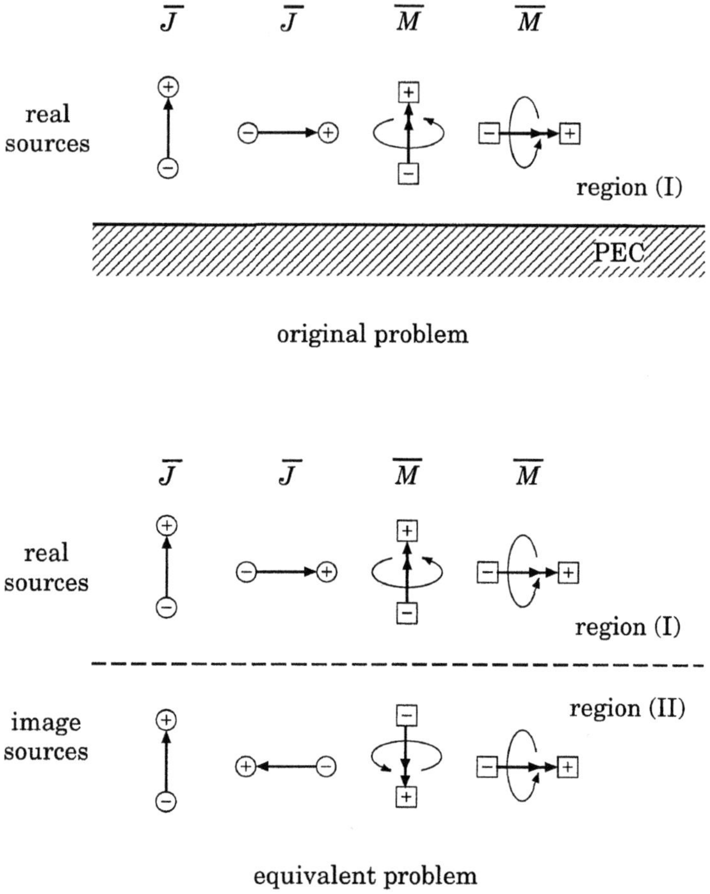

Fig. 6.2. Image method with PEC (perfect electric conductor) plane. ⊕: positive electric charge, ⊖: negative electric charge, →: electric current density $\overline{J}$, ⊞: positive magnetic charge, ⊟: negative magnetic charge, ↠: magnetic current density $\overline{M}$.

consisting of two opposite electric point charges. Similarly, the real source $\overline{M}$ represents a small magnetic current density, which is a small magnetic dipole consisting of two opposite magnetic point charges. Since the outside field produced by a small magnetic dipole is identical to the outside field produced by a small electric current loop, a small magnetic dipole is considered a small electric current loop. The total field in region (I) is the sum of the real source contribution in the absence of a boundary and the reflection from the boundary. The reflection is regarded as the response of an image source that is placed in free space without a PEC boundary. The image sources must be constructed so as to satisfy the tangential electric field continuity condition requiring a zero tangential electric field at the boundary. Figure 6.2 illustrates that the equivalent problem gives the same field in region (I) as the original problem. The image method indicates that the total field in region (I) is the response due to the real and imaginary sources.

As a dual problem, consider the real sources $\overline{J}$ and $\overline{M}$ that are placed above the PMC boundary, as shown in Fig. 6.3. The image sources must also be constructed so as to satisfy the tangential magnetic field continuity requiring a zero tangential magnetic field at the boundary.

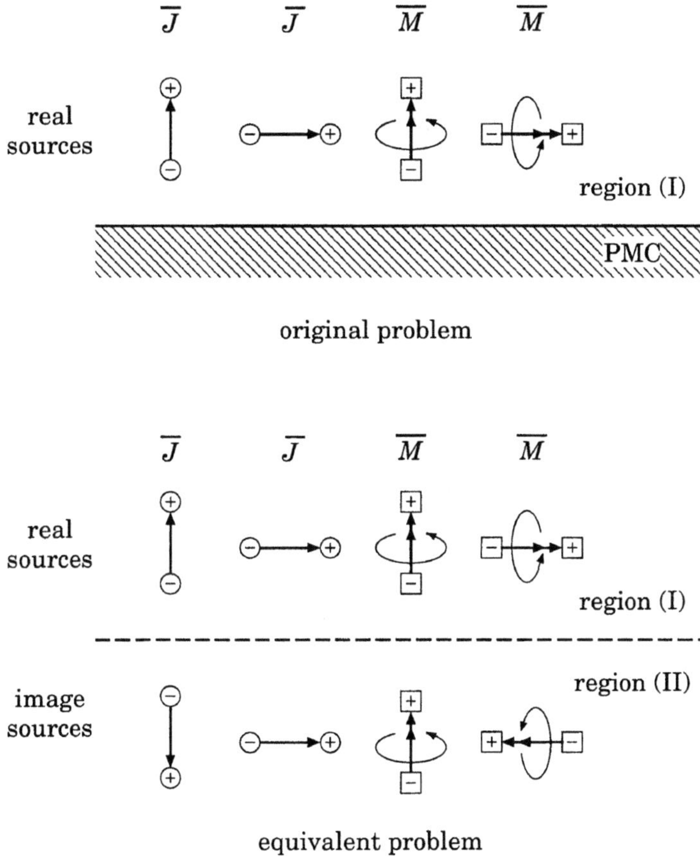

Fig. 6.3. Image theorem with PMC (perfect magnetic conductor) plane. ⊕: positive electric charge, ⊖: negative electric charge, →: electric current density $\overline{J}$, ⊞: positive magnetic charge, ⊟: negative magnetic charge, ↠: magnetic current density $\overline{M}$.

6.2.2 Current Above Perfect Electric Conducting Plane

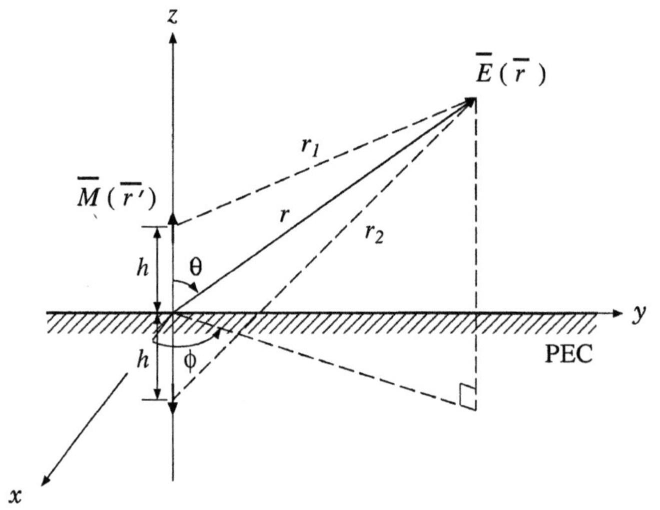

Fig. 6.4. Small current above perfect electric conducting plane.

Consider an infinitesimally small magnetic current source $\overline{M}(\overline{r}')$

$$\overline{M}(\overline{r}') = \hat{z}M\delta(x')\delta(y')\delta(z'-h) \tag{6.14}$$

located above a perfect electric conductor, as shown in Fig. 6.4. Let us evaluate the fields $\overline{E}(\overline{r})$ and $\overline{H}(\overline{r})$ radiating from $\overline{M}(\overline{r}')$ when $r \gg h$. Based on the image method, the electric vector potential $\overline{F}(\overline{r})$ is the sum of the contributions from the real and image sources that are placed in free space. Substituting $\overline{M} = \hat{z}M\delta(x')\delta(y')\big[\delta(z'-h) - \delta(z'+h)\big]$ into a free-space solution (1.102) gives

$$\begin{aligned}\overline{F}(\overline{r}) &= \frac{\epsilon}{4\pi}\int_{V'} \overline{M}(\overline{r}')\frac{e^{ik|\overline{r}-\overline{r}'|}}{|\overline{r}-\overline{r}'|}dv' \\ &= \hat{z}\frac{\epsilon M}{4\pi}\left(\frac{e^{ikr_1}}{r_1} - \frac{e^{ikr_2}}{r_2}\right)\end{aligned} \tag{6.15}$$

where ϵ and k are the medium permittivity and wavenumber, respectively. Note that r_1 and r_2 are the distances from the real and image sources to the observing point $\overline{r}$, respectively. $\overline{F}(\overline{r})$ is expediently expressed using spherical coordinates (r, θ, ϕ). Substituting the approximation in the radiation field at $r \gg h$

$$r_1 \approx r - h\cos\theta \tag{6.16}$$

$$r_2 \approx r + h\cos\theta \tag{6.17}$$

into $\overline{F}(\overline{r})$ with $\hat{z} = \hat{r}\cos\theta - \hat{\theta}\sin\theta$ yields

$$\overline{F}(\overline{r}) = -\frac{i\epsilon M e^{ikr}}{2\pi r}\left(\hat{r}\cos\theta - \hat{\theta}\sin\theta\right)\sin(kh\cos\theta)\,. \tag{6.18}$$

Substituting $\overline{F}(\overline{r})$ into

$$\overline{E} = -\frac{1}{\epsilon}\nabla \times \overline{F}(\overline{r}) \tag{6.19}$$

and collecting the $\frac{1}{r}$ terms gives

$$\overline{E} = \hat{\phi}\underbrace{\frac{Mke^{ikr}}{2\pi r}\sin\theta\sin(kh\cos\theta)}_{E_\phi}\,. \tag{6.20}$$

Subsequently, substituting $\overline{F}(\overline{r})$ into

$$\overline{H} = i\omega\overline{F}(\overline{r}) + \frac{i}{\omega\mu\epsilon}\nabla\left[\nabla\cdot\overline{F}(\overline{r})\right] \tag{6.21}$$

and collecting the $\frac{1}{r}$ terms gives

$$\overline{H} = -\hat{\theta}\sqrt{\frac{\epsilon}{\mu}}E_\phi\,. \tag{6.22}$$

6.3 Equivalence Principle

The equivalence principle is a useful tool for solving boundary-value problems in electromagnetic scattering and radiation. The equivalence principle allows the actual sources to be replaced by equivalent, fictitious sources that produce the same field within a region of interest. Some relevant discussion on the equivalence principle is available in [7]. This section introduces Love's equivalence principle and its application to the problem of transmission through a circular aperture.

6.3.1 Love's Equivalence Principle

Let us consider problem (1) in Fig. 6.5, where the current sources $\overline{J}$ and $\overline{M}$ radiate the fields $\left(\overline{E}, \overline{H}\right)$ throughout regions (I) and (II). In problem (2), the

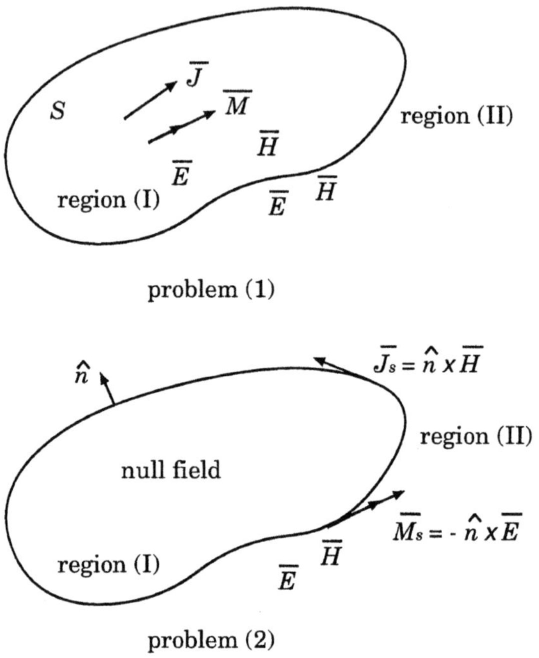

Fig. 6.5. Equivalent problems.

sources $\overline{J}$ and $\overline{M}$ are removed and the fields are assumed to be null in region (I) and $(\overline{E}, \overline{H})$ in region (II), respectively. To retain a null field in region (I) and $(\overline{E}, \overline{H})$ in region (II), the surface currents $\overline{J}_s$ and $\overline{M}_s$ must be placed on the surface boundary

$$\overline{J}_s = \hat{n} \times \overline{H} \tag{6.23}$$

$$\overline{M}_s = -\hat{n} \times \overline{E} \,. \tag{6.24}$$

Then, the surface current densities $\overline{J}_s$ and $\overline{M}_s$ radiate $(\overline{E}, \overline{H})$ in region (II), while maintaining a null field in region (I). Problem (2) is thus equivalent to problem (1) as far as region (II) is concerned.

6.3.2 Transmission Through Circular Aperture

Consider electromagnetic transmission through a circular aperture in a perfectly conducting plane, as shown in Fig. 6.6. The incident field impinges on the aperture from below a perfectly conducting plane. Let us evaluate the far-zone transmitted field for $z > 0$. For simplicity, the aperture electric field

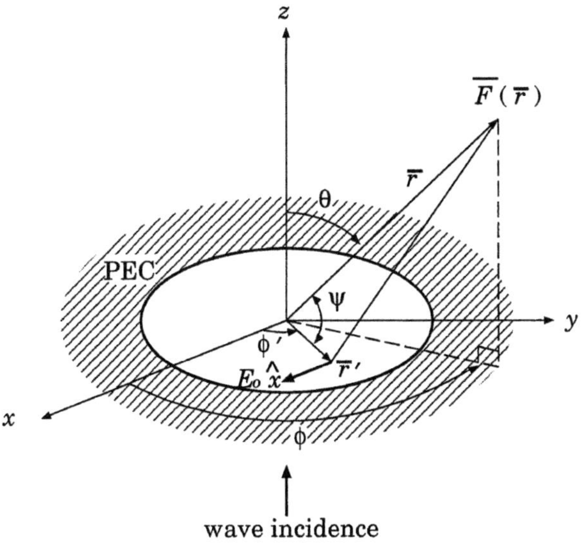

Fig. 6.6. Circular aperture in conducting plane.

is assumed to be $\overline{E}_{ap} = \hat{x}E_0$. It is convenient to transform the original problem into an equivalent problem (b), as shown in Fig. 6.7, where the surface magnetic current is $\overline{M}_s(\overline{r}') = -\hat{n} \times \overline{E}_{ap}$. The equivalent problem (b) in Fig. 6.7 can then be transformed into the equivalent problem (c) in Fig. 6.7, where a PEC sheet is placed beneath the boundary S. The surface electric current $\overline{J}_s(\overline{r}')$ is short-circuited by the PEC sheet beneath the boundary S, and the field $(\overline{E}, \overline{H})$ in region (II) is solely due to the surface magnetic current $\overline{M}_s(\overline{r}')$. According to the image method, the surface magnetic current $\overline{M}_s(\overline{r}')$ on an infinitely extended PEC can be considered as $2\overline{M}_s(\overline{r}')$ in free space, and the electric vector potential for the radiated field for $z > 0$ is

$$\begin{aligned}\overline{F}(\overline{r}) &= \frac{\epsilon}{4\pi}\int_{S'} 2\overline{M}_s(\overline{r}')\frac{e^{ik|\overline{r}-\overline{r}'|}}{|\overline{r}-\overline{r}'|}\,ds' \\ &= \frac{\epsilon}{4\pi}\int_{S'} \left(-2\hat{n} \times \overline{E}_{ap}\right)\frac{e^{ik|\overline{r}-\overline{r}'|}}{|\overline{r}-\overline{r}'|}\,ds' \\ &= -\hat{y}\frac{\epsilon E_0}{2\pi}\int_{S'} \frac{e^{ik|\overline{r}-\overline{r}'|}}{|\overline{r}-\overline{r}'|}\,ds' \ . \end{aligned} \tag{6.25}$$

Let us evaluate the electric vector potential $\overline{F}(\overline{r})$ when $r \gg r'$. Note

$$\begin{aligned}|\,\overline{r}-\overline{r}'\,| &\approx r - r'\cos\psi \\ &= r - r'\sin\theta\cos(\phi' - \phi) \ . \end{aligned} \tag{6.26}$$

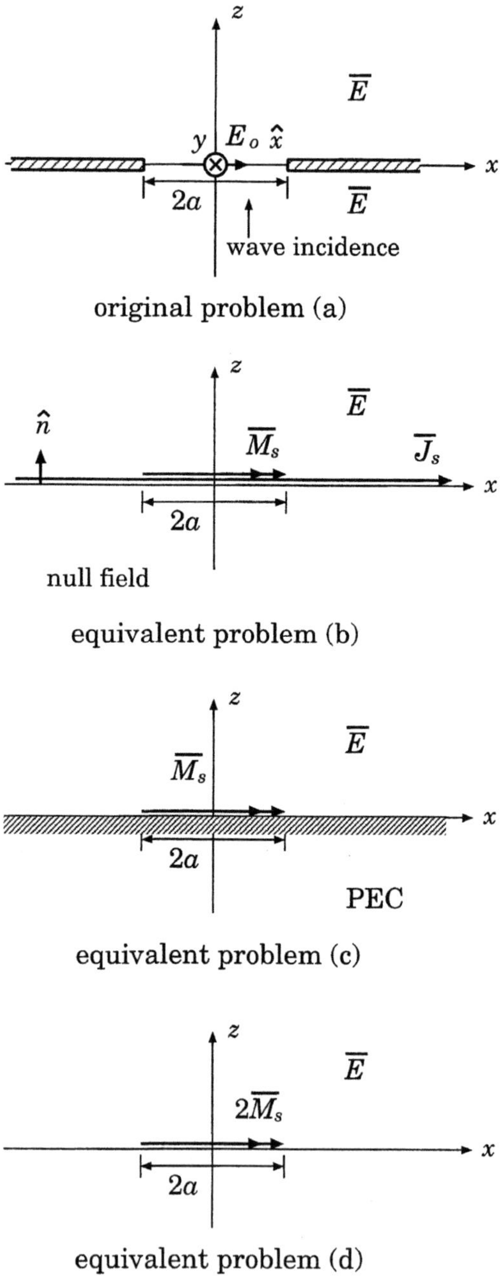

Fig. 6.7. Equivalent problems for circular aperture in conducting plane.

The electric vector potential $\overline{F}(\overline{r})$ is

$$\overline{F}(\overline{r}) \approx -\hat{y}\frac{\epsilon E_0}{2\pi}\int_{S'}\frac{e^{ikr}}{r}\exp\left[-ikr'\sin\theta\cos(\phi'-\phi)\right]\,ds'$$
$$= -\hat{y}\frac{\epsilon E_0}{2\pi}\frac{e^{ikr}}{r}\int_0^a\int_0^{2\pi}\exp\left[-ikr'\sin\theta\cos(\phi'-\phi)\right]r'\,d\phi'\,dr' \;. \quad (6.27)$$

Substituting the identity

$$\exp\left[-ikr'\sin\theta\cos(\phi'-\phi)\right] = \sum_{n=-\infty}^{\infty} i^n J_n(-kr'\sin\theta)e^{in(\phi'-\phi)} \quad (6.28)$$

into $\overline{F}(\overline{r})$ results in

$$\overline{F}(\overline{r}) = -\hat{y}\epsilon E_0\frac{e^{ikr}}{r}\int_0^a J_0(-kr'\sin\theta)r'\,dr' \;. \quad (6.29)$$

Since

$$\hat{y} = \hat{r}\sin\theta\sin\phi + \hat{\theta}\cos\theta\sin\phi + \hat{\phi}\cos\phi \quad (6.30)$$

$$\int J_0(u)u\,du = uJ_1(u) \quad (6.31)$$

$\overline{F}(\overline{r})$ can be written as

$$\overline{F}(\overline{r}) = -\left(\hat{r}\sin\theta\sin\phi + \hat{\theta}\cos\theta\sin\phi + \hat{\phi}\cos\phi\right)$$
$$\cdot\epsilon E_0\frac{e^{ikr}}{r}\frac{a^2 J_1(ka\sin\theta)}{ka\sin\theta}$$
$$\equiv \hat{r}F_r + \hat{\theta}F_\theta + \hat{\phi}F_\phi \;. \quad (6.32)$$

Let us evaluate $\overline{E}$ and $\overline{H}$ in the far zone. Substituting $\overline{F}(\overline{r})$ into $\overline{E} = -\frac{1}{\epsilon}\nabla\times\overline{F}(\overline{r})$ and collecting the $\frac{1}{r}$ terms gives

$$\overline{E} = -\frac{ik}{\epsilon}\left(-\hat{\theta}F_\phi + \hat{\phi}F_\theta\right) \;. \quad (6.33)$$

Similarly, substituting $\overline{F}(\overline{r})$ into $\overline{H} = i\omega\overline{F}(\overline{r}) + \frac{i}{\omega\mu\epsilon}\nabla\left[\nabla\cdot\overline{F}(\overline{r})\right]$ and collecting the $\frac{1}{r}$ terms gives

$$\overline{H} = i\omega\left(\hat{\theta}F_\theta + \hat{\phi}F_\phi\right) \;. \quad (6.34)$$

6.4 Induction Theorem

The induction theorem is often used to solve electromagnetic scattering problems. This section first introduces the induction theorem, then considers its application to scattering from a conducting rectangular plate.

6.4.1 Equivalence Based on Induction Theorem

- Consider an incident wave $\left(\overline{E}^i, \overline{H}^i\right)$ that impinges on a dielectric scatterer, as shown in the original problem (1) in Fig. 6.8. The sources $\left(\overline{J}, \overline{M}\right)$ produce the incident field $\left(\overline{E}^i, \overline{H}^i\right)$ in the absence of the scatterer. The total field in region (II) consists of the incident and scattered components $\left(\overline{E}^i, \overline{H}^i\right)$ and $\left(\overline{E}^s, \overline{H}^s\right)$, respectively. The transmitted field in region (I) is $\left(\overline{E}^t, \overline{H}^t\right)$.

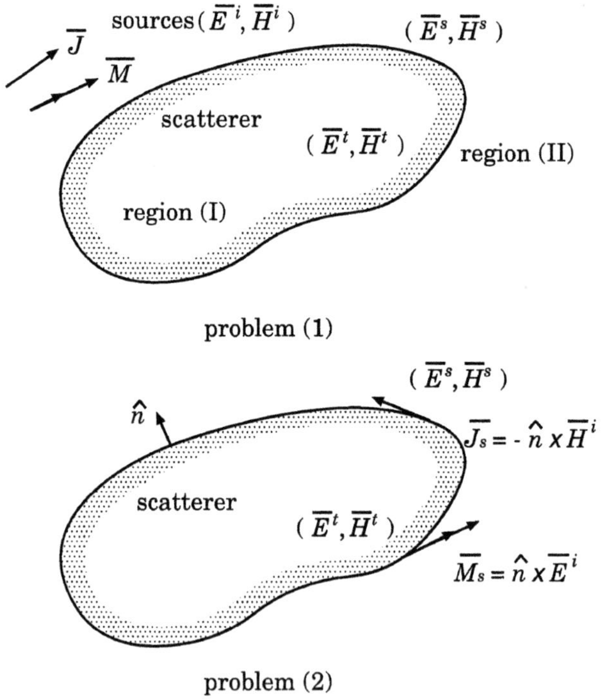

Fig. 6.8. Equivalent problems.

- Based on the induction theorem, the original problem (1) can be transformed into the equivalent problem (2), where the field $\left(\overline{E}^t, \overline{H}^t\right)$ is assumed in region (I) and the field $\left(\overline{E}^s, \overline{H}^s\right)$ is assumed in region (II). At the boundary between regions (I) and (II), the surface currents $\overline{J}_s = \hat{n} \times \left(\overline{H}^s - \overline{H}^t\right) = -\hat{n} \times \overline{H}^i$ and $\overline{M}_s = -\hat{n} \times \left(\overline{E}^s - \overline{E}^t\right) = \hat{n} \times \overline{E}^i$ must be impressed in order to satisfy the boundary conditions.
- It is of interest to consider what happens when region (I) becomes a PEC scatterer. Note that $\overline{E}^t = \overline{H}^t = 0$ in region (I) and $\hat{n} \times \left(\overline{E}^i + \overline{E}^s\right) = 0$ between regions (I) and (II). The impressed surface electric current $\overline{J}_s = -\hat{n} \times \overline{H}^i$ is short-circuited on the PEC surface, and the field $\left(\overline{E}^s, \overline{H}^s\right)$ in region (II) is due solely to the impressed surface magnetic current $\overline{M}_s = \hat{n} \times \overline{E}^i$.

6.4.2 Scattering from Conducting Rectangular Plate

Let us consider the problem of scattering from a perfectly conducting rectangular plate based on the induction theorem. A uniform plane wave

$$\overline{E}^i(x, y, z) = \hat{x} E_0 \exp\left[i(ky \sin\theta_i - kz\cos\theta_i)\right] \tag{6.35}$$

is incident on an $(a \times b)$ perfectly conducting rectangular plate, as shown in Fig. 6.9. Let us evaluate the far-zone scattered field for $z > 0$. If the size of the plate is relatively large compared to the wavelength $(a, b \gg \lambda)$, it is possible to assume that the plate is infinitely extended. According to the image method, the surface magnetic current $\overline{M}_s$ on an infinitely extended PEC can be considered as $2\overline{M}_s$ in free space. The electric vector potential for $z > 0$ is given by

$$\begin{aligned}
\overline{F}(\overline{r}) &= \frac{\epsilon}{4\pi} \int_{S'} 2\hat{n} \times \overline{E}^i (x', y', 0^+) \frac{e^{ik|\overline{r}-\overline{r}'|}}{|\overline{r}-\overline{r}'|} \, ds' \\
&\approx \frac{\epsilon}{4\pi} \int_{S'} 2\hat{n} \times \overline{E}^i (x', y', 0^+) \frac{1}{r} \exp\left[ik(r - \overline{r}' \cdot \hat{r})\right] ds' \\
&= \frac{\epsilon e^{ikr}}{4\pi r} \int_{-b/2}^{b/2} \int_{-a/2}^{a/2} \\
&\qquad 2\hat{n} \times \overline{E}^i (x', y', 0^+) \exp\left(-ik\overline{r}' \cdot \hat{r}\right) dx' \, dy' \, .
\end{aligned} \tag{6.36}$$

Let us evaluate the electric vector potential using spherical coordinates (r, θ, ϕ). Substituting

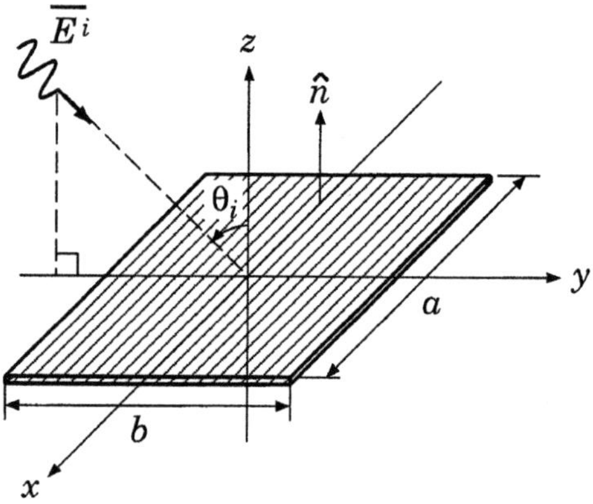

Fig. 6.9. Perfectly conducting rectangular plate of size $(a \times b)$.

$$\hat{n} \times \overline{E}^i \left(x', y', 0^+\right) = \hat{y} E_0 \exp(iky' \sin\theta_i) \tag{6.37}$$

$$\begin{aligned} \overline{r}' \cdot \hat{r} &= (\hat{x}x' + \hat{y}y') \cdot (\hat{x}\sin\theta\cos\phi + \hat{y}\sin\theta\sin\phi + \hat{z}\cos\theta) \\ &= x'\sin\theta\cos\phi + y'\sin\theta\sin\phi \end{aligned} \tag{6.38}$$

into $\overline{F}(\overline{r})$ and performing integration gives

$$\begin{aligned} \overline{F}(\overline{r}) = &\left(\hat{r}\sin\theta\sin\phi + \hat{\theta}\cos\theta\sin\phi + \hat{\phi}\cos\phi\right) \\ &\cdot \frac{\epsilon e^{ikr}}{2\pi r} E_0 ab \left(\frac{\sin\alpha}{\alpha}\right)\left(\frac{\sin\beta}{\beta}\right) \\ \equiv &\, \hat{r}F_r + \hat{\theta}F_\theta + \hat{\phi}F_\phi \end{aligned} \tag{6.39}$$

where

$$\alpha = \frac{ka}{2}\sin\theta\cos\phi \tag{6.40}$$

$$\beta = \frac{kb}{2}(\sin\theta\sin\phi - \sin\theta_i) \ . \tag{6.41}$$

The far-zone scattered fields $\overline{E}$ and $\overline{H}$ are given by

$$\overline{E} = -\frac{ik}{\epsilon}\left(-\hat{\theta}F_\phi + \hat{\phi}F_\theta\right) \tag{6.42}$$

$$\overline{H} = i\omega\left(\hat{\theta}F_\theta + \hat{\phi}F_\phi\right) . \tag{6.43}$$

The far-zone fields (6.42) and (6.43) are good approximations of the exact solution as long as (θ_i and θ) $\ll \frac{\pi}{2}$ and (a and b) $\gg \lambda$.

6.5 Duality Theorem

The duality theorem utilizes the symmetric property of Maxwell's equations. Maxwell's equations for the electric source $\left(\overline{J} \neq 0 \text{ and } \overline{M} = 0\right)$ are given as

$$\nabla \times \overline{E}_e = i\omega\mu\overline{H}_e \tag{6.44}$$

$$\nabla \times \overline{H}_e = -i\omega\epsilon\overline{E}_e + \overline{J} \tag{6.45}$$

$$\nabla \cdot \epsilon\overline{E}_e = \rho_e \tag{6.46}$$

$$\nabla \cdot \mu\overline{H}_e = 0 . \tag{6.47}$$

Similarly, Maxwell's equations for the magnetic source $\left(\overline{M} \neq 0 \text{ and } \overline{J} = 0\right)$ are

$$\nabla \times \overline{H}_m = -i\omega\epsilon\overline{E}_m \tag{6.48}$$

$$-\nabla \times \overline{E}_m = -i\omega\mu\overline{H}_m + \overline{M} \tag{6.49}$$

$$\nabla \cdot \mu\overline{H}_m = \rho_m \tag{6.50}$$

$$\nabla \cdot \epsilon\overline{E}_m = 0 . \tag{6.51}$$

A dual set for the electric and magnetic sources is shown in Fig. 6.10. Maxwell's equations (6.48) through (6.51) can be constructed from (6.44) through (6.47) by interchanging the quantities, as shown in Table 6.1 $\left(\overline{E}_e \to \overline{H}_m, \overline{H}_e \to -\overline{E}_m, \cdots\right)$. Equations (6.44) through (6.47) and (6.48) through (6.51) are termed dual to each other since they take a symmetric mathematical form. If the solution $\left(\overline{E}_e \text{ and } \overline{H}_e\right)$ to (6.44) through (6.47) is known,

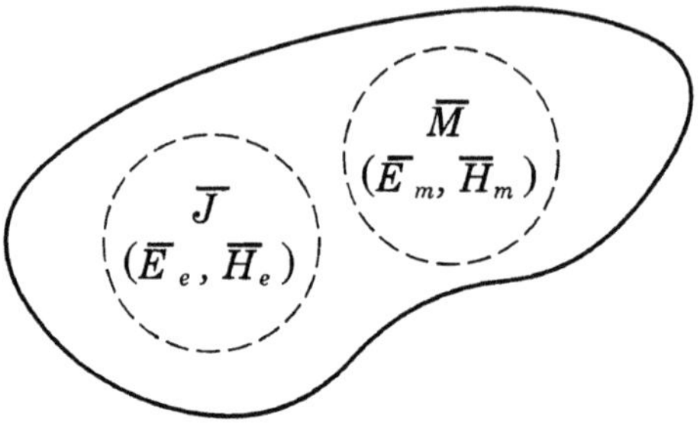

Fig. 6.10. Dual set.

Table 6.1. Dual quantities

Electric source	Magnetic source
$\overline{E}_e$	$\overline{H}_m$
$\overline{H}_e$	$-\overline{E}_m$
$\overline{J}$	$\overline{M}$
ρ_e	ρ_m
μ	ϵ
ϵ	μ
$\overline{A}$	$\overline{F}$

the solution to (6.48) through (6.51) can immediately be constructed by interchanging the quantities, as shown in Table 6.1 $\left(\overline{E}_e \to \overline{H}_m,\ \overline{H}_e \to -\overline{E}_m,\ \cdots\right)$.

Consider radiation from a small electric current $\overline{J}$ above a perfect magnetic conductor plane, as shown in Fig. 6.11. The real current $\overline{J}$ above the PMC surface is equivalent to the sum of the real $\overline{J}$ and its image source in a medium of infinite extent. The problem considered in Fig. 6.11 is dual to the problem considered in Fig. 6.4 in Section 6.2, and the radiation field in the far zone

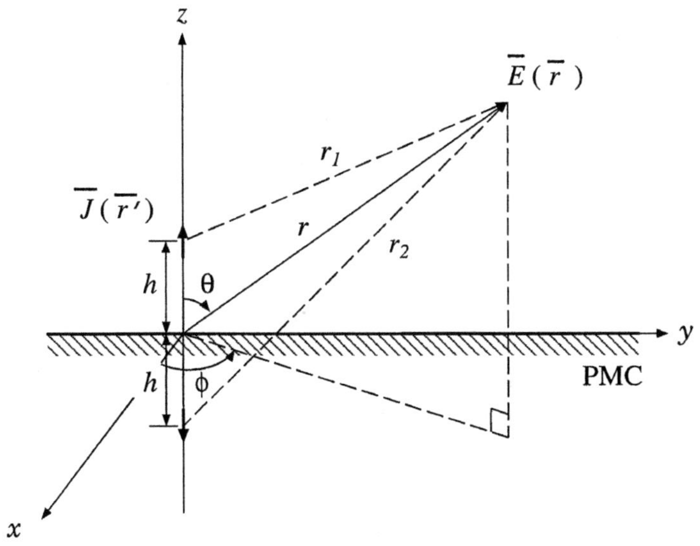

Fig. 6.11. Small electric current above perfect magnetic conductor plane.

is immediately given from the duality theorem. The solution to the problem in Fig. 6.4 is given by (6.20) and (6.22). The solution to the dual problem in Fig. 6.11 is thus

$$\overline{H} = -\hat{\phi} J \frac{ke^{ikr}}{2\pi r} \sin\theta \sin(kh\cos\theta) \tag{6.52}$$

$$\overline{E} = -\hat{\theta}\eta J \frac{ke^{ikr}}{2\pi r} \sin\theta \sin(kh\cos\theta) \tag{6.53}$$

where $\eta = \sqrt{\frac{\mu}{\epsilon}}$.

6.6 Reciprocity Theorem

In circuit theory, the reciprocity theorem holds for any linear network. In electromagnetic theory, the Lorentz reciprocity theorem holds for any linear medium. The reciprocity theorem in circuit theory is a special form of the Lorentz reciprocity theorem in electromagnetic theory. This section derives the Lorentz reciprocity theorem, starting from Maxwell's equations.

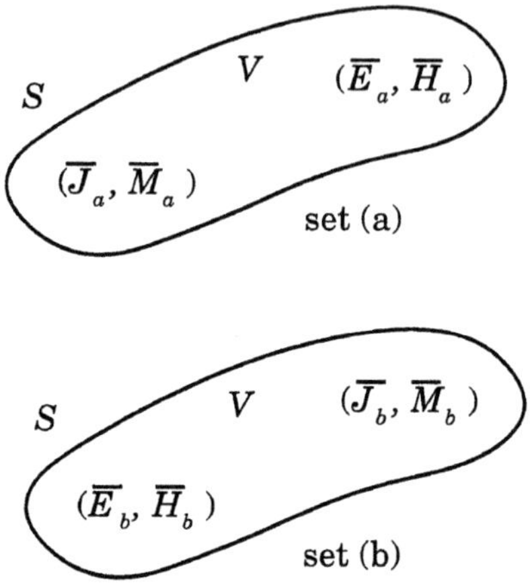

Fig. 6.12. Two different sources and fields.

6.6.1 Lorentz Reciprocity Theorem

Consider two sets of Maxwell's equations in the same linear medium with different sources. The problem geometry is shown in Fig. 6.12. Set (a) is given by

$$\nabla \times \overline{E}_a = i\omega\mu\overline{H}_a - \overline{M}_a \tag{6.54}$$

$$\nabla \times \overline{H}_a = -i\omega\epsilon\overline{E}_a + \overline{J}_a \ . \tag{6.55}$$

Set (b) is

$$\nabla \times \overline{E}_b = i\omega\mu\overline{H}_b - \overline{M}_b \tag{6.56}$$

$$\nabla \times \overline{H}_b = -i\omega\epsilon\overline{E}_b + \overline{J}_b \ . \tag{6.57}$$

The vector operations $\nabla \cdot \left(\overline{E}_a \times \overline{H}_b\right)$ and $\nabla \cdot \left(\overline{E}_b \times \overline{H}_a\right)$ give

$$\begin{aligned} \nabla \cdot \left(\overline{E}_a \times \overline{H}_b\right) &= \overline{H}_b \cdot \nabla \times \overline{E}_a - \overline{E}_a \cdot \nabla \times \overline{H}_b \\ &= \overline{H}_b \cdot \left(i\omega\mu\overline{H}_a - \overline{M}_a\right) - \overline{E}_a \cdot \left(-i\omega\epsilon\overline{E}_b + \overline{J}_b\right) \end{aligned} \tag{6.58}$$

$$\begin{aligned} \nabla \cdot \left(\overline{E}_b \times \overline{H}_a\right) &= \overline{H}_a \cdot \nabla \times \overline{E}_b - \overline{E}_b \cdot \nabla \times \overline{H}_a \\ &= \overline{H}_a \cdot \left(i\omega\mu\overline{H}_b - \overline{M}_b\right) - \overline{E}_b \cdot \left(-i\omega\epsilon\overline{E}_a + \overline{J}_a\right) \ . \end{aligned} \tag{6.59}$$

Subtracting (6.59) from (6.58) leads to

$$\nabla \cdot \left(\overline{E}_a \times \overline{H}_b - \overline{E}_b \times \overline{H}_a\right) =$$

$$- \left(\overline{E}_a \cdot \overline{J}_b - \overline{H}_a \cdot \overline{M}_b\right) + \left(\overline{E}_b \cdot \overline{J}_a - \overline{H}_b \cdot \overline{M}_a\right) . \qquad (6.60)$$

The expression (6.60) is called the Lorentz reciprocity theorem in differential form. Taking volume integration of (6.60) and utilizing the divergence theorem on the left-hand side yields the Lorentz reciprocity theorem in integral form as

$$\oint_S \left(\overline{E}_a \times \overline{H}_b - \overline{E}_b \times \overline{H}_a\right) \cdot d\overline{s} =$$

$$\int_V \left[- \left(\overline{E}_a \cdot \overline{J}_b - \overline{H}_a \cdot \overline{M}_b\right) + \left(\overline{E}_b \cdot \overline{J}_a - \overline{H}_b \cdot \overline{M}_a\right)\right] dv . \qquad (6.61)$$

Two special cases of (6.61) are of practical interest. If region V is free of sources $\left(\overline{J}_a = \overline{M}_a = \overline{J}_b = \overline{M}_b = 0\right)$, (6.61) reduces to

$$\oint_S \left(\overline{E}_a \times \overline{H}_b - \overline{E}_b \times \overline{H}_a\right) \cdot d\overline{s} = 0 . \qquad (6.62)$$

If region V contains sources and V extends to infinity, the fields $\left(\overline{E}_a, \overline{E}_b, \overline{H}_a,\right.$ and $\left.\overline{H}_b\right)$ take the asymptotic form $\frac{e^{ikr}}{r}$ at $r \to \infty$. The fields have the θ and ϕ components that satisfy the relations, $E_\theta = \eta H_\phi$ and $E_\phi = -\eta H_\theta$. Thus, the left-hand side terms of (6.61) all vanish at $r \to \infty$, and (6.61) reduces to

$$\int_V \left(\overline{E}_a \cdot \overline{J}_b - \overline{H}_a \cdot \overline{M}_b\right) dv = \int_V \left(\overline{E}_b \cdot \overline{J}_a - \overline{H}_b \cdot \overline{M}_a\right) dv . \qquad (6.63)$$

6.6.2 Reciprocity for Antennas

Let us apply (6.63) to the problem of two small dipole antennas in free space, as shown in Fig. 6.13. Assume that the electric current density $\overline{J}_a$ at antenna 1 produces $\overline{E}_a$, which is the electric field at antenna 2. Similarly, the electric current density $\overline{J}_b$ at antenna 2 produces $\overline{E}_b$, which is the electric field at antenna 1. Therefore, two antennas satisfy the Lorentz reciprocity theorem as

$$\int_V \overline{E}_a \cdot \overline{J}_b \, dv = \int_V \overline{E}_b \cdot \overline{J}_a \, dv . \qquad (6.64)$$

Let us reinterpret (6.64) in terms of the voltages and currents. The voltages and currents are given as

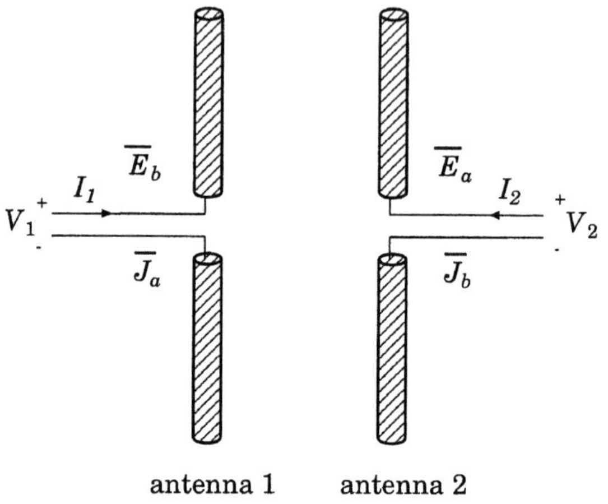

Fig. 6.13. Two small dipole antennas.

$$V_1 = -\int \overline{E}_b \cdot d\overline{l} \tag{6.65}$$

$$I_1 = \int \overline{J}_a \cdot d\overline{s} \tag{6.66}$$

$$V_2 = -\int \overline{E}_a \cdot d\overline{l} \tag{6.67}$$

$$I_2 = \int \overline{J}_b \cdot d\overline{s} \, . \tag{6.68}$$

Therefore, the Lorentz reciprocity theorem (6.64) can be rewritten as

$$V_2 I_2 = V_1 I_1 \tag{6.69}$$

which amounts to the reciprocity theorem in circuit theory.

Problems for Chapter 6

1. Consider a small dipole electric current placed within a conducting parallel-plate waveguide, as shown in Fig. 6.14. Construct the equivalent problem based on the image method.
2. A uniform plane wave is normally incident on a thin PMC circular plate of radius a. Assume $a \gg \lambda$. Evaluate the backscattered far-zone field using the induction theorem.

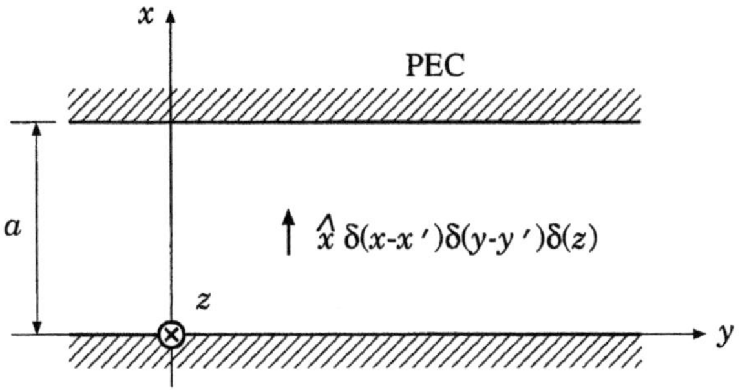

Fig. 6.14. Small dipole within conducting parallel-plate waveguide.

3. Consider radiation from a small electric current $\overline{J}$ above a perfect magnetic conductor plane, as shown in Fig. 6.11. Derive the solution (6.52) and (6.53) without recourse to the duality theorem.

7

Wave Scattering

7.1 Dielectric Circular Cylinder

The problem of scattering from a dielectric circular cylinder is a canonical boundary-value problem that can be solved with the separation of variables technique. This section investigates TM and TE scattering from a long dielectric circular cylinder under the illumination of a uniform plane wave.

7.1.1 TM Scattering

Assume that an incident TM wave ($H_z = 0$ and $E_z \neq 0$) impinges on a cylinder that is infinitely long in the z-direction, as shown in Fig. 7.1. The incident magnetic vector potential $A_z^i = e^{-ikx} = e^{-ik\rho\cos\phi}$ is chosen to yield the incident electric field $E_z^i = E_0 e^{-ikx}$. The incident TM wave has nonzero E_z, H_ρ, and H_ϕ components. No field variation is assumed in the z-direction $\left(\dfrac{\partial}{\partial z} = 0\right)$ so that the geometry of the scattering problem becomes two-dimensional.

The incident electric field is a periodic function of ϕ with period of 2π. Therefore, the incident magnetic vector potential $A_z^i(\rho, \phi)$ can be expanded in a Fourier series as

$$A_z^i(\rho, \phi) = \sum_{n=-\infty}^{\infty} B_n e^{in\phi} \tag{7.1}$$

where

$$\begin{aligned} B_n &= \frac{1}{2\pi} \int_0^{2\pi} e^{-ik\rho\cos\phi - in\phi}\, d\phi \\ &= i^{-n} J_n(k\rho)\ . \end{aligned} \tag{7.2}$$

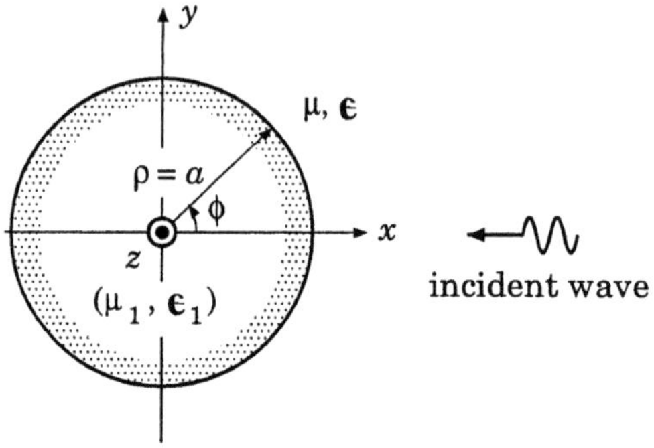

Fig. 7.1. Cross section of dielectric circular cylinder.

Hence, $A_z^i(\rho,\phi)$ is given by

$$A_z^i(\rho,\phi) = \sum_{n=-\infty}^{\infty} i^{-n} J_n(k\rho) e^{in\phi} . \tag{7.3}$$

Scattering from a circular cylinder generates the scattered magnetic vector potential A_z^s for $\rho \geq a$ $\left(k = \omega\sqrt{\mu\epsilon}\right)$ and the transmitted magnetic vector potential A_z^t for $\rho \leq a$ $\left(k_1 = \omega\sqrt{\mu_1\epsilon_1}\right)$, respectively. The scattered magnetic vector potential $A_z^s(\rho,\phi)$ satisfies the Helmholtz equation using cylindrical coordinates

$$\left[\frac{1}{\rho}\frac{\partial}{\partial\rho}\left(\rho\frac{\partial}{\partial\rho}\right) + \frac{1}{\rho^2}\frac{\partial^2}{\partial\phi^2} + k^2\right] A_z^s(\rho,\phi) = 0 . \tag{7.4}$$

Based on the separation of variables technique, it is possible to solve (7.4), where $A_z^s(\rho,\phi)$ takes the form of a Hankel function of the first kind

$$A_z^s(\rho,\phi) = \sum_{n=-\infty}^{\infty} i^{-n} a_n H_n^{(1)}(k\rho) e^{in\phi} \tag{7.5}$$

representing an outward-traveling wave that vanishes at infinity.

Similarly, the transmitted magnetic vector potential can be written as

$$A_z^t(\rho,\phi) = \sum_{n=-\infty}^{\infty} i^{-n} b_n J_n(k_1\rho) e^{in\phi} . \tag{7.6}$$

To determine the unknown coefficients a_n and b_n, the boundary conditions for the E_z and H_ϕ field continuities must be enforced at $\rho = a$. The field components are obtained by substituting $A_z^i(\rho,\phi)$, $A_z^s(\rho,\phi)$, and $A_z^t(\rho,\phi)$ into Table 3.2 in Section 3.1. The continuity of the tangential electric field at $\rho = a$

$$E_z^i(a,\phi) + E_z^s(a,\phi) = E_z^t(a,\phi) \tag{7.7}$$

requires

$$J_n(ka) + a_n H_n^{(1)}(ka) = b_n J_n(k_1 a) \ . \tag{7.8}$$

The continuity of the tangential magnetic field at $\rho = a$

$$H_\phi^i(a,\phi) + H_\phi^s(a,\phi) = H_\phi^t(a,\phi) \tag{7.9}$$

yields

$$\sqrt{\frac{\epsilon}{\mu}}\left[J_n'(ka) + a_n H_n^{(1)\prime}(ka)\right] = \sqrt{\frac{\epsilon_1}{\mu_1}}\left[b_n J_n'(k_1 a)\right] \tag{7.10}$$

where the symbol $\prime$ denotes differentiation with respect to the argument $J_n'(ka) = \left.\frac{dJ_n(k\rho)}{d(k\rho)}\right|_{\rho=a}$.

Solving (7.8) and (7.10) for a_n and b_n yields

$$a_n = \frac{\sqrt{\frac{\epsilon}{\mu}} J_n(k_1 a) J_n'(ka) - \sqrt{\frac{\epsilon_1}{\mu_1}} J_n'(k_1 a) J_n(ka)}{\sqrt{\frac{\epsilon_1}{\mu_1}} J_n'(k_1 a) H_n^{(1)}(ka) - \sqrt{\frac{\epsilon}{\mu}} J_n(k_1 a) H_n^{(1)\prime}(ka)} \tag{7.11}$$

$$\begin{aligned} b_n &= \frac{\sqrt{\frac{\epsilon}{\mu}} H_n^{(1)}(ka) J_n'(ka) - \sqrt{\frac{\epsilon}{\mu}} H_n^{(1)\prime}(ka) J_n(ka)}{\sqrt{\frac{\epsilon_1}{\mu_1}} J_n'(k_1 a) H_n^{(1)}(ka) - \sqrt{\frac{\epsilon}{\mu}} J_n(k_1 a) H_n^{(1)\prime}(ka)} \\ &= \frac{-\frac{2i}{\pi\omega\mu a}}{\sqrt{\frac{\epsilon_1}{\mu_1}} J_n'(k_1 a) H_n^{(1)}(ka) - \sqrt{\frac{\epsilon}{\mu}} J_n(k_1 a) H_n^{(1)\prime}(ka)} \end{aligned} \tag{7.12}$$

where the Bessel function Wronskian

$$\sqrt{\frac{\epsilon}{\mu}} H_n^{(1)}(ka) J_n'(ka) - \sqrt{\frac{\epsilon}{\mu}} H_n^{(1)\prime}(ka) J_n(ka) = -\frac{2i}{\pi\omega\mu a} \tag{7.13}$$

is used to simplify b_n.

The scattered field in the far zone ($k\rho \gg 1$) is written in terms of an asymptotic expression for $H_n^{(1)}(k\rho)$ as

$$E_z^s \underset{k\rho\to\infty}{\sim} E_0 \sum_{n=-\infty}^{\infty} \sqrt{\frac{2}{\pi k\rho}} a_n e^{ik\rho + in(\phi-\pi) - i\pi/4} \ . \tag{7.14}$$

When a circular cylinder is a perfect electric conductor (PEC, $\epsilon_1 \to \infty$), an incident wave cannot penetrate the cylinder. Thus, b_n=0 and

$$a_n = -\frac{J_n(ka)}{H_n^{(1)}(ka)} . \tag{7.15}$$

The electric current density that is induced on the cylindrical surface at $\rho = a$ is

$$\begin{aligned} \overline{J} &= \hat{n} \times \overline{H}\Big|_{\rho=a} \\ &= \hat{\rho} \times \left(\overline{H}^i + \overline{H}^s\right)\Big|_{\rho=a} \\ &= \hat{z}\left[H_\phi^i(a,\phi) + H_\phi^s(a,\phi)\right] \\ &= \hat{z}\frac{2i}{\mu\pi a}\sum_{n=-\infty}^{\infty} i^{-n}\frac{e^{in\phi}}{H_n^{(1)}(ka)} . \end{aligned} \tag{7.16}$$

7.1.2 TE Scattering

Assume that an incident TE wave impinges on a cylinder that is infinitely long in the z-direction. The incident TE wave has nonzero H_z, E_ρ, and E_ϕ components. The analysis for TE scattering is somewhat similar to that of the previous TM scattering case.

The incident electric vector potential is written as

$$\begin{aligned} F_z^i(\rho,\phi) &= e^{-ikx} \\ &= e^{-ik\rho\cos\phi} \\ &= \sum_{n=-\infty}^{\infty} i^{-n} J_n(k\rho) e^{in\phi} . \end{aligned} \tag{7.17}$$

Similarly, the scattered and transmitted electric vector potentials are, respectively

$$F_z^s(\rho,\phi) = \sum_{n=-\infty}^{\infty} i^{-n} c_n H_n^{(1)}(k\rho) e^{in\phi} \tag{7.18}$$

$$F_z^t(\rho,\phi) = \sum_{n=-\infty}^{\infty} i^{-n} d_n J_n(k_1\rho) e^{in\phi} . \tag{7.19}$$

The continuity of the tangential magnetic field at $\rho = a$

$$H_z^i(a,\phi) + H_z^s(a,\phi) = H_z^t(a,\phi) \tag{7.20}$$

yields

$$J_n(ka) + c_n H_n^{(1)}(ka) = d_n J_n(k_1 a) \, . \tag{7.21}$$

The continuity of the tangential electric field at $\rho = a$

$$E^i_\phi(a,\phi) + E^s_\phi(a,\phi) = E^t_\phi(a,\phi) \tag{7.22}$$

yields

$$\sqrt{\frac{\mu}{\epsilon}} \left[J_n'(ka) + c_n H_n^{(1)\prime}(ka) \right] = \sqrt{\frac{\mu_1}{\epsilon_1}} \left[d_n J_n'(k_1 a) \right] \tag{7.23}$$

where the symbol $\prime$ denotes differentiation with respect to the argument. Solving (7.21) and (7.23) for c_n and d_n yields

$$c_n = \frac{\sqrt{\frac{\mu}{\epsilon}} J_n(k_1 a) J_n'(ka) - \sqrt{\frac{\mu_1}{\epsilon_1}} J_n'(k_1 a) J_n(ka)}{\sqrt{\frac{\mu_1}{\epsilon_1}} J_n'(k_1 a) H_n^{(1)}(ka) - \sqrt{\frac{\mu}{\epsilon}} J_n(k_1 a) H_n^{(1)\prime}(ka)} \tag{7.24}$$

$$\begin{aligned} d_n &= \frac{\sqrt{\frac{\mu}{\epsilon}} H_n^{(1)}(ka) J_n'(ka) - \sqrt{\frac{\mu}{\epsilon}} H_n^{(1)\prime}(ka) J_n(ka)}{\sqrt{\frac{\mu_1}{\epsilon_1}} J_n'(k_1 a) H_n^{(1)}(ka) - \sqrt{\frac{\mu}{\epsilon}} J_n(k_1 a) H_n^{(1)\prime}(ka)} \\ &= \frac{-\frac{2i}{\pi \omega \epsilon a}}{\sqrt{\frac{\mu_1}{\epsilon_1}} J_n'(k_1 a) H_n^{(1)}(ka) - \sqrt{\frac{\mu}{\epsilon}} J_n(k_1 a) H_n^{(1)\prime}(ka)} \, . \end{aligned} \tag{7.25}$$

Note that the modal coefficients c_n and d_n are a dual set of a_n and b_n since c_n and d_n can be obtained from (7.11) and (7.12) (a_n and b_n) by interchanging μ and ϵ.

7.1.3 Electrostatic Fields

Let us consider the field distribution due to a circular cylinder when the applied field is static $\left(k = 0 \text{ and } E^a_y = E_0\right)$. It is convenient to work with the electrostatic potential ϕ_e using Laplace's equation

$$\nabla^2 \phi_e = 0 \tag{7.26}$$

which is rewritten as

$$\rho \frac{\partial}{\partial \rho} \left(\rho \frac{\partial \phi_e}{\partial \rho} \right) + \frac{\partial^2 \phi_e}{\partial \phi^2} = 0 \, . \tag{7.27}$$

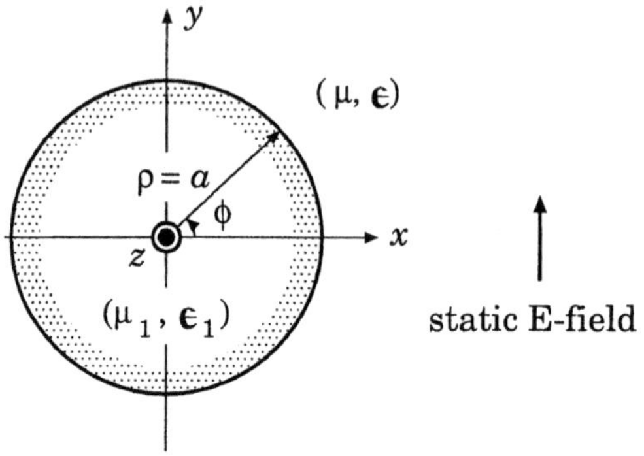

Fig. 7.2. Cross section of circular cylindrical dielectric cylinder.

Note that there is no field variation in the z-direction $\left(\frac{\partial \phi_e}{\partial z} = 0\right)$. Laplace's equation is a special form of Poisson's equation in Section 1.6, where the electric charge is assumed to be zero ($\rho_e = 0$). A solution is available from the separation of variables technique, as was discussed in Section 3.4. Substituting the product form for $\phi_e(\rho, \phi)$

$$\phi_e(\rho, \phi) = R(\rho)\, \Phi(\phi) \tag{7.28}$$

into Laplace's equation and dividing by $R(\rho)\, \Phi(\phi)$ yields

$$\underbrace{\frac{1}{R(\rho)} \left[\rho^2 \frac{d^2 R(\rho)}{d\rho^2} + \rho \frac{dR(\rho)}{d\rho}\right]}_{m^2} + \underbrace{\frac{1}{\Phi(\phi)} \frac{d^2 \Phi(\phi)}{d\phi^2}}_{-m^2} = 0\,. \tag{7.29}$$

Thus

$$\rho^2 \frac{d^2 R(\rho)}{d\rho^2} + \rho \frac{dR(\rho)}{d\rho} - m^2 R(\rho) = 0 \tag{7.30}$$

$$\frac{d^2 \Phi(\phi)}{d\phi^2} + m^2 \Phi(\phi) = 0\,. \tag{7.31}$$

Equation (7.30) is a type of Euler-Cauchy equation. Substituting the assumption

$$R(\rho) = \rho^p \tag{7.32}$$

into (7.30) yields

$$p^2 - m^2 = 0\,. \tag{7.33}$$

Therefore, $p = \pm m$ and

$$R(\rho) = A\rho^m + B\rho^{-m} . \tag{7.34}$$

Since $\Phi(\phi)$ is periodic in ϕ with 2π periodicity, $m = 1, 2, 3, \cdots$, and the solution to (7.31) is

$$\Phi(\phi) = C \cos m\phi + D \sin m\phi \tag{7.35}$$

for every integer $m \geq 1$.

The total potential for the region $\rho \geq a$ consists of the applied and scattered potentials $\phi_e^a(\rho, \phi)$ and $\phi_e^s(\rho, \phi)$, respectively,

$$\begin{aligned} \phi_e^a(\rho, \phi) &= -E_0 y \\ &= -E_0 \rho \sin \phi \end{aligned} \tag{7.36}$$

$$\phi_e^s(\rho, \phi) = \sum_{m=1}^{\infty} B_m \rho^{-m} \sin m\phi . \tag{7.37}$$

The transmitted potential for $\rho \leq a$ is

$$\phi_e^t(\rho, \phi) = \sum_{m=1}^{\infty} A_m \rho^m \sin m\phi . \tag{7.38}$$

To determine the unknown modal coefficients A_m and B_m, the boundary conditions must be enforced. The continuity boundary conditions

$$\phi_e^a(a, \phi) + \phi_e^s(a, \phi) = \phi_e^t(a, \phi) \tag{7.39}$$

$$-\epsilon \left\{ \frac{\partial \left[\phi_e^a(\rho, \phi) + \phi_e^s(\rho, \phi)\right]}{\partial \rho} \right\}_{\rho=a} = -\epsilon_1 \left[\frac{\partial \phi_e^t(\rho, \phi)}{\partial \rho} \right]_{\rho=a} \tag{7.40}$$

yield

$$-E_0 a + B_1 a^{-1} = A_1 a \tag{7.41}$$

$$\epsilon(E_0 + B_1 a^{-2}) = -\epsilon_1 A_1 \tag{7.42}$$

and all other $A_m = B_m = 0$ for $m \geq 2$. Hence

$$A_1 = -\frac{2\epsilon E_0}{\epsilon_1 + \epsilon} \tag{7.43}$$

$$B_1 = \frac{\epsilon_1 - \epsilon}{\epsilon_1 + \epsilon} a^2 E_0 . \tag{7.44}$$

Therefore, the electric field can be obtained from

$$\begin{aligned}\overline{E}^s &= -\nabla\phi_e^s \\ &= \left(\hat{\rho}\sin\phi - \hat{\phi}\cos\phi\right) E_0 \frac{\epsilon_1 - \epsilon}{\epsilon_1 + \epsilon} \frac{a^2}{\rho^2}\end{aligned} \tag{7.45}$$

$$\begin{aligned}\overline{E}^t &= -\nabla\phi_e^t \\ &= \left(\hat{\rho}\sin\phi + \hat{\phi}\cos\phi\right) E_0 \frac{2\epsilon}{\epsilon_1 + \epsilon} \\ &= \hat{y} E_0 \frac{2\epsilon}{\epsilon_1 + \epsilon} .\end{aligned} \tag{7.46}$$

Note that the direction of the transmitted field $\overline{E}^t$, $\hat{y}$, coincides with the direction of the applied field $\overline{E}^a$.

7.2 Dielectric Sphere

The boundary-value problem of electromagnetic wave scattering from a dielectric sphere was first solved by Mie; thus, its solution is known as Mie scattering. The Mie scattering solution has been widely used in practical applications for electromagnetic wave propagation in a medium embedded with spherical particles. This section derives the Mie scattering solution using magnetic and electric vector potentials.

7.2.1 Electromagnetic Case

Let us consider scattering when the uniform plane wave $\overline{E}^i = \hat{x} E_0 e^{ikr\cos\theta}$ is incident on a dielectric sphere of radius a, as shown in Fig. 7.3. The wavenumbers for $r > a$ and $r < a$ are $k = \omega\sqrt{\mu\epsilon}$ and $k_1 = \omega\sqrt{\mu_1\epsilon_1}$, respectively. In view of the field analysis in Section 4.3, the vector potentials, $\overline{A} = \hat{r} A_r$ and $\overline{F} = \hat{r} F_r$, can be separated into the components TM and TE to the $\hat{r}$-direction. When $\overline{A} = \hat{r} A_r$ and $\overline{F} = \hat{r} F_r$ are chosen, $\frac{A_r}{r}$ and $\frac{F_r}{r}$ satisfy the scalar Helmholtz equations

$$(\nabla^2 + k^2) \left\{ \begin{array}{c} \dfrac{A_r(r,\theta,\phi)}{r} \\[2ex] \dfrac{F_r(r,\theta,\phi)}{r} \end{array} \right\} = 0 . \tag{7.47}$$

It is possible to solve (7.47) using the separation of variables technique, as in Section 4.3. Since the scattering geometry is a sphere, it is expedient to rewrite the incident plane wave using spherical coordinates (r, θ, ϕ). The incident wave is represented in terms of the $\hat{r}$ components of the vector potentials $\overline{A}^i$ and $\overline{F}^i$, where the $\hat{r}$ component of the incident wave is

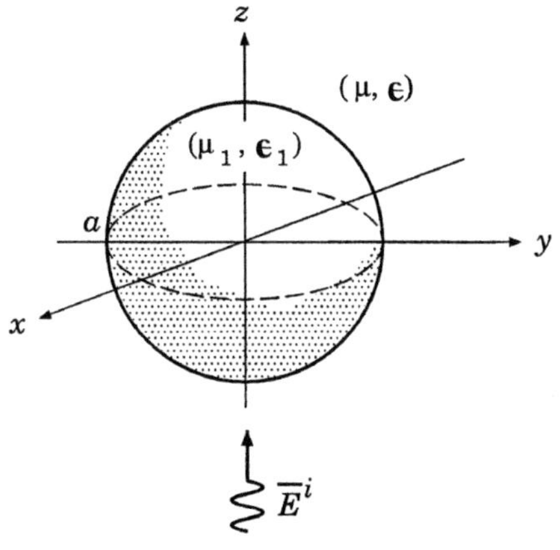

Fig. 7.3. Dielectric sphere with radius a.

$$
\begin{aligned}
E_r^i &= \cos\phi \sin\theta E_0 e^{ikr\cos\theta} \\
&= \cos\phi \frac{E_0}{-ikr}\frac{\partial}{\partial\theta}\left(e^{ikr\cos\theta}\right) . \qquad (7.48)
\end{aligned}
$$

Since the generating function for the spherical Bessel function (see Appendix B) gives

$$
\begin{aligned}
e^{ikr\cos\theta} &= \sqrt{\frac{\pi}{2kr}} \sum_{n=0}^{\infty}(2n+1)i^n J_{n+\frac{1}{2}}(kr)P_n(\cos\theta) \\
&= \sum_{n=0}^{\infty} i^n(2n+1)\frac{\hat{J}_n(kr)}{kr}P_n(\cos\theta) \qquad (7.49)
\end{aligned}
$$

E_r^i is written as

$$
\begin{aligned}
E_r^i &= \cos\phi \frac{E_0}{-i(kr)^2}\sum_{n=0}^{\infty} i^n(2n+1)\hat{J}_n(kr)\frac{d}{d\theta}\left[P_n(\cos\theta)\right] \\
&= \cos\phi \frac{E_0}{-i(kr)^2}\sum_{n=0}^{\infty} i^n(2n+1)\hat{J}_n(kr)\left[-P_n^1(\cos\theta)\right] \\
&= -i\cos\phi \frac{E_0}{(kr)^2}\sum_{n=1}^{\infty} i^n(2n+1)\hat{J}_n(kr)P_n^1(\cos\theta) \qquad (7.50)
\end{aligned}
$$

where $P_0^1(\cos\theta) = 0$. The magnetic vector potential A_r^i generates E_r^i, while the electric vector potential F_r^i does not. Substituting

$$A_r^i = \frac{E_0 \cos\phi}{\omega} \sum_{n=1}^{\infty} a_n \hat{J}_n(kr) P_n^1(\cos\theta) \tag{7.51}$$

into

$$E_r^i = \frac{i}{\omega\mu\epsilon}\left(\frac{\partial^2}{\partial r^2} + k^2\right) A_r^i$$

$$= \frac{i}{\omega\mu\epsilon} \sum_{n=1}^{\infty} \frac{n(n+1)}{r^2} \frac{E_0 \cos\phi}{\omega} a_n \hat{J}_n(kr) P_n^1(\cos\theta) \tag{7.52}$$

yields

$$a_n = -\frac{i^n(2n+1)}{n(n+1)} \,. \tag{7.53}$$

Similarly, from H_r^i, the incident electric potential F_r^i takes the form of

$$F_r^i = \frac{\epsilon}{k} E_0 \sin\phi \sum_{n=1}^{\infty} a_n \hat{J}_n(kr) P_n^1(\cos\theta) \,. \tag{7.54}$$

In view of A_r^i and F_r^i, the scattered potentials A_r^s and F_r^s $(r \geq a)$ can be written as

$$A_r^s = \frac{E_0 \cos\phi}{\omega} \sum_{n=1}^{\infty} b_n \hat{H}_n^{(1)}(kr) P_n^1(\cos\theta) \tag{7.55}$$

$$F_r^s = \frac{\epsilon}{k} E_0 \sin\phi \sum_{n=1}^{\infty} c_n \hat{H}_n^{(1)}(kr) P_n^1(\cos\theta) \,. \tag{7.56}$$

The transmitted potentials A_r^t and F_r^t $(r \leq a)$ take the form of

$$A_r^t = \frac{E_0 \cos\phi}{\omega} \sum_{n=1}^{\infty} d_n \hat{J}_n(k_1 r) P_n^1(\cos\theta) \tag{7.57}$$

$$F_r^t = \frac{\epsilon_1}{k_1} E_0 \sin\phi \sum_{n=1}^{\infty} e_n \hat{J}_n(k_1 r) P_n^1(\cos\theta) \,. \tag{7.58}$$

Four unknown modal coefficients–b_n, c_n, d_n, and e_n–are introduced to represent the scattered and transmitted vector potentials. To determine the unknown coefficients–b_n, c_n, d_n, and e_n, the boundary conditions for the field continuities

$$\left(E_\theta^i + E_\theta^s\right)\Big|_{r=a} = E_\theta^t\Big|_{r=a} \tag{7.59}$$

$$\left(E_\phi^i + E_\phi^s\right)\Big|_{r=a} = E_\phi^t\Big|_{r=a} \tag{7.60}$$

$$\left(H_\theta^i + H_\theta^s\right)\Big|_{r=a} = H_\theta^t\Big|_{r=a} \tag{7.61}$$

$$\left(H_\phi^i + H_\phi^s\right)\Big|_{r=a} = H_\phi^t\Big|_{r=a} \tag{7.62}$$

must be enforced, thereby resulting in

$$\frac{1}{\mu}(A_r^i + A_r^s)\Big|_{r=a} = \frac{1}{\mu_1}A_r^t\Big|_{r=a} \tag{7.63}$$

$$\frac{1}{\mu\epsilon}\frac{\partial(A_r^i + A_r^s)}{\partial r}\Big|_{r=a} = \frac{1}{\mu_1\epsilon_1}\frac{\partial A_r^t}{\partial r}\Big|_{r=a} \tag{7.64}$$

$$\frac{1}{\epsilon}(F_r^i + F_r^s)\Big|_{r=a} = \frac{1}{\epsilon_1}F_r^t\Big|_{r=a} \tag{7.65}$$

$$\frac{1}{\mu\epsilon}\frac{\partial(F_r^i + F_r^s)}{\partial r}\Big|_{r=a} = \frac{1}{\mu_1\epsilon_1}\frac{\partial F_r^t}{\partial r}\Big|_{r=a} . \tag{7.66}$$

Rewriting (7.63) through (7.66) explicitly gives

$$\frac{1}{\mu}\left[a_n\hat{J}_n(u) + b_n\hat{H}_n^{(1)}(u)\right] = \frac{1}{\mu_1}d_n\hat{J}_n(u_1) \tag{7.67}$$

$$\frac{k}{\mu\epsilon}\left[a_n\hat{J}_n'(u) + b_n\hat{H}_n^{(1)\prime}(u)\right] = \frac{k_1}{\mu_1\epsilon_1}d_n\hat{J}_n'(u_1) \tag{7.68}$$

$$\frac{1}{k}\left[a_n\hat{J}_n(u) + c_n\hat{H}_n^{(1)}(u)\right] = \frac{1}{k_1}e_n\hat{J}_n(u_1) \tag{7.69}$$

$$\frac{1}{\mu}\left[a_n\hat{J}_n'(u) + c_n\hat{H}_n^{(1)\prime}(u)\right] = \frac{1}{\mu_1}e_n\hat{J}_n'(u_1) \tag{7.70}$$

where $ka = u$ and $k_1a = u_1$. Here, the prime denotes differentiation with respect to the argument. Solving (7.67) and (7.68) for b_n and d_n gives

$$b_n = \frac{-\sqrt{\mu\epsilon_1}\hat{J}_n'(u)\hat{J}_n(u_1) + \sqrt{\mu_1\epsilon}\hat{J}_n(u)\hat{J}_n'(u_1)}{\sqrt{\mu\epsilon_1}\hat{H}_n^{(1)\prime}(u)\hat{J}_n(u_1) - \sqrt{\mu_1\epsilon}\hat{H}_n^{(1)}(u)\hat{J}_n'(u_1)}a_n \tag{7.71}$$

$$d_n = \frac{\mu_1\sqrt{\dfrac{\epsilon_1}{\mu}}\left[\hat{H}_n^{(1)\prime}(u)\hat{J}_n(u) - \hat{H}_n^{(1)}(u)\hat{J}_n'(u)\right]}{\sqrt{\mu\epsilon_1}\hat{H}_n^{(1)\prime}(u)\hat{J}_n(u_1) - \sqrt{\mu_1\epsilon}\hat{H}_n^{(1)}(u)\hat{J}_n'(u_1)}a_n . \tag{7.72}$$

Since

$$\begin{aligned}
&\hat{H}_n^{(1)\prime}(u)\hat{J}_n(u) - \hat{H}_n^{(1)}(u)\hat{J}_n'(u) \\
&= \frac{\pi u}{2}\left[H_\nu^{(1)\prime}(u)J_\nu(u) - H_\nu^{(1)}(u)J_\nu'(u)\right] \\
&= \frac{\pi u}{2} i\left[N_\nu'(u)J_\nu(u) - N_\nu(u)J_\nu'(u)\right] \\
&= i
\end{aligned} \tag{7.73}$$

where $\nu = n + \dfrac{1}{2}$, the modal coefficient d_n is simplified to

$$d_n = \frac{\mu_1\sqrt{\dfrac{\epsilon_1}{\mu}}i}{\sqrt{\mu\epsilon_1}\hat{H}_n^{(1)\prime}(u)\hat{J}_n(u_1) - \sqrt{\mu_1\epsilon}\hat{H}_n^{(1)}(u)\hat{J}_n'(u_1)} a_n \, . \tag{7.74}$$

Similarly, (7.69) and (7.70) produces

$$c_n = \frac{-\sqrt{\mu\epsilon_1}\hat{J}_n(u)\hat{J}_n'(u_1) + \sqrt{\mu_1\epsilon}\hat{J}_n'(u)\hat{J}_n(u_1)}{\sqrt{\mu\epsilon_1}\hat{H}_n^{(1)}(u)\hat{J}_n'(u_1) - \sqrt{\mu_1\epsilon}\hat{H}_n^{(1)\prime}(u)\hat{J}_n(u_1)} a_n \tag{7.75}$$

$$\begin{aligned}
e_n &= \frac{-\mu_1\sqrt{\dfrac{\epsilon_1}{\mu}}\left[\hat{H}_n^{(1)\prime}(u)\hat{J}_n(u) - \hat{H}_n^{(1)}(u)\hat{J}_n'(u)\right]}{\sqrt{\mu\epsilon_1}\hat{H}_n^{(1)}(u)\hat{J}_n'(u_1) - \sqrt{\mu_1\epsilon}\hat{H}_n^{(1)\prime}(u)\hat{J}_n(u_1)} a_n \\
&= \frac{-\mu_1\sqrt{\dfrac{\epsilon_1}{\mu}}i}{\sqrt{\mu\epsilon_1}\hat{H}_n^{(1)}(u)\hat{J}_n'(u_1) - \sqrt{\mu_1\epsilon}\hat{H}_n^{(1)\prime}(u)\hat{J}_n(u_1)} a_n \, .
\end{aligned} \tag{7.76}$$

The scattered TM and TE fields can be evaluated by substituting A_r^s and F_r^s into Table 4.1 in Section 4.3. Within the low-frequency limit $k_1 a \ll 1$, the Mie scattering solution reduces to a much simpler form of Rayleigh scattering. Using the low-frequency approximation $\hat{H}_1^{(1)}(kr) \approx -\dfrac{i}{kr}$ and $\hat{J}_1(kr) \approx \dfrac{(kr)^2}{3}$ yields the Rayleigh scattering solution $b_n = c_n = 0 \quad (n \geq 2)$ and

$$b_1 = (ka)^3 \left(\frac{\epsilon_1 - \epsilon}{\epsilon_1 + 2\epsilon}\right) \tag{7.77}$$

$$c_1 = (ka)^3 \left(\frac{\mu_1 - \mu}{\mu_1 + 2\mu}\right) \, . \tag{7.78}$$

When sunlight impinges upon small particles that are present in the sky, blue light (higher k) tends to scatter more than red light (lower k), thus causing the sky to appear blue. Within the static limit $k \to 0$

$$A_r^s = \frac{E_0 \cos\phi}{\omega} b_1 \hat{H}_1^{(1)}(kr) P_1^1(\cos\theta) \tag{7.79}$$

where $P_1^1(\cos\theta) = \sin\theta$. Substituting the vector potentials into Table 4.1 in Section 4.3 yields

$$\overline{E}^s = (\hat{r} 2 \sin\theta\cos\phi - \hat{\theta}\cos\theta\cos\phi + \hat{\phi}\sin\phi) E_0 \left(\frac{\epsilon_1 - \epsilon}{\epsilon_1 + 2\epsilon}\right) \frac{a^3}{r^3} \tag{7.80}$$

$$\begin{aligned} \overline{E}^t &= \underbrace{(\hat{r}\sin\theta\cos\phi + \hat{\theta}\cos\theta\cos\phi - \hat{\phi}\sin\phi)}_{\hat{x}} E_0 \frac{3\epsilon}{\epsilon_1 + 2\epsilon} \\ &= \hat{x} E_0 \left(\frac{3\epsilon}{\epsilon_1 + 2\epsilon}\right) . \end{aligned} \tag{7.81}$$

The direction of the transmitted field $\overline{E}^t$, $\hat{x}$, coincides with the direction of the incident field $\overline{E}^i$.

7.2.2 Electrostatic Case

Let us consider the field distribution due to a dielectric sphere of radius $r = a$ when the applied field $E_x^a = E_0$ is static ($k = 0$). It is convenient to work with the electrostatic potential ϕ_e using Laplace's equation

$$\nabla^2 \phi_e = 0 \tag{7.82}$$

where a solution is available from the separation of variables technique, as discussed in Section 4.3. Substituting

$$\phi_e(r, \theta, \phi) = \psi_1(r)\psi_2(\theta)\psi_3(\phi) \tag{7.83}$$

into Laplace's equation yields the following three ordinary differential equations

$$\frac{d^2\psi_1}{dr^2} + \frac{2}{r}\frac{d\psi_1}{dr} - \frac{n(n+1)}{r^2}\psi_1 = 0 \tag{7.84}$$

$$\frac{1}{\sin\theta}\frac{d}{d\theta}\left(\sin\theta\frac{d\psi_2}{d\theta}\right) + \left[n(n+1) - \frac{m^2}{\sin^2\theta}\right]\psi_2 = 0 \tag{7.85}$$

$$\frac{d^2\psi_3}{d\phi^2} + m^2\psi_3 = 0 \tag{7.86}$$

where the solutions for every integer $n \geq 0$ and $m \geq 0$ are

$$\psi_1(r) = Ar^n + Br^{-(n+1)} \tag{7.87}$$

$$\psi_2(\theta) = CP_n^m(\cos\theta) \tag{7.88}$$

$$\psi_3(\phi) = D\cos m\phi + E\sin m\phi\,. \tag{7.89}$$

The total potential for the region $r \geq a$ consists of the applied and scattered potentials $\phi_e^a(r,\theta,\phi)$ and $\phi_e^s(r,\theta,\phi)$, respectively, whereas the potential for the region $r \leq a$ is given by the transmitted potential $\phi_e^t(r,\theta,\phi)$. The continuity boundary conditions require

$$\phi_e^a(a,\theta,\phi) + \phi_e^s(a,\theta,\phi) = \phi_e^t(a,\theta,\phi) \tag{7.90}$$

$$\epsilon\frac{\partial}{\partial r}\left[\phi_e^a(r,\theta,\phi) + \phi_e^s(r,\theta,\phi)\right]_{r=a} = \epsilon_1\left[\frac{\partial\phi_e^t(r,\theta,\phi)}{\partial r}\right]_{r=a}\,. \tag{7.91}$$

Since the applied potential is in the form of

$$\phi_e^a(r,\theta,\phi) = -E_0 x = -E_0 r\sin\theta\cos\phi \tag{7.92}$$

the scattered potential takes the form of

$$\phi_e^s(r,\theta,\phi) = \sum_{n=0}^{\infty}\left[A_n r^n + B_n r^{-(n+1)}\right] P_n^1(\cos\theta)\cos\phi\,. \tag{7.93}$$

In view of the associated Legendre polynomials

$$P_0^1(\cos\theta) = 0 \tag{7.94}$$

$$P_1^1(\cos\theta) = \sin\theta \tag{7.95}$$

$$P_2^1(\cos\theta) = 3\sin\theta\cos\theta \tag{7.96}$$

$$P_3^1(\cos\theta) = -\frac{3}{2}\sin\theta\left(1 - 5\cos^2\theta\right) \tag{7.97}$$

the scattered potential can be written as

$$\phi_e^s(r,\theta,\phi) = \left(A_1 r + B_1 r^{-2}\right) P_1^1(\cos\theta)\cos\phi$$

$$= \left(A_1 r + B_1 r^{-2}\right)\sin\theta\cos\phi\,. \tag{7.98}$$

Furthermore, since the scattered potential vanishes at infinity ($r \to \infty$), $A_1 = 0$ and

$$\phi_e^s(r,\theta,\phi) = B_1 r^{-2} \sin\theta \cos\phi \,. \tag{7.99}$$

Similarly, the transmitted potential for $r \leq a$ can be written as

$$\begin{aligned}
\phi_e^t(r,\theta,\phi) &= \sum_{n=0}^{\infty} C_n r^n P_n^1(\cos\theta)\cos\phi \\
&= C_1 r P_1^1(\cos\theta)\cos\phi \\
&= C_1 r \sin\theta\cos\phi \,.
\end{aligned} \tag{7.100}$$

Substituting $\phi_e^a(r,\theta,\phi)$, $\phi_e^s(r,\theta,\phi)$, and $\phi_e^t(r,\theta,\phi)$ into (7.90) and (7.91) yields

$$B_1 = \left(\frac{\epsilon_1 - \epsilon}{\epsilon_1 + 2\epsilon}\right) a^3 E_0 \tag{7.101}$$

$$C_1 = -\left(\frac{3\epsilon}{\epsilon_1 + 2\epsilon}\right) E_0 \,. \tag{7.102}$$

Hence

$$\phi_e^s(r,\theta,\phi) = E_0 \left(\frac{\epsilon_1 - \epsilon}{\epsilon_1 + 2\epsilon}\right) \frac{a^3}{r^2} \sin\theta\cos\phi \tag{7.103}$$

$$\phi_e^t(r,\theta,\phi) = -E_0 \left(\frac{3\epsilon}{\epsilon_1 + 2\epsilon}\right) r \sin\theta\cos\phi \,. \tag{7.104}$$

Note that the electric fields from

$$\overline{E}^s = -\nabla\phi_e^s(r,\theta,\phi) \tag{7.105}$$

$$\overline{E}^t = -\nabla\phi_e^t(r,\theta,\phi) \tag{7.106}$$

are shown to be identical with the expressions (7.80) and (7.81).

7.3 Step in Parallel-Plate Waveguide

Consider a step between two perfectly conducting parallel-plate waveguides, as shown in Fig. 7.4. Assume that a TE wave $E_y^i(x,z)$ impinges on a step between perfectly conducting parallel-plate waveguides. When an incident wave $E_y^i(x,z)$ impinges on the step at $z = 0$, scattering takes place due to a step discontinuity. No field variation is assumed in the y-direction $\left(\dfrac{\partial}{\partial y} = 0\right)$ and nonzero field components are E_y, H_x, and H_z. The electric field $E_y(x,z)$ should satisfy the Helmholtz equation

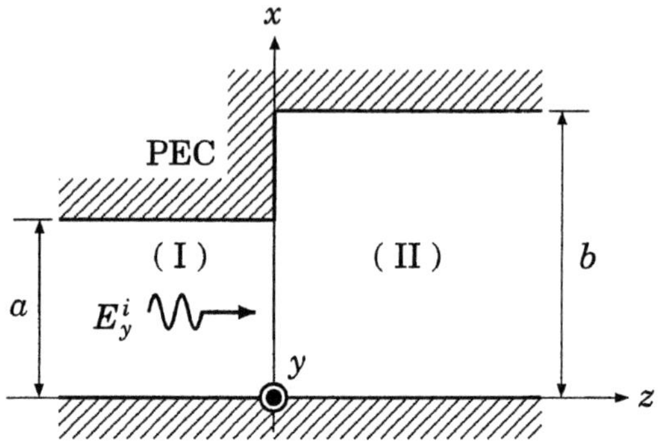

Fig. 7.4. Step between two perfect conducting parallel-plate waveguides.

$$\left(\nabla^2 + k^2\right) E_y(x, z) = 0 \tag{7.107}$$

where $k \left(= \omega\sqrt{\mu\epsilon}\right)$ is the wavenumber. This section uses a mode-matching technique to analyze scattering.

The field in region (I) ($z \le 0$) consists of the incident and reflected components

$$E_y^i(x, z) = \sin(a_n x) \exp(ik_{z1n} z) \tag{7.108}$$

$$E_y^r(x, z) = \sum_{m=1}^{\infty} R_m \sin(a_m x) \exp(-ik_{z1m} z) \tag{7.109}$$

where $k_{z1m} = \sqrt{k^2 - a_m^2}$ and $a_m = \dfrac{m\pi}{a}$. Note that $E_y^i(x, z)$ and $E_y^r(x, z)$ are chosen to satisfy the boundary conditions

$$E_y^i(x, z)\Big|_{x=0,a} = E_y^r(x, z)\Big|_{x=0,a} = 0\,. \tag{7.110}$$

Similarly, the transmitted field $E_y^t(x, z)$ in region (II) ($z \ge 0$) is set as

$$E_y^t(x, z) = \sum_{m=1}^{\infty} T_m \sin(b_m x) \exp(ik_{z2m} z) \tag{7.111}$$

where $k_{z2m} = \sqrt{k^2 - b_m^2}$ and $b_m = \dfrac{m\pi}{b}$.

To determine the reflection and transmission coefficients R_m and T_m, respectively, the boundary conditions for the tangential field continuities must be enforced. The tangential electric field continuity at $z = 0$ is

$$\sum_{m=1}^{\infty} T_m \sin(b_m x) = \begin{cases} \sin(a_n x) + \sum_{m=1}^{\infty} R_m \sin(a_m x) & \text{for } 0 \le x \le a \\ 0 & \text{for } a \le x \le b\,. \end{cases} \tag{7.112}$$

The orthogonality property of the sinusoidal function $\sin(b_m x)$ can be used for further simplification. Multiplying (7.112) by $\sin(b_p x)$ and integrating with respect to x for $0 \le x \le b$ yields

$$T_p = \frac{2}{b}\left(G_{np} + \sum_{m=1}^{\infty} R_m G_{mp}\right) \tag{7.113}$$

where

$$\int_0^a \sin(a_m x)\sin(b_p x)\,dx = \underbrace{\frac{(-1)^m a_m \sin(b_p a)}{b_p^2 - a_m^2}}_{G_{mp}}\,. \tag{7.114}$$

The tangential magnetic field is given by

$$H_x(x,z) = \frac{i}{\omega\mu}\frac{\partial E_y(x,z)}{\partial z}\,. \tag{7.115}$$

Similarly, multiplying the tangential magnetic field continuity at $z = 0$ and $0 \le x \le a$

$$k_{z1n}\sin(a_n x) - \sum_{m=1}^{\infty} k_{z1m} R_m \sin(a_m x) = \sum_{m=1}^{\infty} k_{z2m} T_m \sin(b_m x) \tag{7.116}$$

by $\sin(a_p x)$ and integrating with respect to x for $0 \le x \le a$ yields

$$k_{z1p}(\delta_{np} - R_p) = \frac{2}{a}\sum_{m=1}^{\infty} k_{z2m} T_m G_{pm} \tag{7.117}$$

where δ_{np} is the Kronecker delta.

Equations (7.113) and (7.117) constitute a set of simultaneous systems for the unknown coefficients T_m and R_m. Once T_m and R_m are determined, it is possible to evaluate the reflected and transmitted powers. The reflected magnetic field is given by

$$H_x^r(x,z) = \frac{i}{\omega\mu}\frac{\partial E_y^r(x,z)}{\partial z}$$
$$= \frac{1}{\omega\mu}\sum_{m=1}^{\infty} k_{z1m} R_m \sin(a_m x)\exp(-ik_{z1m}z) . \tag{7.118}$$

A Poynting vector gives the reflected time-average power

$$P_r = \frac{1}{2}\mathrm{Re}\int_0^a E_y^r(x,z) H_x^{r*}(x,z)\,dx$$
$$= \frac{a}{4\omega\mu}\sum_{m=1}^{M_1} |R_m|^2 k_{z1m} \tag{7.119}$$

and the transmitted time-average power

$$P_t = \frac{b}{4\omega\mu}\sum_{m=1}^{M_2} |T_m|^2 k_{z2m} \tag{7.120}$$

where M_1 and M_2 denote the numbers of propagation modes as $M_1 < \frac{ka}{\pi}$ and $M_2 < \frac{kb}{\pi}$, respectively.

7.4 Slit in Conducting Plane

Electromagnetic scattering from a slit in a conducting plane finds many applications in antenna and electromagnetic interference problems. One of the basic radiating elements for aperture antennas is a slit in a conducting plane. An analysis of scattering from a slit in a thick conducting plane is available in [8] based on the mode matching and Fourier transform. Meanwhile, this section investigates TE wave scattering from a slit in a conducting plane using the mode matching and Fourier transform.

7.4.1 Field Analysis

Consider a TE wave $E_z^i(x,y)$ impinging on a thick slit in a perfectly conducting plane. The problem geometry is shown in Fig. 7.5 where the wavenumbers in regions (I), (II), and (III) are k_0 $(=\omega\sqrt{\mu_0\epsilon_0})$, k_1 $(=\omega\sqrt{\mu_1\epsilon_1})$, and k_2 $(=\omega\sqrt{\mu_2\epsilon_2})$, respectively. No field variation is assumed in the z-direction $\left(\frac{\partial}{\partial z}=0\right)$ and field components are $E_z(x,y)$, $H_x(x,y)$, and $H_y(x,y)$.

The field in region (I) $(x \le 0)$ consists of the incident, reflected, and scattered components, although the inclusion of the reflected component in region (I) is arbitrary. The incident and reflected fields are assumed to be

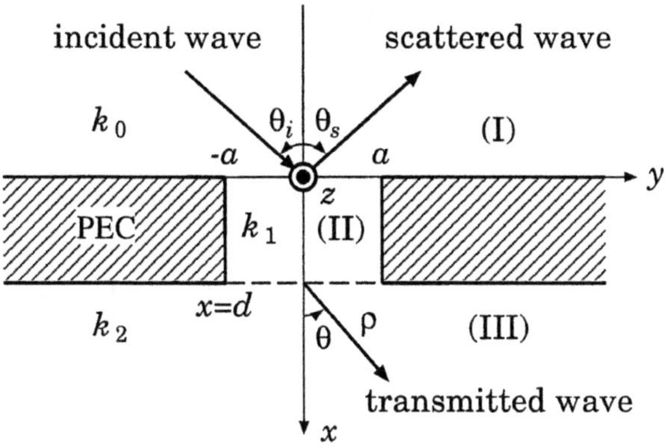

Fig. 7.5. Thick slit in conducting plane.

$$E_z^i(x,y) = \exp(ik_x x + ik_y y) \tag{7.121}$$

$$E_z^r(x,y) = -\exp(-ik_x x + ik_y y) \tag{7.122}$$

where $k_x = k_0 \cos\theta_i$ and $k_y = k_0 \sin\theta_i$. It is possible to represent the scattered field in region (I) in terms of a continuous spectrum. Let the scattered field $E_z^s(x,y)$ be represented in terms of the Fourier transform

$$E_z^s(x,y) = \int_{-\infty}^{\infty} \tilde{E}_z^s(x,\zeta) e^{-i\zeta y}\, d\zeta \tag{7.123}$$

where its inverse is

$$\tilde{E}_z^s(x,\zeta) = \frac{1}{2\pi}\int_{-\infty}^{\infty} E_z^s(x,y) e^{i\zeta y}\, dy\ . \tag{7.124}$$

Substituting $E_z^s(x,y)$ into the Helmholtz equation

$$\left(\nabla^2 + k_0^2\right) E_z^s(x,y) = 0 \tag{7.125}$$

gives

$$\left(\frac{d^2}{dx^2} + \kappa_0^2\right) \tilde{E}_z^s(x,\zeta) = 0 \tag{7.126}$$

where $\kappa_0 = \sqrt{k_0^2 - \zeta^2}$. The solution that satisfies the radiation condition $\tilde{E}_z^s(-\infty,\zeta) \to 0$ is given by

$$\tilde{E}_z^s(x,\zeta) = \tilde{E}_z^s(\zeta) e^{-i\kappa_0 x}\ . \tag{7.127}$$

Substituting $\tilde{E}_z^s(x,\zeta)$ into (7.123) finally gives the representation for $E_z^s(x,y)$

$$E_z^s(x,y) = \int_{-\infty}^{\infty} \tilde{E}_z^s(\zeta) e^{-i\zeta y - i\kappa_0 x}\, d\zeta\,. \tag{7.128}$$

The fields, $E_z^d(x,y)$ and $E_z^t(x,y)$, in regions (II) and (III) respectively, are

$$E_z^d(x,y) = \sum_{m=1}^{\infty} (b_m \cos\xi_m x + c_m \sin\xi_m x) \sin a_m(y+a) \tag{7.129}$$

$$E_z^t(x,y) = \int_{-\infty}^{\infty} \tilde{E}_z^t(\zeta) e^{i\kappa_2(x-d)} e^{-i\zeta y}\, d\zeta \tag{7.130}$$

where $a_m = \dfrac{m\pi}{2a}$, $\xi_m = \sqrt{k_1^2 - a_m^2}$, and $\kappa_2 = \sqrt{k_2^2 - \zeta^2}$. Note that $E_z^d(x,y)$ is constructed so as to satisfy the boundary condition $E_z^d(x,\pm a) = 0$.

Equations (7.128), (7.129), and (7.130) represent the scattered fields in regions (I), (II), and (III), respectively, in terms of the four unknown modal coefficients–namely, $\tilde{E}_z^s(\zeta)$, b_m, c_m, and $\tilde{E}_z^t(\zeta)$. The boundary conditions must be used to obtain simultaneous equations for the modal coefficients. Let us enforce the four boundary conditions for the continuities of the tangential fields–$E_z(0,y)$, $H_y(0,y)$, $E_z(d,y)$, and $H_y(d,y)$. Applying the inverse Fourier transform to the tangential electric field continuity at $x=0$

$$E_z^s(0,y) = \begin{cases} E_z^d(0,y) & \text{for } |y| < a \\ 0 & \text{otherwise} \end{cases} \tag{7.131}$$

gives

$$\tilde{E}_z^s(\zeta) = \sum_{m=1}^{\infty} \frac{b_m a_m a^2}{2\pi} F_m(\zeta a) \tag{7.132}$$

where $F_m(u) = \dfrac{\left[e^{iu}(-1)^m - e^{-iu}\right]}{u^2 - \left(\dfrac{m\pi}{2}\right)^2}$.

The continuity of the tangential magnetic field at $x = 0$ and $|y| < a$ requires

$$H_y^i(0,y) + H_y^r(0,y) + H_y^s(0,y) = H_y^d(0,y)\,. \tag{7.133}$$

Since

$$H_y(x,y) = \frac{i}{\omega\mu} \frac{\partial E_z(x,y)}{\partial x} \tag{7.134}$$

(7.133) can be rewritten as

$$
-\frac{2k_x}{\mu_0}e^{ik_y y} + \frac{1}{\mu_0}\int_{-\infty}^{\infty} \tilde{E}_z^s(\zeta)\kappa_0 e^{-i\zeta y}\, d\zeta
$$
$$
= \frac{i}{\mu_1}\sum_{m=1}^{\infty} \xi_m c_m \sin a_m(y+a) \, . \tag{7.135}
$$

The orthogonality property of the sinusoidal function $\sin a_m(y+a)$ can be used for further simplification. Multiplying (7.135) by $\sin a_n(y+a)$ and integrating with respect to y for $|y| < a$ yields

$$
-\frac{2k_x a_n}{\mu_0}a^2 F_n(k_y a) + \frac{a_n}{2\pi\mu_0}\sum_{m=1}^{\infty} b_m a_m a^2 I(k_0) = \frac{ic_n\xi_n a}{\mu_1} \tag{7.136}
$$

where

$$
I(k_0) = \int_{-\infty}^{\infty} a^2 F_m(\zeta a)F_n(-\zeta a)\kappa_0 \, d\zeta \, . \tag{7.137}
$$

An evaluation of $I(k_0)$ is given in Section 7.4.3 using residue calculus.

Let us enforce the boundary conditions at $x = d$. Applying the inverse Fourier transform to the tangential electric field continuity at $x = d$

$$
E_z^t(d,y) = \begin{cases} E_z^d(d,y) & \text{for } |y| < a \\ 0 & \text{otherwise} \end{cases} \tag{7.138}
$$

gives

$$
\tilde{E}_z^t(\zeta) = \sum_{m=1}^{\infty}(b_m \cos\xi_m d + c_m \sin\xi_m d)\frac{a_m a^2}{2\pi}F_m(\zeta a) \, . \tag{7.139}
$$

Similarly, the continuity of the tangential magnetic field at $x = d$ and $|y| < a$

$$
H_y^t(d,y) = H_y^d(d,y) \tag{7.140}
$$

results in

$$
-i\frac{\mu_2}{\mu_1}(-b_n \sin\xi_n d + c_n \cos\xi_n d)\xi_n a
$$
$$
= \frac{a_n}{2\pi}\sum_{m=1}^{\infty} a_m(b_m \cos\xi_m d + c_m \sin\xi_m d)a^2 I(k_2) \tag{7.141}
$$

where

$$
I(k_2) = \int_{-\infty}^{\infty} a^2 F_m(\zeta a)F_n(-\zeta a)\kappa_2 \, d\zeta \, . \tag{7.142}
$$

Equations (7.136) and (7.141) constitute a set of simultaneous equations for the discrete modal coefficients b_m and c_m. Once (7.136) and (7.141) are numerically solved, it is possible to evaluate the scattered and transmitted fields, $E_z^s(x,y)$ and $E_z^t(x,y)$, respectively. While the scattering solution requires the evaluation of (7.137) and (7.142), an approximate simple solution is available within the high-frequency limit [8]. When $k_0 a \to \infty$,

$$I(k_0) \to \frac{2\pi p_m}{a a_m^2} \delta_{mn} \tag{7.143}$$

$$I(k_2) \to \frac{2\pi q_m}{a a_m^2} \delta_{mn} \tag{7.144}$$

where $p_m = \sqrt{k_0^2 - a_m^2}$ and $q_m = \sqrt{k_2^2 - a_m^2}$. Substituting (7.143) and (7.144) into (7.136) and (7.141) yields an approximate high-frequency solution for b_n and c_n in closed form

$$\begin{pmatrix} p_n & -i\frac{\mu_0}{\mu_1}\xi_n \\ i\frac{\mu_2}{\mu_1}\tan(\xi_n d)\xi_n - q_n & -i\frac{\mu_2}{\mu_1}\xi_n - \tan(\xi_n d) q_n \end{pmatrix} \begin{pmatrix} b_n \\ c_n \end{pmatrix}$$

$$= \begin{pmatrix} 2k_x a_n a F_n(k_y a) \\ 0 \end{pmatrix} . \tag{7.145}$$

7.4.2 Far Field and Transmission Coefficient

It is possible to obtain the far-zone transmitted field ($k_2\rho \gg 1$) based on the saddle-point method [9]. Substituting the relations

$$x - d = \rho \cos\theta \tag{7.146}$$

$$y = \rho \sin\theta \tag{7.147}$$

$$\zeta = k_2 \sin\psi \tag{7.148}$$

$$\kappa_2 = \sqrt{k_2^2 - \zeta^2} = k_2 \cos\psi \tag{7.149}$$

into the transmitted field (7.130) gives

$$E_z^t = \int \tilde{E}_z^t(k_2 \sin\psi) \exp\left[ik_2\rho \underbrace{\cos(\psi+\theta)}_{f(\psi)}\right] k_2 \cos\psi \, d\psi . \tag{7.150}$$

The condition $f'(\psi_0) = 0$ yields the saddle point $\psi_0 = -\theta$. Hence, when $k_2\rho \gg 1$,

$$E_z^t \approx \tilde{E}_z^t(-k_2 \sin\theta)\sqrt{\frac{2\pi}{k_2\rho}} \exp\left[i\left(k_2\rho - \frac{\pi}{4}\right)\right] k_2 \cos\theta \,. \tag{7.151}$$

Substituting (7.139) into (7.151) finally yields the far-zone transmitted field

$$E_z^t \approx \exp\left[i\left(k_2\rho - \frac{\pi}{4}\right)\right]\sqrt{\frac{k_2}{2\pi\rho}}\cos\theta$$

$$\sum_{m=1}^{\infty}(b_m \cos\xi_m d + c_m \sin\xi_m d) a_m a^2 F_m(-k_2 a \sin\theta) \,. \tag{7.152}$$

The transmission coefficient (τ) is defined as the ratio of the transmitted power P_t to the incident power P_i in a slit $\left(\tau = \frac{P_t}{P_i}\right)$. The incident power is given in terms of a Poynting vector as

$$\begin{aligned}
P_i &= \int_{-a}^{a} \frac{1}{2} \mathrm{Re}\left[\overline{E}^i(0,y) \times \overline{H}^i(0,y)^*\right] \cdot \hat{x}\, dy \\
&= \int_{-a}^{a} \frac{1}{2} \mathrm{Re}\left[-E_z^i(0,y) H_y^i(0,y)^*\right] dy \\
&= \frac{a k_x}{\omega\mu_0} \,.
\end{aligned} \tag{7.153}$$

Similarly, the power transmitted through a slit at $x = d$ is

$$P_t = \int_{-a}^{a} \frac{1}{2} \mathrm{Re}\left[-E_z^d(d,y) H_y^d(d,y)^*\right] dy \,. \tag{7.154}$$

Since

$$E_z^d(d,y) = \sum_{m=1}^{\infty}(b_m \cos\xi_m d + c_m \sin\xi_m d)\sin a_m(y+a) \tag{7.155}$$

$$\begin{aligned}
H_y^d(d,y) &= \frac{i}{\omega\mu_1}\frac{\partial E_z^d(x,y)}{\partial x}\bigg|_{x=d} \\
&= \frac{i}{\omega\mu_1}\sum_{m=1}^{\infty}\xi_m\left(-b_m \sin\xi_m d + c_m \cos\xi_m d\right) \\
&\quad \cdot \sin a_m(y+a)
\end{aligned} \tag{7.156}$$

the transmitted power P_t becomes

$$\begin{aligned}
P_t = \mathrm{Im}\Bigg\{ &-\frac{a}{2\omega\mu_1}\sum_{m=1}^{\infty}\xi_m^*\Big[-|b_m|^2 \cos\xi_m d(\sin\xi_m d)^* \\
&+ b_m c_m^* |\cos\xi_m d|^2 - b_m^* c_m |\sin\xi_m d|^2 \\
&+ |c_m|^2 \sin\xi_m d(\cos\xi_m d)^*\Big]\Bigg\} \,.
\end{aligned} \tag{7.157}$$

The transmission coefficient $\left(\tau = \frac{P_t}{P_i}\right)$ can be readily obtained from (7.153) and (7.157).

7.4.3 Residue Calculus for $I(k_0)$

Consider the integral

$$I(k_0) = \int_{-\infty}^{\infty} \underbrace{a^2 F_m(\zeta a) F_n(-\zeta a)\kappa_0}_{f(\zeta)} \, d\zeta \, . \tag{7.158}$$

When $m+n$ is odd, $f(\zeta)$ is an odd function of ζ. Therefore, $I(k_0) = 0$. When $m+n$ is even, $I(k_0)$ can be rewritten as

$$I(k_0) = \int_{-\infty}^{\infty} \underbrace{2\left[\frac{1-(-1)^n e^{i2\zeta a}}{(\zeta^2 - a_m^2)(\zeta^2 - a_n^2)a^2}\right]\kappa_0}_{f(\zeta)} \, d\zeta \, . \tag{7.159}$$

It is expedient to evaluate $I(k_0)$ in the complex ζ-plane using residue calculus and contour integration. The integrand contains branch points at $\zeta = \pm k_0$ and simple poles at $\zeta = \pm a_m$ when $m = n$, as shown in Fig. 7.6. Assume $R \to \infty$. For analytic convenience, the medium wavenumber $k_0 = k_r + ik_i$ is assumed to have an infinitesimally small positive imaginary part ($k_i > 0$). Considering the contour Γ, which consists of Γ_1, Γ_2, $\cdots$, Γ_7 in the upper half-plane, the residue theorem gives

$$\oint f(\zeta)\, d\zeta = \int_\Gamma f(\zeta)\, d\zeta = 0 \, . \tag{7.160}$$

The paths Γ_2 and Γ_3 denote infinitesimal semicircles and the path Γ_6 denotes an infinitesimal circle. The path Γ_4 denotes an infinite semicircle. Note

$$\int_{\Gamma_1} f(\zeta)\, d\zeta = I(k_0) \tag{7.161}$$

$$\int_{\Gamma_2} f(\zeta)\, d\zeta = -\pi i \operatorname*{Res}_{\zeta=-a_m} f(\zeta) = -\frac{\pi\sqrt{k_0^2 - a_m^2}}{a a_m^2}\delta_{mn} \tag{7.162}$$

$$\int_{\Gamma_3} f(\zeta)\, d\zeta = -\pi i \operatorname*{Res}_{\zeta=a_m} f(\zeta) = -\frac{\pi\sqrt{k_0^2 - a_m^2}}{a a_m^2}\delta_{mn} \tag{7.163}$$

$$\int_{\Gamma_4} f(\zeta)\, d\zeta = 0 \tag{7.164}$$

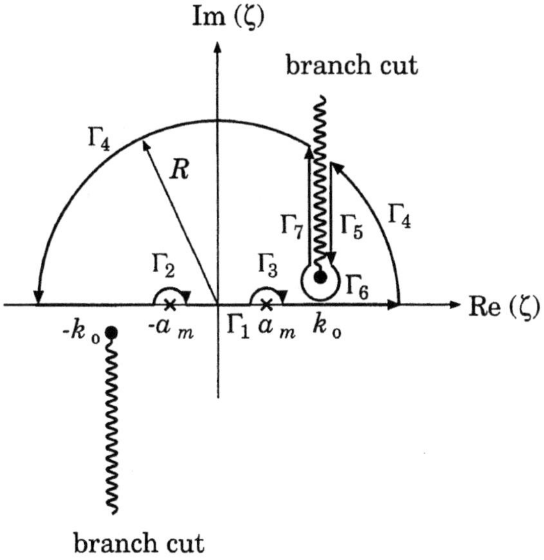

Fig. 7.6. Complex ζ-plane.

$$\int_{\Gamma_5} f(\zeta)\, d\zeta = K_1 \tag{7.165}$$

$$\int_{\Gamma_6} f(\zeta)\, d\zeta = 0 \tag{7.166}$$

$$\int_{\Gamma_7} f(\zeta)\, d\zeta = K_2\,. \tag{7.167}$$

Hence

$$I(k_0) = \frac{2\pi\sqrt{k_0^2 - a_m^2}}{a a_m^2}\delta_{mn} - (K_1 + K_2) \tag{7.168}$$

where K_1 and K_2 are integrations along the branch cut Γ_5 and Γ_7 associated with the branch point k_0.

In order to evaluate K_1 and K_2, consider $\kappa_0 = \sqrt{k_0^2 - \zeta^2}$. Introducing the polar forms

$$\zeta - k_0 = r_1 e^{i\theta_1} \tag{7.169}$$

$$\zeta + k_0 = r_2 e^{i\theta_2} \tag{7.170}$$

leads to

$$\kappa_0 = \sqrt{r_1 r_2} \exp\Big(\underbrace{i\,\frac{\theta_1+\theta_2+\pi}{2}}_{\psi}\Big) . \tag{7.171}$$

The corresponding phase diagram is illustrated in Fig. 7.7. The phase variations along the branch cut path A_1, A_2, A_3, and A_4 are in Table 7.1, showing that the difference in ψ at A_1 and A_4 (or at A_2 and A_3) is π. Therefore

$$\kappa_0\Big|_{\Gamma_5} = -\kappa_0\Big|_{\Gamma_7} \tag{7.172}$$

$$K_1 = K_2 . \tag{7.173}$$

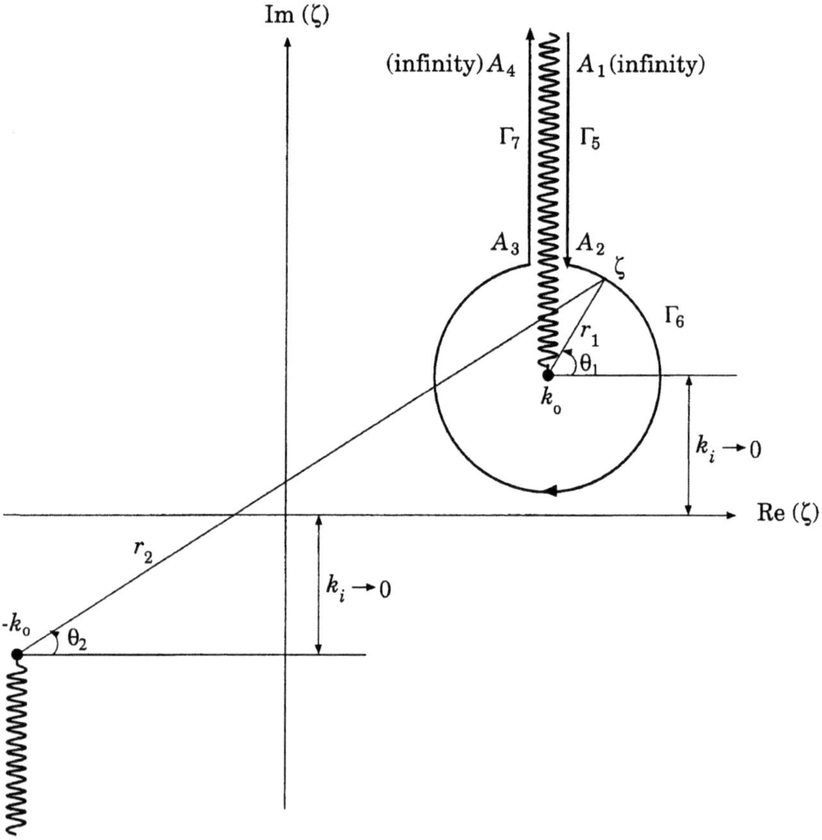

Fig. 7.7. Phase diagram.

Table 7.1. Phase variations along branch cut

	θ_1	θ_2	ψ
A_1	$\frac{\pi}{2}$	$\frac{\pi}{2}$	π
A_2	$\frac{\pi}{2}$	0	$\frac{3\pi}{4}$
A_3	$-\frac{3\pi}{2}$	0	$-\frac{\pi}{4}$
A_4	$-\frac{3\pi}{2}$	$\frac{\pi}{2}$	0

7.4.4 Thin Slit Within High-Frequency Limit

The previous subsections performed a rigorous analysis of the scattering from a slit in a conducting plane. While a rigorous scattering analysis provides an exact solution, it requires further tedious computations to obtain the scattering characteristics. This subsection presents an alternative approximate solution to the problem based on Love's equivalence principle and image method. Consider an incident wave $E_z^i(x,y)$ that impinges on an infinitesimally thin slit ($d=0$), as shown in Fig. 7.8 (a). Assume that the medium wavenumber is $k(=\omega\sqrt{\mu\epsilon})$ throughout. Let the transmitted ($x \geq 0$) wave be denoted by $\overline{E}^t(x,y)$.

Based on Love's equivalence principle, the original problem (a) can be transformed into an equivalent problem (b). In problem (b), the surface impressed currents $\overline{J}_s = \hat{n} \times \overline{H}^t$ and $\overline{M}_s = -\hat{n} \times \overline{E}^t$ are placed at $x=0$, thereby radiating $\overline{E}^t$ and $\overline{H}^t$ for $x>0$ and the null field for $x<0$. When a perfect electric conductor is placed at $x=0$ beneath the impressed current in problem (c), the impressed electric current $\overline{J}_s$ becomes short-circuited on the PEC. The impressed current $\overline{M}_s$ on the infinite PEC plane in problem (c) is equivalent to the impressed current

$$2\overline{M}_s = -2\hat{x} \times \overline{E}^t(0,y) \tag{7.174}$$

in a medium of infinite extent in problem (d), as far as the field $\overline{E}^t$ and $\overline{H}^t$ for $x>0$ is concerned.

To determine $\left[-2\hat{x} \times \overline{E}^t(0,y)\right]$, the tangential field continuities at $x=0$ must be used. The continuity of the tangential electric field for $|y|<a$ requires

$$E_z^i(0,y) + E_z^r(0,y) + E_z^s(0,y) = E_z^t(0,y)\ . \tag{7.175}$$

When the operating frequency is high enough that $ka \gg 1$, the condition

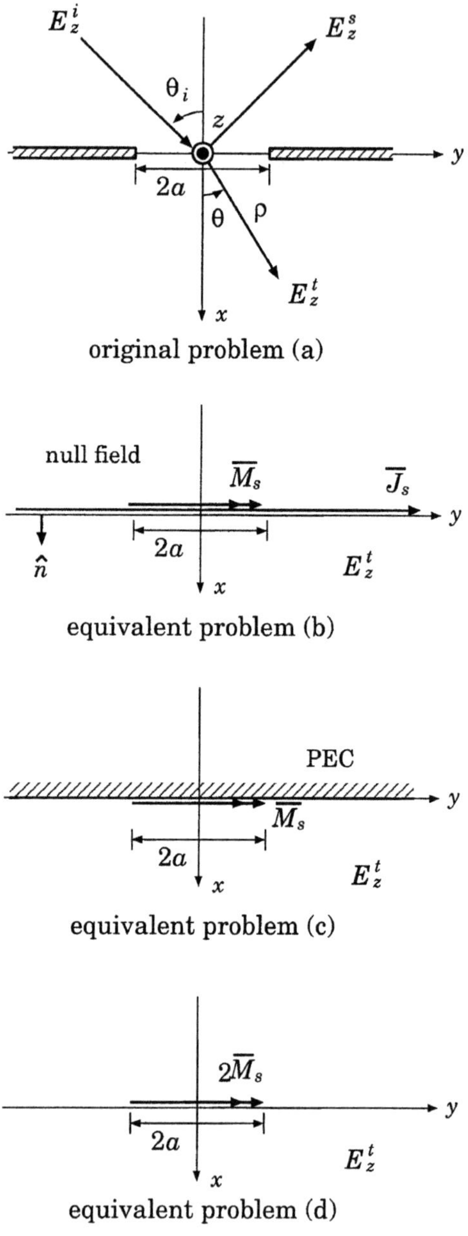

Fig. 7.8. Thin slit in conducting plane.

$$|E_z^r(0,y) + E_z^s(0,y)| \ll \left|E_z^i(0,y)\right| \tag{7.176}$$

applies for $-a < y < a$. Therefore

$$E_z^t(0,y) \approx E_z^i(0,y) \tag{7.177}$$

$$2\overline{M}_s(0,y') = \hat{y}' 2\overline{E}^i(0,y') \; . \tag{7.178}$$

The electric vector potential $\overline{F}(\overline{\rho})$ due to 2 $\overline{M}_s(0,y')$ in a medium of infinite extent is given in terms of the free-space Green's function in Section 1.5.2. The electric vector potential $\overline{F}(\overline{\rho})$ is thus written as

$$\begin{aligned} \overline{F}(\overline{\rho}) &= \epsilon \int_{-a}^{a} 2\overline{M}_s(0,y') \frac{i}{4} H_0^{(1)}\left(k|\overline{\rho} - \overline{\rho}'|\right) dy' \\ &= \frac{i\epsilon}{2} \int_{-a}^{a} \hat{y}' e^{ik_y y'} H_0^{(1)}\left(k|\overline{\rho} - \overline{\rho}'|\right) dy' \end{aligned} \tag{7.179}$$

where $\frac{i}{4} H_0^{(1)}\left(k|\overline{\rho} - \overline{\rho}'|\right)$ is the two-dimensional free-space Green's function. Substituting the asymptotic form

$$H_0^{(1)}(k|\overline{\rho} - \overline{\rho}'|) \rightarrow \sqrt{\frac{2}{\pi k \rho}} \exp\left[i\left(k\rho - ky' \sin\theta - \frac{\pi}{4}\right)\right] \tag{7.180}$$

for the far zone $\rho \gg \rho'$ and $k\rho \gg 1$ into $\overline{F}(\overline{\rho})$ gives

$$\overline{F}(\overline{\rho}) = \hat{y} \underbrace{i\epsilon \sqrt{\frac{2}{\pi k \rho}} \exp\left[i\left(k\rho - \frac{\pi}{4}\right)\right] \frac{\sin\left[k(\sin\theta - \sin\theta_i)a\right]}{k(\sin\theta - \sin\theta_i)}}_{F_y} \; . \tag{7.181}$$

Since $\hat{y} = \hat{\rho}\sin\theta + \hat{\theta}\cos\theta$, the electric vector potential $\overline{F}(\overline{\rho})$ using cylindrical coordinates (ρ, θ, z) is

$$\overline{F}(\overline{\rho}) = \hat{\rho} F_y \sin\theta + \hat{\theta} F_y \cos\theta \; . \tag{7.182}$$

In the far zone $k\rho \gg 1$, the radiation field is of the $\frac{1}{\sqrt{\rho}}$ order. Taking the $\frac{1}{\sqrt{\rho}}$ order terms from

$$\overline{E}^t = -\frac{1}{\epsilon} \nabla \times \overline{F} \tag{7.183}$$

gives the far-zone field

$$\begin{aligned} E_z^t &\approx -\frac{1}{\epsilon} ik F_\theta \\ &= \cos\theta \sqrt{\frac{2}{\pi k \rho}} \exp\left[i\left(k\rho - \frac{\pi}{4}\right)\right] \frac{\sin\left[k\left(\sin\theta - \sin\theta_i\right)a\right]}{\left(\sin\theta - \sin\theta_i\right)} \; . \end{aligned} \tag{7.184}$$

7.5 Circular Aperture: Electrostatic Case

The subject of wave penetration into a circular aperture in a thick conducting plane is important in electromagnetic interference problems. When the wave frequency is relatively low, low-frequency fields can approximately be described in terms of the static potentials. This section investigates electrostatic potential penetration into a circular aperture in a thick perfectly conducting plane, as shown in Fig. 7.9. A similar discussion is available in [10].

Regions (I) ($z \geq 0$), (II) ($-d \leq z \leq 0$ and $\rho \leq a$), and (III) ($z \leq -d$) represent the upper half-space, circular aperture, and lower half-space, respectively. In region (I) ($z \geq 0$) the primary potential $\phi_e^p(\rho, z)$ is applied to a circular aperture with radius a and depth d in a thick perfectly conducting plane at a zero potential. The electrostatic potential ϕ_e is governed by Laplace's equation

$$\nabla^2 \phi_e(\rho, z) = 0 \,. \tag{7.185}$$

Our aim is to solve Laplace's equation subject to the boundary conditions. Assume that the primary potential is

$$\phi_e^p(\rho, z) = z \,. \tag{7.186}$$

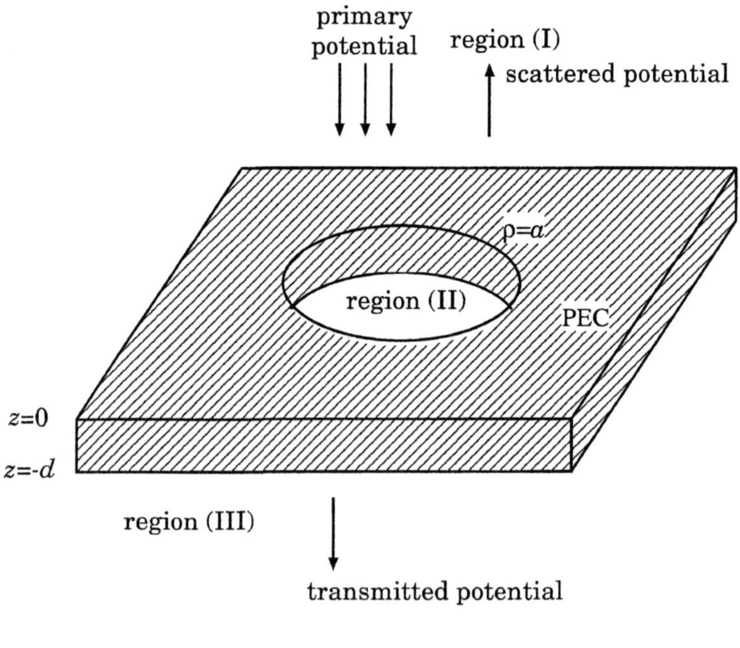

Fig. 7.9. Thick circular aperture in conducting plane.

Since the primary potential is assumed to be azimuthally symmetrical, Laplace's equation is independent of ϕ $\left(\frac{\partial}{\partial \phi} = 0\right)$

$$\left[\frac{1}{\rho}\frac{\partial}{\partial \rho}\left(\rho\frac{\partial}{\partial \rho}\right) + \frac{\partial^2}{\partial z^2}\right]\phi_e(\rho, z) = 0 \,. \tag{7.187}$$

Let us first consider the scattered potential in region (I). Since the scattering domain in region (I) is open ($\rho \to \infty$) in the radial direction ρ, it is expedient to use the Hankel transform representation

$$\tilde{f}(\zeta) = \int_0^\infty f(\rho)J_0(\zeta\rho)\rho \, d\rho \tag{7.188}$$

$$f(\rho) = \int_0^\infty \tilde{f}(\zeta)J_0(\zeta\rho)\zeta \, d\zeta \,. \tag{7.189}$$

Substituting the scattered potential based on the Hankel transform

$$\phi_e^s(\rho, z) = \int_0^\infty \tilde{\phi}_e^s(\zeta, z)J_0(\zeta\rho)\zeta \, d\zeta \tag{7.190}$$

into (7.187) gives

$$\frac{d^2\tilde{\phi}_e^s(\zeta, z)}{dz^2} - \zeta^2\tilde{\phi}_e^s(\zeta, z) = 0 \tag{7.191}$$

where the solution

$$\tilde{\phi}_e^s(\zeta, z) = \tilde{\phi}_e^s(\zeta)e^{-\zeta z} \tag{7.192}$$

is chosen to satisfy the condition $\tilde{\phi}_e^s(\zeta, \infty) = 0$. Hence

$$\phi_e^s(\rho, z) = \int_0^\infty \tilde{\phi}_e^s(\zeta)J_0(\zeta\rho)e^{-\zeta z}\zeta \, d\zeta \,. \tag{7.193}$$

The total electrostatic potential in region (I) is the sum of the primary and scattered components, namely, $\phi_e^p(\rho, z) + \phi_e^s(\rho, z)$.

Let us represent the potential within the circular aperture. Since the scattering domain is closed ($\rho \le a$ and $-d \le z \le 0$), it is expedient to represent the scattered potential in terms of discrete eigenfunctions. The total electrostatic potential in region (II) is

$$\phi_e^d(\rho, z) = \sum_{n=1}^{\infty}\left[b_n \sinh k_n(z + d) + c_n \cosh k_n(z + d)\right]J_0(k_n\rho) \tag{7.194}$$

where the boundary condition

$$\phi_e^d(a, z) = 0 \tag{7.195}$$

yields the characteristic equation $J_0(k_n a) = 0$, which determines the eigenvalues k_n.

The representation of the transmitted potential in region (III) is somewhat similar to that in region (I). The transmitted electrostatic potential in region (III) is

$$\phi_e^t(\rho, z) = \int_0^\infty \tilde{\phi}_e^t(\zeta) J_0(\zeta\rho) e^{\zeta(z+d)} \zeta \, d\zeta \,. \tag{7.196}$$

To represent the potentials, the four unknown modal coefficients–$\tilde{\phi}_e^s(\zeta)$, $\tilde{\phi}_e^t(\zeta)$, b_n, and c_n–are introduced, thus requiring four boundary conditions at $z = 0$ and $-d$. The boundary condition for the continuity of the potential at $z = 0$ requires

$$\phi_e^p(\rho, 0) + \phi_e^s(\rho, 0) = \begin{cases} \phi_e^d(\rho, 0) & \text{for } \rho < a \\ 0 & \text{otherwise} \,. \end{cases} \tag{7.197}$$

For arbitrary Bessel functions Z_p and B_p [11, page 634],

$$\int \rho Z_p(\alpha\rho) B_p(\beta\rho) \, d\rho = \frac{\beta\rho Z_p(\alpha\rho) B_{p-1}(\beta\rho) - \alpha\rho Z_{p-1}(\alpha\rho) B_p(\beta\rho)}{\alpha^2 - \beta^2} \,. \tag{7.198}$$

Applying the Hankel transform $\int_0^\infty (\cdot) J_0(\zeta\rho)\rho \, d\rho$ to (7.197) yields

$$\tilde{\phi}_e^s(\zeta) = \sum_{n=1}^\infty (b_n \sinh k_n d + c_n \cosh k_n d) \cdot \left[\frac{-a k_n J_0(\zeta a) J_1(k_n a)}{\zeta^2 - k_n^2} \right] \,. \tag{7.199}$$

Let us substitute $\tilde{\phi}_e^s(\zeta)$ into the other boundary condition at $z = 0$ for $\rho < a$

$$\frac{\partial}{\partial z} \left[\phi_e^p(\rho, z) + \phi_e^s(\rho, z) \right]_{z=0} = \frac{\partial}{\partial z} \left[\phi_e^d(\rho, z) \right]_{z=0} \,. \tag{7.200}$$

To simplify (7.200), it is necessary to utilize the orthogonality property of Bessel functions

$$\int_0^a J_0(k_p\rho) J_0(k_n\rho)\rho \, d\rho = \begin{cases} \dfrac{a^2}{2} J_1^2(k_p a) & \text{for } n = p \\ 0 & \text{for } n \neq p \,. \end{cases} \tag{7.201}$$

Multiplying (7.200) by $J_0(k_p\rho)\rho$ and integrating with respect to ρ from 0 to a yields

$$\frac{a}{k_p} - \sum_{n=1}^{\infty}(b_n \sinh k_n d + c_n \cosh k_n d)a^2 k_n k_p J_1(k_n a) I_{np}$$
$$= \frac{a^2}{2} k_p J_1(k_p a)(b_p \cosh k_p d + c_p \sinh k_p d) \tag{7.202}$$

where

$$I_{np} = \int_0^{\infty} \frac{J_0^2(\zeta a)\zeta^2}{(\zeta^2 - k_n^2)(\zeta^2 - k_p^2)}\, d\zeta\,. \tag{7.203}$$

Similarly, the boundary conditions at $z = -d$

$$\phi_e^t(\rho, -d) = \begin{cases} \phi_e^d(\rho, -d) & \text{for } \rho < a \\ 0 & \text{otherwise} \end{cases} \tag{7.204}$$

$$\frac{\partial}{\partial z}\left[\phi_e^t(\rho, z)\right]_{z=-d} = \frac{\partial}{\partial z}\left[\phi_e^d(\rho, z)\right]_{z=-d} \quad \text{for } \rho < a \tag{7.205}$$

finally give

$$\sum_{n=1}^{\infty} c_n k_n J_1(k_n a) I_{np} = \frac{b_p}{2} J_1(k_p a)\,. \tag{7.206}$$

To evaluate the scattered and transmitted potentials, the simultaneous equations (7.202) and (7.206) must be solved for the unknown modal coefficients b_n and c_n.

The behavior of the aperture field is often discussed in terms of the electric polarizability $\chi(z)$. The electric polarizability is represented in terms of b_n and c_n as

$$\chi(z) \equiv 2\pi \int_0^a \phi_e^d(\rho, z)\rho\, d\rho$$
$$= 2\pi a \sum_{n=1}^{\infty} \left[b_n \sinh k_n(z+d) + c_n \cosh k_n(z+d)\right] \frac{J_1(k_n a)}{k_n}\,. \tag{7.207}$$

Problems for Chapter 7

1. Derive (7.120).
2. Show that $I(k_0)$ is given by (7.143) within the high-frequency limit.

3. Consider electromagnetic scattering from the thick PEC slit, as shown in Fig. 7.5. Assume that the TM uniform plane wave $H_z^i(x,y) = \exp(ik_x x + ik_y y)$ is incident on the slit. Using the Fourier transform and mode matching, obtain a set of simultaneous equations for the modal coefficients. Using the high-frequency approximation and equivalence principle, evaluate approximately the transmitted ($x > 0$) field for the thin slit ($d = 0$).
4. Derive (7.206).

8

Green's Functions: Fundamentals

8.1 Delta Function and Sturm-Liouville Equation

The delta function is an important concept in dealing with point charges and currents. The Sturm-Liouville equation also plays an important role in the study of wave propagation and radiation. For instance, Bessel's equation is of the Sturm-Liouville equation type. This section introduces the fundamentals of the delta function and Sturm-Liouville equation, which are useful in radiation and scattering formulation.

8.1.1 Delta Function

The delta function $\delta(\bar{r}-\bar{r}')$ is defined as

$$\delta(\bar{r}-\bar{r}') = 0 \qquad \text{when } \bar{r} \neq \bar{r}' \tag{8.1}$$

$$\int_V \delta(\bar{r}-\bar{r}')\, dv = 1 \qquad \text{when } V \text{ contains } \bar{r}' . \tag{8.2}$$

The delta function has the sifting property

$$\int_V f(\bar{r})\delta(\bar{r}-\bar{r}')\, dv = f(\bar{r}') . \tag{8.3}$$

Let us represent the delta function using rectangular, cylindrical, and spherical coordinates, respectively.

- Consider the delta function $\delta(\bar{r}-\bar{r}')$ where the source is located at the position $x = x'$. Then

$$\delta(\bar{r}-\bar{r}') = \delta(x-x') . \tag{8.4}$$

Similarly, in two- and three-dimensional rectangular coordinates

$$\delta(\overline{r}-\overline{r}') = \delta(x-x')\delta(y-y') \tag{8.5}$$

$$\delta(\overline{r}-\overline{r}') = \delta(x-x')\delta(y-y')\delta(z-z') \,. \tag{8.6}$$

- Consider the delta function $\delta(\overline{r}-\overline{r}')$ where the source is located at the position (ρ', ϕ') using two-dimensional polar coordinates (ρ, ϕ). Let us express the delta function as

$$\delta(\overline{r}-\overline{r}') = A_1\delta(\rho-\rho')\delta(\phi-\phi') \,. \tag{8.7}$$

Substituting (8.7) into (8.2) yields

$$\int_0^{2\pi}\int_0^{\infty} A_1\delta(\rho-\rho')\delta(\phi-\phi')\rho\, d\rho\, d\phi = 1 \,. \tag{8.8}$$

Therefore, $A_1 = \dfrac{1}{\rho}$ and

$$\delta(\overline{r}-\overline{r}') = \frac{1}{\rho}\delta(\rho-\rho')\delta(\phi-\phi') \,. \tag{8.9}$$

Consider the delta source $\delta(\overline{r}-\overline{r}') = A_2\delta(\rho-\rho')\delta(\phi-\phi')\delta(z-z')$ using three-dimensional cylindrical coordinates. In view of (8.9),

$$\delta(\overline{r}-\overline{r}') = \frac{1}{\rho}\delta(\rho-\rho')\delta(\phi-\phi')\delta(z-z') \,. \tag{8.10}$$

- Consider the delta source $\delta(\overline{r}-\overline{r}') = A_3\delta(r-r')\delta(\theta-\theta')\delta(\phi-\phi')$ using three-dimensional spherical coordinates. Similarly, substituting $\delta(\overline{r}-\overline{r}')$ into (8.2) gives

$$\int_0^{2\pi}\int_0^{\pi}\int_0^{\infty} A_3\delta(r-r')\delta(\theta-\theta')\delta(\phi-\phi')r^2\sin\theta\, dr\, d\theta\, d\phi = 1 \,. \tag{8.11}$$

Therefore, $A_3 = \dfrac{1}{r^2\sin\theta}$ and

$$\delta(\overline{r}-\overline{r}') = \frac{1}{r^2\sin\theta}\delta(r-r')\delta(\theta-\theta')\delta(\phi-\phi') \,. \tag{8.12}$$

It is possible to express the delta function in the spectral domain based on transform pairs. Consider the Fourier transform pair

$$\tilde{f}(\zeta) = \int_{-\infty}^{\infty} f(x)e^{-i\zeta x}\, dx \tag{8.13}$$

$$f(x) = \frac{1}{2\pi}\int_{-\infty}^{\infty} \tilde{f}(\zeta)e^{i\zeta x}\, d\zeta \,. \tag{8.14}$$

Substituting $\tilde{f}(\zeta)$ into $f(x)$ gives

$$f(x) = \frac{1}{2\pi}\int_{-\infty}^{\infty}\left[\int_{-\infty}^{\infty} f(x')e^{-i\zeta x'}\,dx'\right]e^{i\zeta x}\,d\zeta\,. \tag{8.15}$$

Since

$$f(x) = \int_{-\infty}^{\infty}\delta(x-x')f(x')\,dx' \tag{8.16}$$

the delta function can be written as

$$\delta(x-x') = \frac{1}{2\pi}\int_{-\infty}^{\infty} e^{i\zeta(x-x')}\,d\zeta\,. \tag{8.17}$$

8.1.2 Sturm-Liouville Equation

Consider the differential equation

$$\left\{\frac{d}{dx}\left[p(x)\frac{d}{dx}\right] + q(x) + \lambda_n r(x)\right\}\psi_n(x) = 0 \tag{8.18}$$

on the interval $a \le x \le b$ subject to the boundary conditions

$$\left[\alpha_1\psi_n(x) + \beta_1\frac{d\psi_n(x)}{dx}\right]_{x=a} = 0 \tag{8.19}$$

$$\left[\alpha_2\psi_n(x) + \beta_2\frac{d\psi_n(x)}{dx}\right]_{x=b} = 0 \tag{8.20}$$

where the functions $p(x)$, $q(x)$, and $r(x)$ as well as the parameters α_1, α_2, β_1, and β_2 are real. Equation (8.18) is called a Sturm-Liouville equation with the eigenfunction $\psi_n(x)$ and eigenvalue λ_n.

- When $p(x) = 1$, $q(x) = 0$, $\lambda_n = k_n^2$, and $r(x) = 1$, (8.18) becomes the differential equation of simple harmonic motion

$$\left(\frac{d^2}{dx^2} + k_n^2\right)\psi_n(x) = 0\,. \tag{8.21}$$

- When $p(x) = x$, $q(x) = -\dfrac{m^2}{x}$, $\lambda_n = k_n^2$, and $r(x) = x$, (8.18) becomes Bessel's equation

$$x^2\frac{d^2\psi_n(x)}{dx^2} + x\frac{d\psi_n(x)}{dx} + \left(k_n^2x^2 - m^2\right)\psi_n(x) = 0\,. \tag{8.22}$$

- When $p(x) = 1 - x^2$, $q(x) = -\dfrac{m^2}{1-x^2}$, $\lambda_n = n(n+1)$, and $r(x) = 1$, (8.18) becomes the associated Legendre equation

$$\left(1-x^2\right)\frac{d^2\psi_n(x)}{dx^2} - 2x\frac{d\psi_n(x)}{dx} + \left[n(n+1) - \frac{m^2}{1-x^2}\right]\psi_n(x) = 0 \,. \tag{8.23}$$

Let us show that the eigenfunctions $\psi_n(x)$ are orthogonal to each other. Multiplying (8.18) by $\psi_m^*(x)$ and integrating from a to b gives

$$\int_a^b \psi_m^*(x)\frac{d}{dx}\left[p(x)\frac{d\psi_n(x)}{dx}\right] dx + \int_a^b q(x)\psi_m^*(x)\psi_n(x)\,dx + \lambda_n \int_a^b r(x)\psi_m^*(x)\psi_n(x)\,dx = 0 \,. \tag{8.24}$$

Integration by parts gives

$$\begin{aligned}&\int_a^b \psi_m^*(x)\frac{d}{dx}\left[p(x)\frac{d\psi_n(x)}{dx}\right] dx \\ &= \left[\psi_m^*(x)p(x)\frac{d\psi_n(x)}{dx}\right]_a^b - \int_a^b \frac{d\psi_m^*(x)}{dx}p(x)\frac{d\psi_n(x)}{dx}\,dx \\ &= p(a)\frac{\alpha_1}{\beta_1}\psi_m^*(a)\psi_n(a) - p(b)\frac{\alpha_2}{\beta_2}\psi_m^*(b)\psi_n(b) \\ &\quad - \int_a^b \frac{d\psi_m^*(x)}{dx}p(x)\frac{d\psi_n(x)}{dx}\,dx \,.\end{aligned} \tag{8.25}$$

Substituting (8.25) into (8.24) gives

$$\begin{aligned}&p(a)\frac{\alpha_1}{\beta_1}\psi_m^*(a)\psi_n(a) - p(b)\frac{\alpha_2}{\beta_2}\psi_m^*(b)\psi_n(b) \\ &- \int_a^b \frac{d\psi_m^*(x)}{dx}p(x)\frac{d\psi_n(x)}{dx}\,dx + \int_a^b q(x)\psi_m^*(x)\psi_n(x)\,dx \\ &+\lambda_n \int_a^b r(x)\psi_m^*(x)\psi_n(x)\,dx = 0 \,.\end{aligned} \tag{8.26}$$

Interchanging the indices m and n and taking the complex conjugate of (8.26) gives

$$\begin{aligned}&p(a)\frac{\alpha_1}{\beta_1}\psi_n(a)\psi_m^*(a) - p(b)\frac{\alpha_2}{\beta_2}\psi_n(b)\psi_m^*(b) \\ &- \int_a^b \frac{d\psi_n(x)}{dx}p(x)\frac{d\psi_m^*(x)}{dx}\,dx + \int_a^b q(x)\psi_n(x)\psi_m^*(x)\,dx \\ &+\lambda_m^* \int_a^b r(x)\psi_n(x)\psi_m^*(x)\,dx = 0 \,.\end{aligned} \tag{8.27}$$

Subtracting (8.27) from (8.26) finally yields

$$(\lambda_n - \lambda_m^*) \int_a^b r(x)\psi_n(x)\psi_m^*(x)\, dx = 0\,. \tag{8.28}$$

Assume that $r(x)$ does not change sign on the interval (a, b). For $m \neq n$ $(\lambda_n \neq \lambda_m^*)$

$$\int_a^b r(x)\psi_n(x)\psi_m^*(x)\, dx = 0\,. \tag{8.29}$$

For $m = n$

$$\int_a^b r(x)\psi_n(x)\psi_n^*(x)\, dx \neq 0 \tag{8.30}$$

$$\lambda_n - \lambda_n^* = 0\,. \tag{8.31}$$

The eigenfunctions $\psi_n(x)$ are normalized as

$$\int_a^b r(x)\psi_n(x)\psi_n^*(x)\, dx = 1\,. \tag{8.32}$$

The condition $\lambda_n = \lambda_n^*$ shows that the eigenvalue λ_n is real. The property (8.29) and (8.32) is referred to as the orthogonality of the eigenfunctions $\psi_n(x)$. The eigenfunctions $\psi_n(x)$ constitute a complete set. This completeness implies that an arbitrary function $f(x)$ can be represented to any desired accuracy as a linear combination of $\psi_n(x)$

$$f(x) = \sum_n f_n\psi_n(x)\,. \tag{8.33}$$

Multiplying $f(x)$ by $r(x)\psi_m^*(x)$ and integrating from a to b yields the expansion coefficient

$$f_n = \int_a^b r(x)f(x)\psi_n^*(x)\, dx\,. \tag{8.34}$$

Consider the Sturm-Liouville equation with a delta source $\delta(x - x')$

$$\left\{\frac{d}{dx}\left[p(x)\frac{d}{dx}\right] + q(x) + \lambda r(x)\right\} g(x; x') = -\delta(x - x') \tag{8.35}$$

on the interval $a \leq (x,\, x') \leq b$, where the boundary conditions are given by

$$\left[\alpha_1 g(x; x') + \beta_1 \frac{dg(x; x')}{dx}\right]_{x=a} = 0 \tag{8.36}$$

$$\left[\alpha_2 g(x; x') + \beta_2 \frac{dg(x; x')}{dx}\right]_{x=b} = 0\,. \tag{8.37}$$

The solution $g(x;x')$, which is the response to the delta source, is usually known as a Green's function. Let us represent the Green's function and delta function by a series of eigenfunctions

$$g(x;x') = \sum_n g_n(x')\psi_n(x) \tag{8.38}$$

$$\frac{\delta(x-x')}{r(x')} = \sum_n \psi_n^*(x')\psi_n(x) \,. \tag{8.39}$$

Substituting (8.38) and (8.39) into (8.35) finally gives

$$g(x;x') = -\sum_n \frac{\psi_n^*(x')\psi_n(x)}{\lambda - \lambda_n} \tag{8.40}$$

which is a Green's function representation based on eigenfunction expansions.

8.2 One-Dimensional Green's Function

This section discusses one-dimensional Green's functions in free space, half space, and closed space, respectively. The one-dimensional Green's functions are often used in scattering and radiation problem formulations. Let us first consider a free-space case.

8.2.1 Free Space

Consider the Sturm-Liouville equation in one-dimensional free space ($-\infty < x < \infty$) with the parameters $p(x) = r(x) = 1$, $q(x) = 0$, and $\lambda = k^2$. The Sturm-Liouville equation is given by

$$\left(\frac{d^2}{dx^2} + k^2\right) g(x;x') = -\delta(x-x') \tag{8.41}$$

subject to the radiation condition $g(\pm\infty;x') = 0$. The response $g(x;x')$ at x is due to the delta source at x', as shown in Fig. 8.1. In the following, two different approaches are introduced to derive $g(x;x')$.

Fig. 8.1. Free-space Green's function.

Approach 1

It is possible to represent $g(x;x')$ in terms of eigenfunctions. Since the domain is open $(-\infty < x < \infty)$, $g(x;x')$ is represented in the inverse Fourier transform

$$g(x;x') = \frac{1}{2\pi}\int_{-\infty}^{\infty} \tilde{g}(\zeta;x')e^{i\zeta x}\,d\zeta\,. \tag{8.42}$$

Substituting (8.42) and the identity

$$\delta(x-x') = \frac{1}{2\pi}\int_{-\infty}^{\infty} e^{i\zeta(x-x')}\,d\zeta \tag{8.43}$$

into (8.41) yields

$$g(x;x') = \int_{-\infty}^{\infty} \underbrace{\frac{1}{2\pi}\frac{e^{i\zeta(x-x')}}{(\zeta^2-k^2)}}_{f(\zeta)}\,d\zeta\,. \tag{8.44}$$

The physical condition requires that the delta source response $g(x;x')$ be an outgoing wave in the form of $e^{\pm ikx}$ that vanishes as $x \to \pm\infty$. This condition can be met if the medium is assumed to be slightly lossy, where the medium wavenumber $k = k_r + ik_i$ has an infinitesimally small positive imaginary part $(k_i > 0)$. For $x - x' > 0$, let us use the residue theorem in the complex ζ-plane, as shown in Fig. 8.2 (a). Performing contour integration along the path Γ_1 and Γ_2 yields

$$\int_{\Gamma_1} f(\zeta)\,d\zeta + \int_{\Gamma_2} f(\zeta)\,d\zeta = 2\pi i \operatorname*{Res}_{\zeta=k} f(\zeta) \tag{8.45}$$

Since $\int_{\Gamma_2} f(\zeta)\,d\zeta \to 0$ and $\int_{\Gamma_1} f(\zeta)\,d\zeta = g(x;x')$ as $R \to \infty$,

$$g(x;x') = \frac{i}{2k}e^{ik(x-x')}\,. \tag{8.46}$$

When $x - x' < 0$, the semicircle in the lower half-plane is chosen for Γ_2 so as to make $\int_{\Gamma_2} f(\zeta)\,d\zeta \to 0$ as $R \to \infty$, as shown in Fig. 8.2 (b). Contour integration gives

$$g(x;x') = \frac{i}{2k}e^{-ik(x-x')}\,. \tag{8.47}$$

The one-dimensional free-space Green's function is, therefore, given by

$$g(x;x') = \frac{i}{2k}e^{ik|x-x'|}\,. \tag{8.48}$$

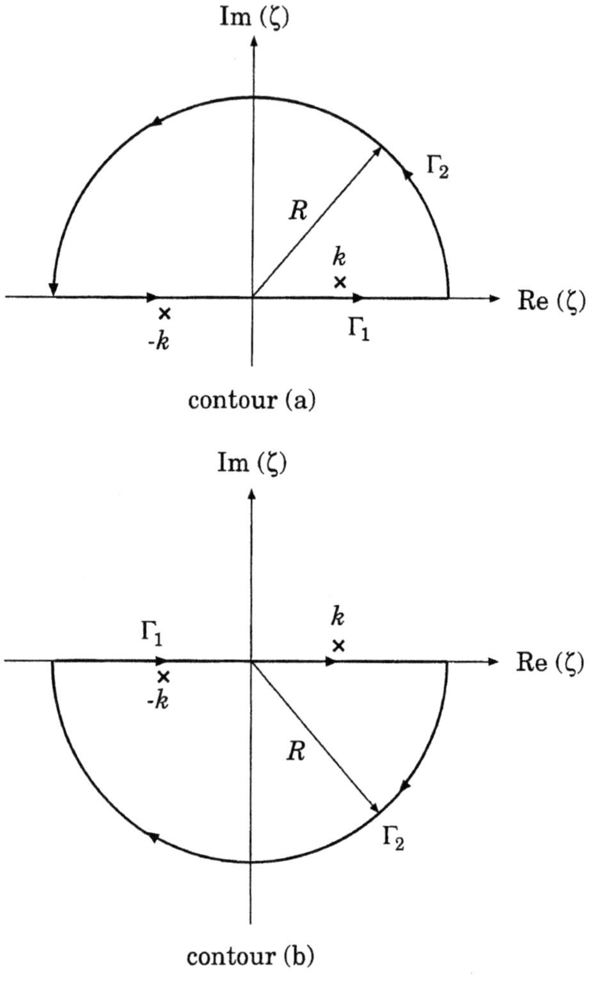

Fig. 8.2. Complex ζ-plane.

Approach 2

The solution $g(x;x')$ can also be obtained by another approach as follows: Consider the homogeneous differential equation

$$\left(\frac{d^2}{dx^2}+k^2\right)g(x;x')=0 \tag{8.49}$$

when $x \neq x'$. The Green's function that satisfies the radiation condition at $x=\pm\infty$ is

$$g(x;x') = \begin{cases} Ae^{ikx} & \text{for } x > x' \\ Be^{-ikx} & \text{for } x < x' \,. \end{cases} \tag{8.50}$$

The unknown constants A and B can be determined by the boundary conditions at $x = x' \pm \Delta$, where Δ denotes an infinitesimally small interval. Integrating (8.41) from $x = x' - \Delta$ to $x = x' + \Delta$ yields

$$\left.\frac{dg(x;x')}{dx}\right|_{x'-\Delta}^{x'+\Delta} + \int_{x'-\Delta}^{x'+\Delta} k^2 g(x;x')\, dx = -1 \,. \tag{8.51}$$

Since $\frac{dg(x;x')}{dx}$ is discontinuous and $g(x;x')$ is continuous at x', the boundary conditions

$$\left.\frac{dg(x;x')}{dx}\right|_{x'-\Delta}^{x'+\Delta} = -1 \tag{8.52}$$

$$\left. g(x;x')\right|_{x'-\Delta}^{x'+\Delta} = 0 \tag{8.53}$$

give

$$ik\left(Ae^{ikx'} + Be^{-ikx'}\right) = -1 \tag{8.54}$$

$$Be^{-ikx'} = Ae^{ikx'} \,. \tag{8.55}$$

Solving (8.54) and (8.55) for A and B finally gives

$$g(x;x') = \begin{cases} \frac{i}{2k} e^{ik(x-x')} & \text{for } x > x' \\ \frac{i}{2k} e^{-ik(x-x')} & \text{for } x < x' \end{cases}$$

$$= \frac{i}{2k} e^{ik|x-x'|} \,. \tag{8.56}$$

8.2.2 Half Space

Let us investigate the Sturm-Liouville equation in one-dimensional half space with the parameters $p(x) = r(x) = 1$, $q(x) = 0$, and $\lambda = k^2$. These types of problems are often encountered in scattering and radiation problems dealing with perfectly conducting boundaries. Consider the half-space Sturm-Liouville equation

$$\left(\frac{d^2}{dx^2} + k^2\right) g(x;x') = -\delta(x - x') \tag{8.57}$$

for $x \geq 0$ subject to the boundary conditions $g(0;x') = 0$ and $g(\infty;x') = 0$, as shown in Fig. 8.3. Let us introduce four different approaches to obtain $g(x;x')$.

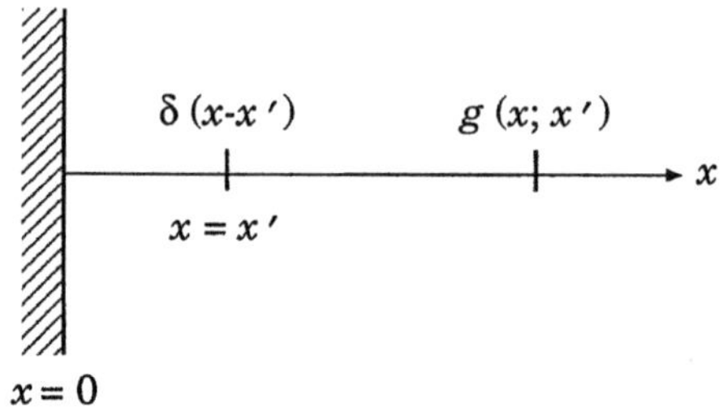

Fig. 8.3. Half-space Green's function.

Approach 1

Due to reflection from the boundary at $x = 0$, $g(x; x')$ takes a standing wave $\sin kx$ for $0 \leq x < x'$ and a traveling wave e^{ikx} for $x > x'$. Assume

$$g(x; x') = \begin{cases} Ae^{ikx} & \text{for } \; x > x' \\ B \sin kx & \text{for } \; 0 \leq x < x' \end{cases} \tag{8.58}$$

which satisfies the boundary conditions $g(\infty; x') = 0$ and $g(0; x') = 0$. Integrating (8.57) gives

$$\left. \frac{dg(x; x')}{dx} \right|_{x'-\Delta}^{x'+\Delta} = -1 \tag{8.59}$$

$$\left. g(x; x') \right|_{x'-\Delta}^{x'+\Delta} = 0 \,. \tag{8.60}$$

Solving (8.59) and (8.60) for A and B yields

$$A = \frac{\sin kx'}{k} \tag{8.61}$$

$$B = \frac{e^{ikx'}}{k} \,. \tag{8.62}$$

Hence, the half-space Green's function is

$$g(x; x') = \begin{cases} \dfrac{\sin kx'}{k} e^{ikx} & \text{for } \; x > x' \\ \dfrac{e^{ikx'}}{k} \sin kx & \text{for } \; 0 \leq x < x' \,. \end{cases} \tag{8.63}$$

Approach 2

It is also possible to obtain $g(x;x')$ using eigenfunction expansions. Since the domain is open for $x > 0$ and $g(0;x') = 0$, it is convenient to represent $g(x;x')$ in terms of the Fourier sine transform. Substituting

$$g(x;x') = \frac{2}{\pi}\int_0^\infty \tilde{g}(\zeta;x')\sin\zeta x\,d\zeta \tag{8.64}$$

$$\delta(x-x') = \frac{2}{\pi}\int_0^\infty \sin\zeta x\sin\zeta x'\,d\zeta\,. \tag{8.65}$$

into (8.57) yields

$$\tilde{g}(\zeta;x') = \frac{\sin\zeta x'}{\zeta^2-k^2}\,. \tag{8.66}$$

Therefore

$$\begin{aligned} g(x;x') &= \frac{2}{\pi}\int_0^\infty \frac{\sin\zeta x\sin\zeta x'}{\zeta^2-k^2}\,d\zeta \\ &= \frac{1}{\pi}\int_{-\infty}^\infty \frac{\sin\zeta x\sin\zeta x'}{\zeta^2-k^2}\,d\zeta \\ &= -\frac{1}{4\pi}\int_{-\infty}^\infty \\ &\quad \left[\frac{e^{i\zeta(x+x')}-e^{i\zeta(x-x')}-e^{i\zeta(-x+x')}+e^{i\zeta(-x-x')}}{\zeta^2-k^2}\right]d\zeta\,. \end{aligned} \tag{8.67}$$

In view of the free-space Green's function representations (8.44) and (8.48), $g(x;x')$ is written as

$$\begin{aligned} g(x;x') &= -\frac{i}{4k}\left(e^{ik|x+x'|}-e^{ik|x-x'|}-e^{ik|-x+x'|}+e^{ik|-x-x'|}\right) \\ &= \begin{cases} \dfrac{\sin kx'}{k}e^{ikx} & \text{for } x > x' \\ \dfrac{e^{ikx'}}{k}\sin kx & \text{for } 0 \le x < x'\,. \end{cases} \end{aligned} \tag{8.68}$$

Approach 3

Let us show an approach based on the image method, as shown in Fig. 8.4, where the image source $-\delta(x+x')$ is added to satisfy the boundary condition $g(0;x') = 0$. Since both $\delta(x-x')$ and $-\delta(x+x')$ are assumed to be in free

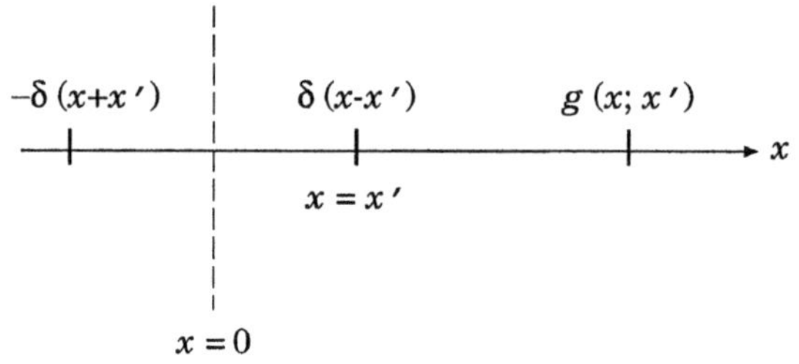

Fig. 8.4. Half-space Green's function based on image theorem.

space, the response is the sum of two free-space Green's functions

$$g(x;x') = \frac{i}{2k}\left(e^{ik|x-x'|} - e^{ik|x+x'|}\right)$$

$$= \begin{cases} \dfrac{\sin kx'}{k} e^{ikx} & \text{for } x > x' \\ \dfrac{e^{ikx'}}{k} \sin kx & \text{for } 0 \le x < x' . \end{cases} \tag{8.69}$$

Approach 4

A solution to (8.57) consists of the incident $[g^i(x;x')]$ and scattered $[g^s(x;x')]$ Green's functions. The incident Green's function is a response to a delta source in free space and the scattered one results from reflection from the boundary at $x = 0$. The incident term is the one-dimensional free-space Green's function and the scattered one is given in terms of the reflected waves as

$$g^i(x;x') = \frac{i}{2k} e^{ik|x-x'|} \tag{8.70}$$

$$g^s(x;x') = Ae^{ikx} . \tag{8.71}$$

Since

$$\left[g^i(x;x') + g^s(x;x')\right]_{x=0} = 0 \tag{8.72}$$

the coefficient A is given by

$$A = -\frac{i}{2k} e^{ikx'} . \tag{8.73}$$

Therefore

$$g(x;x') = g^i(x;x') + g^s(x;x')$$

$$= \frac{i}{2k}\left[e^{ik|x-x'|} - e^{ik(x+x')}\right]$$

$$= \begin{cases} \dfrac{\sin kx'}{k} e^{ikx} & \text{for } x > x' \\ \dfrac{e^{ikx'}}{k} \sin kx & \text{for } 0 \le x < x' . \end{cases} \tag{8.74}$$

8.2.3 Closed Space

This subsection investigates the Sturm-Liouville equation in one-dimensional closed space. Consider

$$\left(\frac{d^2}{dx^2} + k^2\right) g(x;x') = -\delta(x - x') \tag{8.75}$$

on the interval $0 \le x \le a$ subject to the boundary conditions $g(0;x') = g(a;x') = 0$, as shown in Fig. 8.5. The following are three different approaches to derive the solution $g(x;x')$.

Approach 1

Due to the boundaries at $x = 0$ and a, the solution $g(x;x')$ can be represented in terms of standing waves, $\sin kx$ and $\cos kx$. The choice

$$g(x;x') = \begin{cases} A \sin k(x - a) & \text{for } x' < x \le a \\ B \sin kx & \text{for } 0 \le x < x' \end{cases} \tag{8.76}$$

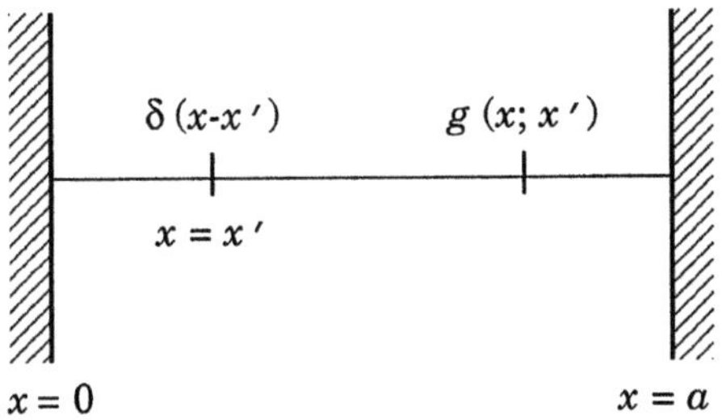

Fig. 8.5. Closed-space Green's function.

automatically satisfies the boundary conditions $g(0;x') = g(a;x') = 0$. Consider the boundary condition near $x = x'$. Integrating (8.75) near $x = x'$ gives

$$\left.\frac{dg(x;x')}{dx}\right|_{x'-\Delta}^{x'+\Delta} = -1 \tag{8.77}$$

$$\left. g(x;x')\right|_{x'-\Delta}^{x'+\Delta} = 0\,. \tag{8.78}$$

Solving (8.77) and (8.78) for A and B yields

$$A = -\frac{\sin kx'}{k \sin ka} \tag{8.79}$$

$$B = -\frac{\sin k(x'-a)}{k \sin ka}\,. \tag{8.80}$$

Hence

$$g(x;x') = \begin{cases} -\dfrac{\sin kx' \sin k(x-a)}{k \sin ka} & \text{for } x' < x \le a \\ -\dfrac{\sin k(x'-a) \sin kx}{k \sin ka} & \text{for } 0 \le x < x'\,. \end{cases} \tag{8.81}$$

Approach 2

It is also possible to obtain $g(x;x')$ using eigenfunction expansions, where the eigenfunction $\psi_m(x)$ is given by

$$\psi_m(x) = \sqrt{\frac{2}{a}} \sin \frac{m\pi}{a} x\,. \tag{8.82}$$

Substituting

$$g(x;x') = \sum_{m=1}^{\infty} A_m(x')\psi_m(x) \tag{8.83}$$

$$\delta(x-x') = \sum_{m=1}^{\infty} \psi_m(x)\psi_m(x') \tag{8.84}$$

into (8.75) yields

$$A_m(x') = -\frac{\psi_m(x')}{k^2 - \left(\frac{m\pi}{a}\right)^2}\,. \tag{8.85}$$

Hence

$$g(x;x') = -\sum_{m=1}^{\infty} \frac{\psi_m(x')\psi_m(x)}{k^2 - \left(\frac{m\pi}{a}\right)^2}$$

$$= -\sum_{m=1}^{\infty} \frac{\left(\frac{2}{a}\right)\sin\left(\frac{m\pi}{a}x\right)\sin\left(\frac{m\pi}{a}x'\right)}{k^2 - \left(\frac{m\pi}{a}\right)^2}$$

$$= \frac{a}{\pi^2}\sum_{m=1}^{\infty} \frac{\cos\frac{m\pi}{a}(x'-x) - \cos\frac{m\pi}{a}(x'+x)}{m^2 - \left(\frac{ka}{\pi}\right)^2} . \qquad (8.86)$$

Using the series formula for $0 \le u \le 2\pi$ [11, page 40]

$$\sum_{m=1}^{\infty} \frac{\cos mu}{m^2-\alpha^2} = \frac{1}{2\alpha^2} - \frac{\pi}{2}\frac{\cos[\alpha(\pi-u)]}{\alpha \sin \alpha\pi} \qquad (8.87)$$

in (8.86) yields (8.81).

Approach 3

A solution to (8.75) consists of the incident $[g^i(x;x')]$ and scattered $[g^s(x;x')]$ Green's functions, where the incident term is a response to a delta source in free space and the scattered one results from multiple reflection from the boundaries at $x = 0$ and a. Hence

$$g^i(x;x') = \frac{i}{2k}e^{ik|x-x'|} \qquad (8.88)$$

$$g^s(x;x') = Ae^{ikx} + Be^{-ikx} . \qquad (8.89)$$

Since

$$\left[g^i(x;x') + g^s(x;x')\right]_{x=0,a} = 0 \qquad (8.90)$$

A and B are given by

$$A = -\frac{i\sin k(a-x')}{2k\sin(ka)} \qquad (8.91)$$

$$B = -\frac{i\sin(kx')e^{ika}}{2k\sin(ka)} . \qquad (8.92)$$

Therefore

$$g(x;x') = g^i(x;x') + g^s(x;x')$$

$$= \frac{i}{2k}\left[e^{ik|x-x'|} - \frac{\sin k(a-x')e^{ikx}}{\sin(ka)} - \frac{\sin(kx')e^{ik(a-x)}}{\sin(ka)}\right]$$

$$= \begin{cases} -\dfrac{\sin kx' \sin k(x-a)}{k \sin ka} & \text{for } x' < x \le a \\ -\dfrac{\sin k(x'-a)\sin kx}{k\sin ka} & \text{for } 0 \le x < x' . \end{cases} \tag{8.93}$$

8.3 Two-Dimensional Green's Function

The Helmholtz equation using cylindrical coordinates (ρ, ϕ) with a delta source (line source at $\overline{\rho} = \overline{\rho}'$) is

$$\left(\nabla^2 + k^2\right) g(\overline{\rho};\overline{\rho}') = -\delta(\overline{\rho}-\overline{\rho}') . \tag{8.94}$$

Its explicit expression is

$$\left[\frac{1}{\rho}\frac{\partial}{\partial\rho}\left(\rho\frac{\partial}{\partial\rho}\right) + \frac{1}{\rho^2}\frac{\partial^2}{\partial\phi^2} + k^2\right] g(\rho,\phi;\rho',\phi')$$

$$= -\frac{\delta(\rho-\rho')}{\rho}\delta(\phi-\phi') . \tag{8.95}$$

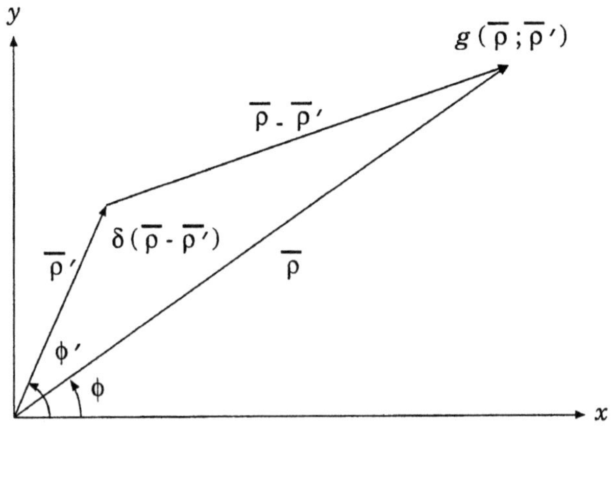

Fig. 8.6. Two-dimensional free-space Green's function.

The problem geometry is shown in Fig. 8.6. Let us find a response at $\overline{\rho}$ due to a delta source at $\overline{\rho'}$ using three different approaches.

8.3.1 Approach 1

Based on the separation of variables technique, the Green's function and delta function are given by

$$g(\rho, \phi; \rho', \phi') = \sum_{m=-\infty}^{\infty} g_m(\rho; \rho', \phi')e^{im\phi} \tag{8.96}$$

$$\delta(\phi - \phi') = \sum_{m=-\infty}^{\infty} \frac{1}{2\pi} e^{im(\phi-\phi')}. \tag{8.97}$$

Substituting $g(\rho, \phi; \rho', \phi')$ and $\delta(\phi - \phi')$ into (8.95) gives

$$\left[\frac{1}{\rho}\frac{d}{d\rho}\left(\rho\frac{d}{d\rho}\right) - \frac{m^2}{\rho^2} + k^2\right] g_m(\rho; \rho', \phi') = -\frac{\delta(\rho - \rho')}{\rho}\frac{e^{-im\phi'}}{2\pi}. \tag{8.98}$$

Let us use an approach based on the cylindrical wave propagation characteristics, as shown in Section 3.4. When $\rho \neq \rho'$, (8.98) reduces to Bessel's equation where its solutions $g_m(\rho; \rho', \phi')$ are the Bessel and Hankel functions

$$g_m(\rho; \rho', \phi') = \begin{cases} A_m J_m(k\rho) & \text{for } \rho < \rho' \\ B_m H_m^{(1)}(k\rho) & \text{for } \rho > \rho' . \end{cases} \tag{8.99}$$

Multiplying (8.98) by ρ and integrating from $\rho = \rho' - \Delta$ to $\rho = \rho' + \Delta$ yields

$$\rho\frac{dg_m(\rho; \rho', \phi')}{d\rho}\bigg|_{\rho'-\Delta}^{\rho'+\Delta} + \int_{\rho'-\Delta}^{\rho'+\Delta}\left(-\frac{m^2}{\rho^2} + k^2\right) g_m(\rho; \rho', \phi')\rho\, d\rho$$
$$= -\frac{e^{-im\phi'}}{2\pi} . \tag{8.100}$$

Since $\dfrac{dg_m(\rho; \rho', \phi')}{d\rho}$ and $g_m(\rho; \rho', \phi')$ are discontinuous and continuous at $\rho = \rho'$, respectively, the boundary conditions are written as

$$\rho'\left[- A_m k J_m'(k\rho') + B_m k H_m^{(1)'}(k\rho')\right] = -\frac{e^{-im\phi'}}{2\pi} \tag{8.101}$$

$$A_m J_m(k\rho') = B_m H_m^{(1)}(k\rho') \tag{8.102}$$

where $J_m'(k\rho') = \dfrac{dJ_m(k\rho)}{d(k\rho)}\bigg|_{\rho=\rho'}$. Solving (8.101) and (8.102) for A_m yields

$$A_m = -\frac{e^{-im\phi'}}{2\pi} \frac{H_m^{(1)}(k\rho')}{k\rho'\left[-J_m'(k\rho')H_m^{(1)}(k\rho') + J_m(k\rho')H_m^{(1)'}(k\rho')\right]} . \tag{8.103}$$

Since

$$-J_m'(k\rho')H_m^{(1)}(k\rho') + J_m(k\rho')H_m^{(1)'}(k\rho')$$

$$= i\left[-J_m'(k\rho')N_m(k\rho') + J_m(k\rho')N_m'(k\rho')\right]$$

$$= i\frac{2}{\pi k\rho'} \tag{8.104}$$

the coefficients A_m and B_m as well as the Green's function $g_m(\rho;\rho',\phi')$ are given by

$$A_m = \frac{i}{4}e^{-im\phi'}H_m^{(1)}(k\rho') \tag{8.105}$$

$$B_m = \frac{i}{4}e^{-im\phi'}J_m(k\rho') \tag{8.106}$$

$$g_m(\rho;\rho',\phi') = \frac{i}{4}\begin{cases} H_m^{(1)}(k\rho')J_m(k\rho)e^{-im\phi'} & \text{for } \rho < \rho' \\ J_m(k\rho')H_m^{(1)}(k\rho)e^{-im\phi'} & \text{for } \rho > \rho' . \end{cases} \tag{8.107}$$

Therefore

$$g(\rho,\phi;\rho',\phi')$$

$$= \sum_{m=-\infty}^{\infty} g_m(\rho;\rho',\phi')e^{im\phi}$$

$$= \frac{i}{4}\begin{cases} \sum_{m=-\infty}^{\infty} H_m^{(1)}(k\rho')J_m(k\rho)e^{im(\phi-\phi')} & \text{for } \rho < \rho' \\ \sum_{m=-\infty}^{\infty} J_m(k\rho')H_m^{(1)}(k\rho)e^{im(\phi-\phi')} & \text{for } \rho > \rho' . \end{cases} \tag{8.108}$$

By the summation theorem for the Hankel function [11, page 979], the two-dimensional free-space Green's function is written as

$$g(\rho,\phi;\rho',\phi') = \frac{i}{4}H_0^{(1)}\left(k|\overline{\rho} - \overline{\rho}'|\right) . \tag{8.109}$$

Let us rewrite $g(\rho,\phi;\rho',\phi')$ in terms of the superposition of plane waves in rectangular coordinates. Rewriting (8.94) in rectangular coordinates (x, y) gives

$$\left(\frac{\partial^2}{\partial x^2} + \frac{\partial^2}{\partial y^2} + k^2\right) g(x, y; x', y') = -\delta(x - x')\delta(y - y') \,. \tag{8.110}$$

Substituting

$$g(x, y; x', y') = \frac{1}{2\pi} \int_{-\infty}^{\infty} \tilde{g}(\zeta) e^{i\zeta(x-x')}\, d\zeta \tag{8.111}$$

into (8.110) results in

$$\left(\frac{d^2}{dy^2} + k^2 - \zeta^2\right) \tilde{g}(\zeta) = -\delta(y - y') \,. \tag{8.112}$$

The solution $\tilde{g}(\zeta)$ is the one-dimensional free-space Green's function

$$\tilde{g}(\zeta) = \frac{i}{2\kappa} e^{i\kappa|y-y'|} \tag{8.113}$$

where $\kappa = \sqrt{k^2 - \zeta^2}$. Substituting $\tilde{g}(\zeta)$ into (8.111) finally gives

$$\frac{i}{4} H_0^{(1)}\left(k|\overline{\rho} - \overline{\rho}'|\right) = \frac{1}{2\pi} \int_{-\infty}^{\infty} \frac{i}{2\kappa} e^{i\zeta(x-x') + i\kappa|y-y'|}\, d\zeta \tag{8.114}$$

which represents the cylindrical wave $H_0^{(1)}\left(k|\overline{\rho} - \overline{\rho}'|\right)$ as the superposition of a continuous spectrum of plane waves.

8.3.2 Approach 2

It is also possible to derive (8.107) based on the residue calculus and Hankel transform approach. Substituting the Hankel transform representations

$$g_m(\rho; \rho', \phi') = \int_0^{\infty} \tilde{g}_m(\zeta) J_m(\zeta\rho) \zeta\, d\zeta \tag{8.115}$$

$$\frac{\delta(\rho - \rho')}{\rho} = \int_0^{\infty} J_m(\zeta\rho') J_m(\zeta\rho) \zeta\, d\zeta \tag{8.116}$$

into (8.98) gives

$$\begin{aligned} &\tilde{g}_m(\zeta) \left[\frac{1}{\rho}\frac{d}{d\rho}\left(\rho\frac{d}{d\rho}\right) - \frac{m^2}{\rho^2} + k^2\right] J_m(\zeta\rho) \\ &= -\frac{e^{-im\phi'}}{2\pi} J_m(\zeta\rho') J_m(\zeta\rho) \,. \end{aligned} \tag{8.117}$$

In view of Bessel's equation

$$\left[\frac{1}{\rho}\frac{d}{d\rho}\left(\rho\frac{d}{d\rho}\right) - \frac{m^2}{\rho^2} + \zeta^2\right] J_m(\zeta\rho) = 0 \tag{8.118}$$

(8.117) can be rewritten as

$$\left(-\zeta^2 + k^2\right) \tilde{g}_m(\zeta) = -\frac{e^{-im\phi'}}{2\pi} J_m(\zeta\rho') \, . \tag{8.119}$$

Substituting $\tilde{g}_m(\zeta)$ of (8.119) into (8.115) gives

$$g_m(\rho; \rho', \phi') = \frac{e^{-im\phi'}}{2\pi} \int_0^\infty \frac{J_m(\zeta\rho) J_m(\zeta\rho')}{(\zeta^2 - k^2)} \zeta \, d\zeta \, . \tag{8.120}$$

Let us compute the integral (8.120) analytically in the complex ζ-plane based on a contour integral technique. Consider two different cases, $\rho' > \rho$ and $\rho' < \rho$. When $\rho' > \rho$, substituting

$$J_m(\zeta\rho') = \frac{1}{2}\left[H_m^{(1)}(\zeta\rho') + H_m^{(2)}(\zeta\rho')\right] \tag{8.121}$$

into (8.120) yields

$$g_m(\rho; \rho', \phi') = \frac{e^{-im\phi'}}{2\pi} \int_0^\infty \frac{J_m(\zeta\rho)\left[H_m^{(1)}(\zeta\rho') + H_m^{(2)}(\zeta\rho')\right]}{2(\zeta^2 - k^2)} \zeta d\zeta \, . \tag{8.122}$$

Since $H_m^{(1)}(-u) = -(-1)^m H_m^{(2)}(u)$,

$$g_m(\rho; \rho', \phi') = \int_{-\infty}^\infty \underbrace{\frac{e^{-im\phi'}}{2\pi} \frac{J_m(\zeta\rho) H_m^{(1)}(\zeta\rho')}{2(\zeta^2 - k^2)} \zeta}_{f(\zeta)} \, d\zeta \, . \tag{8.123}$$

Note that the function $f(\zeta)$ contains simple poles at $\zeta = \pm k$. The Hankel function $H_m^{(1)}(\zeta\rho')$ has the branch point at $\zeta = 0$ and the branch cut is chosen along the negative real ζ-axis, as shown in Fig. 8.7. Performing contour integration in the ζ-plane gives

$$\int_{\Gamma_1} f(\zeta)\, d\zeta + \int_{\Gamma_2} f(\zeta)\, d\zeta + \int_{\Gamma_3} f(\zeta)\, d\zeta = 2\pi i \operatorname*{Res}_{\zeta=k} f(\zeta) \, . \tag{8.124}$$

The asymptotic expressions for $\zeta \to \infty$ (on Γ_2) give

$$J_m(\zeta\rho) \sim \sqrt{\frac{2}{\pi\zeta\rho}} \cos\left(\zeta\rho - \frac{\pi}{4} - \frac{m\pi}{2}\right) \tag{8.125}$$

$$H_m^{(1)}(\zeta\rho') \sim \sqrt{\frac{2}{\pi\zeta\rho'}} \exp\left[i\left(\zeta\rho' - \frac{\pi}{4} - \frac{m\pi}{2}\right)\right] \, . \tag{8.126}$$

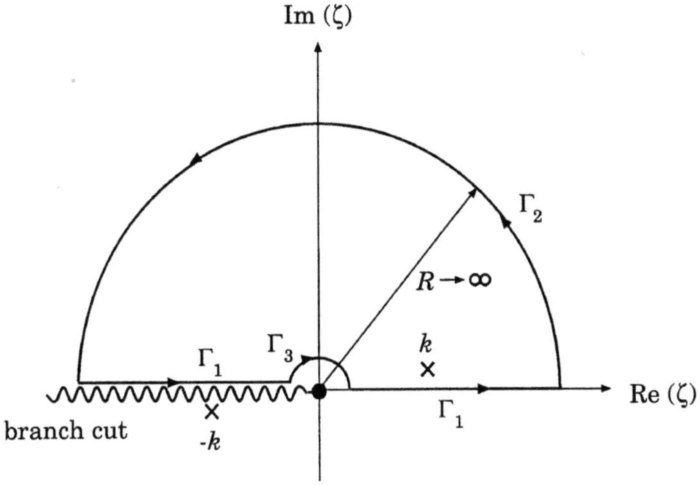

Fig. 8.7. Complex ζ-plane.

It can be shown that $J_m(\zeta\rho)H_m^{(1)}(\zeta\rho') \to 0$ on Γ_2 for $\rho' > \rho$, implying that the integral on path Γ_2 vanishes. It is possible to show that the integral on the infinitesimal semicircular path Γ_3 also vanishes.

Let us consider

$$\int_{\Gamma_1} f(\zeta)\, d\zeta = g_m(\rho;\rho',\phi') \tag{8.127}$$

$$\begin{aligned}\operatorname*{Res}_{\zeta=k} f(\zeta) &= \frac{e^{-im\phi'}}{2\pi}\frac{J_m(\zeta\rho)H_m^{(1)}(\zeta\rho')}{2(\zeta^2-k^2)}\zeta(\zeta-k)\bigg|_{\zeta=k} \\ &= \frac{e^{-im\phi'}}{8\pi}J_m(k\rho)H_m^{(1)}(k\rho')\ .\end{aligned} \tag{8.128}$$

Substituting (8.127) and (8.128) into (8.124) yields the Green's function $g_m(\rho;\rho',\phi')$ as

$$g_m(\rho;\rho',\phi') = \frac{i}{4}e^{-im\phi'}H_m^{(1)}(k\rho')J_m(k\rho)\ . \tag{8.129}$$

Similarly, when $\rho' < \rho$, it is possible to obtain the Green's function $g_m(\rho;\rho',\phi')$ as

$$g_m(\rho;\rho',\phi') = \frac{i}{4}e^{-im\phi'}H_m^{(1)}(k\rho)J_m(k\rho')\ . \tag{8.130}$$

Note that (8.129) and (8.130) are identical with (8.107).

8.3.3 Approach 3

The Helmholtz equation (8.94) is rewritten in rectangular coordinates as

$$\left(\frac{\partial^2}{\partial x^2} + \frac{\partial^2}{\partial y^2} + k^2\right) g(x, y; x', y') = -\delta(x - x')\delta(y - y') . \qquad (8.131)$$

Substituting the two-dimensional Fourier transform representations

$$g(x, y; x', y') = \frac{1}{(2\pi)^2} \int_{-\infty}^{\infty} \int_{-\infty}^{\infty} \tilde{g}(\zeta, \eta) e^{i\zeta(x-x')+i\eta(y-y')} \, d\zeta \, d\eta \qquad (8.132)$$

$$\delta(x - x')\delta(y - y') = \frac{1}{(2\pi)^2} \int_{-\infty}^{\infty} \int_{-\infty}^{\infty} e^{i\zeta(x-x')+i\eta(y-y')} \, d\zeta \, d\eta \qquad (8.133)$$

into (8.131) yields

$$\tilde{g}(\zeta, \eta) = \frac{1}{\zeta^2 + \eta^2 - k^2} . \qquad (8.134)$$

Hence

$$g(x, y; x', y') = \frac{1}{(2\pi)^2} \int_{-\infty}^{\infty} \int_{-\infty}^{\infty} \frac{e^{i\zeta(x-x')+i\eta(y-y')}}{\zeta^2 + \eta^2 - k^2} \, d\zeta \, d\eta . \qquad (8.135)$$

Based on the relations

$$\overline{R} = \hat{\zeta}(x - x') + \hat{\eta}(y - y') \qquad (8.136)$$

$$\overline{K} = \hat{\zeta}\zeta + \hat{\eta}\eta \qquad (8.137)$$

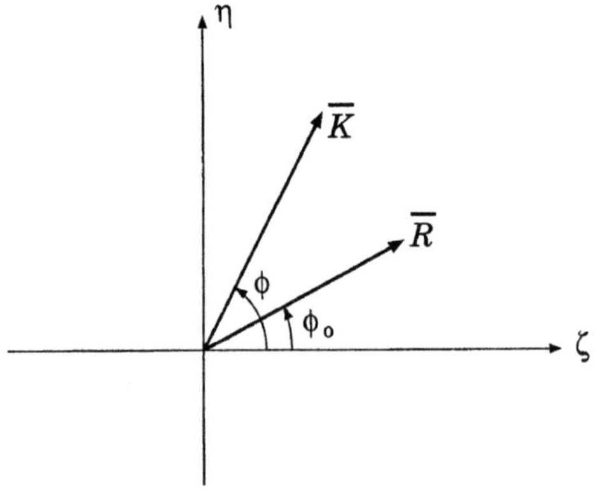

Fig. 8.8. Polar coordinates.

as shown in Fig. 8.8, the Green's function $g(x,y;x',y')$ is rewritten as

$$g(x,y;x',y') = \frac{1}{(2\pi)^2} \int_S \frac{\exp\left(i\overline{K}\cdot\overline{R}\right)}{K^2-k^2}\, ds$$

$$= \frac{1}{(2\pi)^2} \int_0^\infty \int_0^{2\pi} \frac{e^{iKR\cos(\phi-\phi_0)}}{K^2-k^2} K\, d\phi\, dK\,. \qquad (8.138)$$

Substituting

$$e^{iKR\cos(\phi-\phi_0)} = \sum_{n=-\infty}^{\infty} i^n J_n(KR) e^{in(\phi-\phi_0)} \qquad (8.139)$$

into $g(x,y;x',y')$ gives

$$g(x,y;x',y') = \frac{1}{(2\pi)^2} \int_0^\infty \int_0^{2\pi} \frac{1}{K^2-k^2} \cdot \sum_{n=-\infty}^{\infty} i^n J_n(KR) e^{in(\phi-\phi_0)} K\, d\phi\, dK$$

$$= \frac{1}{2\pi} \int_0^\infty \frac{K J_0(KR)}{K^2-k^2}\, dK\,. \qquad (8.140)$$

Since

$$J_0(KR) = \frac{1}{2}\left[H_0^{(1)}(KR) + H_0^{(2)}(KR)\right] \qquad (8.141)$$

the Green's function becomes

$$g(x,y;x',y') = \frac{1}{4\pi} \int_{-\infty}^{\infty} \frac{K H_0^{(1)}(KR)}{(K-k)(K+k)}\, dK\,. \qquad (8.142)$$

It is possible to evaluate $g(x,y;x',y')$ using the residue theorem. Let us integrate along the path in the complex K-plane, as shown in Fig. 8.9. The integrals along the infinite semicircular path $\varGamma_2$ and along the infinitesimal semicircular path $\varGamma_3$ both vanish. The result is

$$g(x,y;x',y') = \frac{1}{4\pi} 2\pi i \operatorname*{Res}_{K=k} \left[\frac{K H_0^{(1)}(KR)}{(K-k)(K+k)}\right]$$

$$= \frac{1}{4\pi} 2\pi i \left.\frac{K H_0^{(1)}(KR)}{(K+k)}\right|_{K\to k}$$

$$= \frac{i}{4} H_0^{(1)}(kR)$$

$$= \frac{i}{4} H_0^{(1)}\left(k|\overline{\rho}-\overline{\rho}'|\right) \qquad (8.143)$$

which is identical with (8.109).

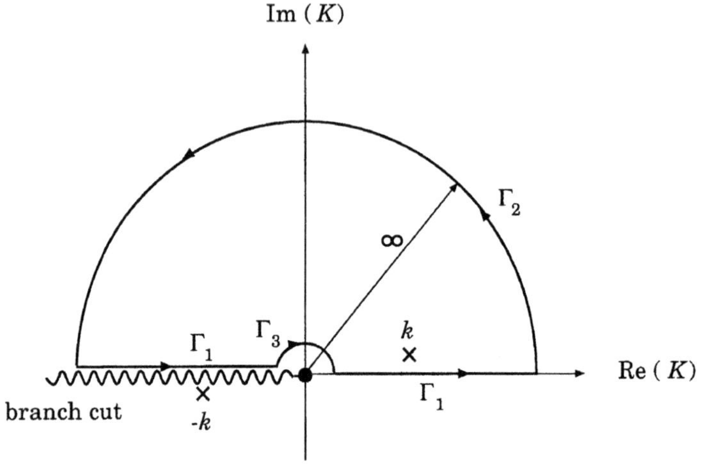

Fig. 8.9. Complex K-plane.

8.4 Three-Dimensional Green's Function

Consider a problem of a three-dimensional Green's function, where the problem geometry is shown in Fig. 8.10. The Helmholtz equation using spherical coordinates (r, θ, ϕ) with the delta source (point source at $\overline{r} = \overline{r}'$) is

$$\left(\nabla^2 + k^2\right) g(\overline{r}; \overline{r}') = -\delta\left(\overline{r} - \overline{r}'\right) \tag{8.144}$$

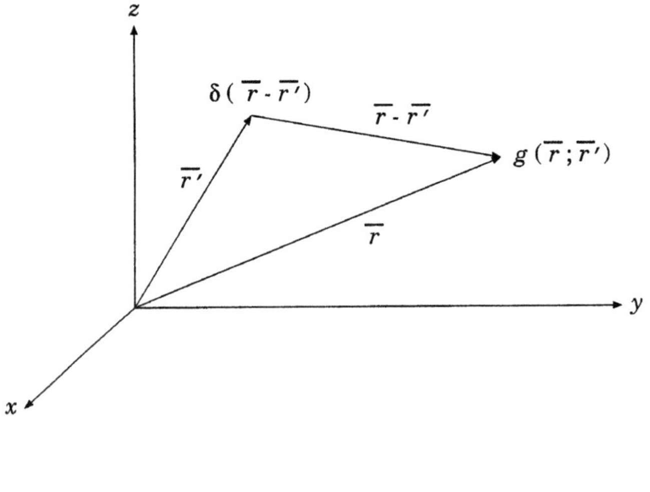

Fig. 8.10. Three-dimensional free-space Green's function.

This section shows three different approaches to derive the solution $g(\overline{r};\overline{r}')$.

8.4.1 Approach 1

Let $\varrho = |\overline{r}-\overline{r}'|$, then $\dfrac{\partial g(\varrho)}{\partial \theta} = \dfrac{\partial g(\varrho)}{\partial \phi} = 0$, and $\delta(\overline{r}-\overline{r}') = \dfrac{\delta(\varrho)}{4\pi\varrho^2}$ using spherical coordinates (ϱ, θ, ϕ). Substituting $g(\varrho) = \dfrac{u(\varrho)}{\varrho}$ into (8.144) gives

$$\left(\frac{d^2}{d\varrho^2} + k^2\right) u(\varrho) = -\frac{\delta(\varrho)}{4\pi\varrho} . \tag{8.145}$$

A solution for $\varrho \neq 0$ is chosen to be the outgoing wave $u(\varrho) = Ae^{ik\varrho}$ and $g(\varrho) = A\dfrac{e^{ik\varrho}}{\varrho}$. The boundary condition for $g(\varrho)$ at $\varrho = 0$ must be enforced to determine A. Integrating

$$\left(\nabla^2 + k^2\right) g(\varrho) = -\frac{\delta(\varrho)}{4\pi\varrho^2} \tag{8.146}$$

over an infinitesimal spherical volume centered at $\varrho = 0$ results in

$$\int_V \nabla^2 g(\varrho)\, dv + \int_V k^2 g(\varrho)\, dv = -1 . \tag{8.147}$$

Since $\nabla^2 g(\varrho)$ is more singular than $g(\varrho)$ at $\varrho = 0$, it is possible to assume

$$\int_V g(\varrho)\, dv = 0 \tag{8.148}$$

$$\int_V \nabla^2 g(\varrho)\, dv = -1 . \tag{8.149}$$

Applying the divergence theorem to (8.149) gives

$$\begin{aligned}
\int_V \underbrace{\nabla^2 g(\varrho)}_{\nabla\cdot\nabla g(\varrho)}\, dv &= \oint_S \nabla g(\varrho) \cdot d\overline{s} \\
&= A\left(\frac{ike^{ik\varrho}\varrho - e^{ik\varrho}}{\varrho^2} 4\pi\varrho^2\right)_{\varrho\to 0} \\
&= -4\pi A .
\end{aligned} \tag{8.150}$$

Hence, $A = \dfrac{1}{4\pi}$, and the three-dimensional free-space Green's function $g(\overline{r};\overline{r}')$ is

$$g(\overline{r};\overline{r}') = \frac{e^{ik|\overline{r}-\overline{r}'|}}{4\pi|\overline{r}-\overline{r}'|} . \tag{8.151}$$

It is useful to express $g(\overline{r};\overline{r}')$ in terms of the superposition of plane waves and cylindrical waves as follows:

- Let us rewrite (8.144) using rectangular coordinates as

$$\left(\frac{\partial^2}{\partial x^2}+\frac{\partial^2}{\partial y^2}+\frac{\partial^2}{\partial z^2}+k^2\right) g(x,y,z;x',y',z')$$
$$= -\delta(x-x')\delta(y-y')\delta(z-z') \,. \tag{8.152}$$

Substituting

$$g(x,y,z;x',y',z')$$
$$= \frac{1}{(2\pi)^2}\int_{-\infty}^{\infty}\int_{-\infty}^{\infty} \tilde{g}(\zeta,\eta)e^{i\zeta(x-x')+i\eta(y-y')}\,d\zeta\,d\eta \tag{8.153}$$

into (8.152) yields

$$\left(\frac{d^2}{dz^2}+k^2-\zeta^2-\eta^2\right)\tilde{g}(\zeta,\eta) = -\delta(z-z') \tag{8.154}$$

where the solution is the one-dimensional free-space Green's function

$$\tilde{g}(\zeta,\eta) = \frac{i}{2\kappa}e^{i\kappa|z-z'|} \tag{8.155}$$

with $\kappa = \sqrt{k^2-\zeta^2-\eta^2}$. Substituting $\tilde{g}(\zeta,\eta)$ into (8.153) finally gives the relation

$$\frac{e^{ik|\overline{r}-\overline{r}'|}}{4\pi|\overline{r}-\overline{r}'|}$$
$$= \frac{1}{(2\pi)^2}\int_{-\infty}^{\infty}\int_{-\infty}^{\infty}\frac{i}{2\kappa}e^{i\kappa|z-z'|}e^{i\zeta(x-x')+i\eta(y-y')}\,d\zeta\,d\eta \tag{8.156}$$

which represents the superposition of a continuous spectrum of plane waves.

- In view of Fig. 8.8, let

$$\overline{R} = \hat{\zeta}(x-x')+\hat{\eta}(y-y') \tag{8.157}$$

$$\overline{K} = \hat{\zeta}\zeta+\hat{\eta}\eta \,. \tag{8.158}$$

Hence

$$\frac{e^{ik|\overline{r}-\overline{r}'|}}{4\pi|\overline{r}-\overline{r}'|}$$
$$= \frac{1}{(2\pi)^2}\int_0^{\infty}\int_0^{2\pi}\frac{i}{2\kappa}e^{i\kappa|z-z'|}e^{iKR\cos(\phi-\phi_0)}K\,d\phi\,dK \tag{8.159}$$

where $R = \sqrt{(x-x')^2 + (y-y')^2}$ and $\kappa = \sqrt{k^2 - K^2}$.
Since

$$e^{iKR\cos(\phi-\phi_0)} = \sum_{n=-\infty}^{\infty} i^n J_n(KR) e^{in(\phi-\phi_0)} \tag{8.160}$$

integration with respect to ϕ gives

$$\frac{e^{ik|\overline{r}-\overline{r}'|}}{4\pi|\overline{r}-\overline{r}'|} = \frac{1}{2\pi}\int_0^\infty \frac{i}{2\kappa} e^{i\kappa|z-z'|} J_0(KR)\, K\, dK \tag{8.161}$$

which represents the three-dimensional free-space Green's function in terms of the cylindrical wave $J_0(KR)$.

8.4.2 Approach 2

The Helmholtz equation (8.144) is explicitly rewritten using spherical coordinates as

$$\frac{1}{r^2}\frac{\partial}{\partial r}\left(r^2\frac{\partial g}{\partial r}\right) + \frac{1}{r^2\sin\theta}\frac{\partial}{\partial\theta}\left(\sin\theta\frac{\partial g}{\partial\theta}\right) + \frac{1}{r^2\sin^2\theta}\frac{\partial^2 g}{\partial\phi^2} + k^2 g =$$

$$-\frac{\delta(r-r')\delta(\theta-\theta')\delta(\phi-\phi')}{r^2\sin\theta} \,. \tag{8.162}$$

It is expedient to represent the Green's function and delta function using spherical harmonics $P_n^m(\cos\theta)e^{im\phi}$, as shown in Appendix E. The Green's function g and the delta function $\dfrac{\delta(\theta-\theta')\delta(\phi-\phi')}{\sin\theta}$ are written as

$$g(r,\theta,\phi;r',\theta',\phi') = \sum_{n=0}^{\infty}\sum_{m=-n}^{n} g_{nm}(r;r',\theta',\phi')P_n^m(\cos\theta)e^{im\phi} \tag{8.163}$$

and

$$\frac{\delta(\theta-\theta')\delta(\phi-\phi')}{\sin\theta} =$$

$$\sum_{n=0}^{\infty}\sum_{m=-n}^{n} A_{nm}P_n^m(\cos\theta')P_n^m(\cos\theta)e^{im(\phi-\phi')} \tag{8.164}$$

where

$$A_{nm} = \frac{2n+1}{4\pi}\frac{(n-m)!}{(n+m)!} \,. \tag{8.165}$$

Substituting $g(r,\theta,\phi;r',\theta',\phi')$ and $\dfrac{\delta(\theta-\theta')\delta(\phi-\phi')}{\sin\theta}$ into (8.162) gives

$$\frac{d^2 g_{nm}}{dr^2} + \frac{2}{r}\frac{dg_{nm}}{dr} + \left[k^2 - \frac{n(n+1)}{r^2}\right] g_{nm} =$$

$$-A_{nm} P_n^m(\cos\theta')e^{-im\phi'}\frac{\delta(r-r')}{r^2} . \tag{8.166}$$

Introducing $\hat{g}_{nm} = kr g_{nm}$ gives

$$\frac{d^2 \hat{g}_{nm}}{dr^2} + \left[k^2 - \frac{n(n+1)}{r^2}\right] \hat{g}_{nm} =$$

$$-A_{nm} P_n^m(\cos\theta')e^{-im\phi'} k\frac{\delta(r-r')}{r} . \tag{8.167}$$

The solution to (8.167) for $r \neq r'$ is the spherical Bessel functions and spherical Hankel functions. Let us assume

$$\hat{g}_{nm} = \begin{cases} B\hat{J}_n(kr) & \text{for } r < r' \\ C\hat{H}_n^{(1)}(kr) & \text{for } r > r' \end{cases} \tag{8.168}$$

where $\hat{J}_n(kr)$ and $\hat{H}_n^{(1)}(kr)$ are the spherical Bessel function of the first kind and the spherical Hankel function of the first kind, respectively. Multiplying (8.167) by r and integrating from $r' - \Delta$ to $r' + \Delta$ gives

$$r\frac{d\hat{g}_{nm}}{dr}\Bigg|_{r'-\Delta}^{r'+\Delta} = -A_{nm} P_n^m(\cos\theta')e^{-im\phi'} k \tag{8.169}$$

$$\hat{g}_{nm}\Bigg|_{r'-\Delta}^{r'+\Delta} = 0 \tag{8.170}$$

which are then rewritten as

$$r'\left[C\hat{H}_n^{(1)'}(kr') - B\hat{J}_n'(kr')\right] = -A_{nm} P_n^m(\cos\theta')e^{-im\phi'} \tag{8.171}$$

$$C\hat{H}_n^{(1)}(kr') - B\hat{J}_n(kr') = 0 \tag{8.172}$$

where $\hat{H}_n^{(1)'}(kr') = \left.\dfrac{d\hat{H}_n^{(1)}(kr)}{d(kr)}\right|_{r=r'}$. Solving (8.171) and (8.172) for B and C gives

$$B = \frac{-A_{nm}\hat{H}_n^{(1)}(kr')P_n^m(\cos\theta')e^{-im\phi'}}{r'\left[\hat{J}_n(kr')\hat{H}_n^{(1)'}(kr') - \hat{J}_n'(kr')\hat{H}_n^{(1)}(kr')\right]}$$

$$= \frac{i}{r'}A_{nm}\hat{H}_n^{(1)}(kr')P_n^m(\cos\theta')e^{-im\phi'} \tag{8.173}$$

$$C = \frac{i}{r'}A_{nm}\hat{J}_n(kr')P_n^m(\cos\theta')e^{-im\phi'} . \tag{8.174}$$

Substituting B and C into (8.168) gives

$$g(r,\theta,\phi;r',\theta',\phi') = \frac{i}{krr'}\sum_{n=0}^{\infty}\sum_{m=-n}^{n} A_{nm}P_n^m(\cos\theta)P_n^m(\cos\theta')e^{im(\phi-\phi')} \cdot \begin{cases} \hat{H}_n^{(1)}(kr')\hat{J}_n(kr) & \text{for } r < r' \\ \hat{J}_n(kr')\hat{H}_n^{(1)}(kr) & \text{for } r > r' . \end{cases} \tag{8.175}$$

The Bessel function summation theorem in [11, page 980] shows that (8.175) reduces to (8.151).

8.4.3 Approach 3

The Helmholtz equation (8.144) is rewritten using rectangular coordinates as

$$\left(\frac{\partial^2}{\partial x^2}+\frac{\partial^2}{\partial y^2}+\frac{\partial^2}{\partial z^2}+k^2\right)g(x,y,z;x',y',z') = -\delta(x-x')\delta(y-y')\delta(z-z') . \tag{8.176}$$

Substituting the three-dimensional Fourier transform representations

$$g(x,y,z;x',y',z') = \frac{1}{(2\pi)^3}\int_{-\infty}^{\infty}\int_{-\infty}^{\infty}\int_{-\infty}^{\infty} \tilde{g}(\zeta,\eta,\xi)e^{i\zeta(x-x')+i\eta(y-y')+i\xi(z-z')}\,d\zeta\,d\eta\,d\xi \tag{8.177}$$

$$\delta(x-x')\delta(y-y')\delta(z-z') = \frac{1}{(2\pi)^3}\int_{-\infty}^{\infty}\int_{-\infty}^{\infty}\int_{-\infty}^{\infty} e^{i\zeta(x-x')+i\eta(y-y')+i\xi(z-z')}\,d\zeta\,d\eta\,d\xi \tag{8.178}$$

into (8.176) results in

$$\tilde{g}(\zeta,\eta,\xi) = \frac{1}{\zeta^2+\eta^2+\xi^2-k^2} \tag{8.179}$$

$$g(x,y,z;x',y',z') =$$

$$\frac{1}{(2\pi)^3}\int_{-\infty}^{\infty}\int_{-\infty}^{\infty}\int_{-\infty}^{\infty}\frac{e^{i\zeta(x-x')+i\eta(y-y')+i\xi(z-z')}}{\zeta^2+\eta^2+\xi^2-k^2}\,d\zeta\,d\eta\,d\xi\,. \quad (8.180)$$

Let

$$\overline{R} = \hat{\zeta}(x-x') + \hat{\eta}(y-y') + \hat{\xi}(z-z') \quad (8.181)$$

$$\overline{K} = \hat{\zeta}\zeta + \hat{\eta}\eta + \hat{\xi}\xi\,. \quad (8.182)$$

Then

$$g(x,y,z;x',y',z') = \frac{1}{(2\pi)^3}\int_V \frac{\exp\left(i\overline{K}\cdot\overline{R}\right)}{K^2-k^2}\,dv\,. \quad (8.183)$$

Let us introduce rectangular coordinates (ζ',η',ξ') whose ξ'-axis coincides with the vector $\overline{R}$, as shown in Fig. 8.11. Therefore

$$\zeta' = K\sin\theta\cos\phi \quad (8.184)$$

$$\eta' = K\sin\theta\sin\phi \quad (8.185)$$

$$\xi' = K\cos\theta\,. \quad (8.186)$$

Using spherical coordinates (K,θ,ϕ) results in

$$\overline{K}\cdot\overline{R} = KR\cos\theta \quad (8.187)$$

$$dv = K^2\sin\theta\,d\phi\,d\theta\,dK\,. \quad (8.188)$$

Hence

$$g(x,y,z;x',y',z') =$$

$$\frac{1}{(2\pi)^3}\int_0^{\infty}\int_0^{\pi}\int_0^{2\pi}\frac{e^{iKR\cos\theta}}{K^2-k^2}K^2\sin\theta\,d\phi\,d\theta\,dK =$$

$$\frac{1}{(2\pi)^2 iR}\int_0^{\infty}\frac{\left(e^{iKR}-e^{-iKR}\right)K}{(K-k)(K+k)}\,dK\,. \quad (8.189)$$

Since

$$\int_0^{\infty}\frac{-e^{-iKR}K}{(K-k)(K+k)}\,dK =$$

$$\int_{-\infty}^{0}\frac{e^{iKR}K}{(K-k)(K+k)}\,dK \quad (8.190)$$

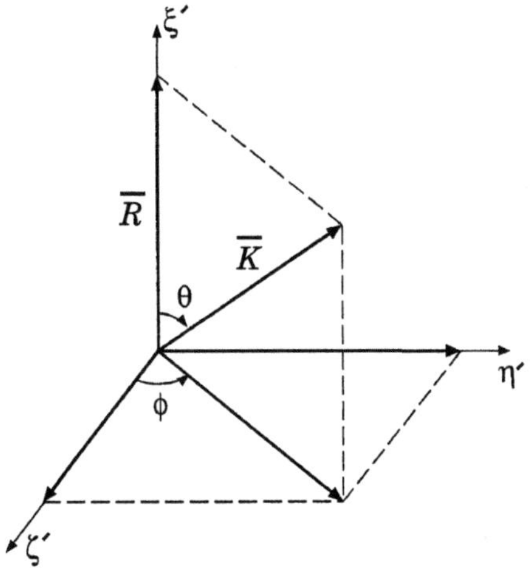

Fig. 8.11. Spherical coordinates.

the Green's function $g(x, y, z; x', y', z')$ is

$$g(x, y, z; x', y', z') = \frac{1}{(2\pi)^2 iR} \int_{-\infty}^{\infty} \frac{e^{iKR} K}{(K-k)(K+k)} \, dK \; . \qquad (8.191)$$

Let us utilize the residue theorem and contour integration to evaluate (8.191). Integrating along the path in the complex K-plane of Fig. 8.12 gives

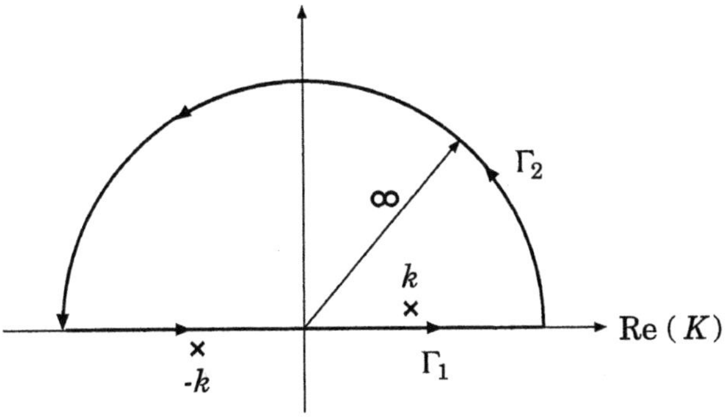

Fig. 8.12. Complex K-plane.

$$g(x,y,z;x',y',z') = \frac{1}{(2\pi)^2 iR} 2\pi i \operatorname*{Res}_{K=k} \left[\frac{e^{iKR}K}{(K-k)(K+k)} \right]$$

$$= \frac{1}{2\pi R} \frac{e^{iKR}K}{(K+k)} \bigg|_{K \to k}$$

$$= \frac{e^{ikR}}{4\pi R}$$

$$= \frac{e^{ik|\overline{r}-\overline{r}'|}}{4\pi|\overline{r}-\overline{r}'|} . \tag{8.192}$$

Problems for Chapter 8

1. Consider the half-space Sturm-Liouville equation

$$\left(\frac{d^2}{dx^2} + k^2 \right) g(x;x') = -\delta(x-x') \tag{8.193}$$

for $x \geq 0$ subject to the boundary conditions $\left. \frac{dg(x;x')}{dx} \right|_{x=0} = 0$ and $g(\infty;x') = 0$. The problem geometry is shown in Fig. 8.13. Derive $g(x;x')$.

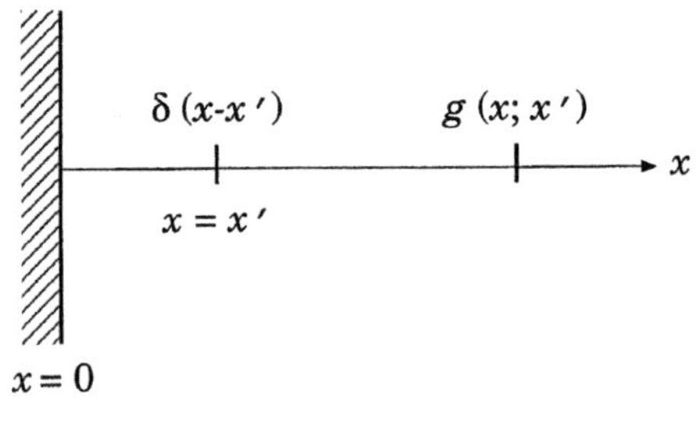

Fig. 8.13. Half-space Green's function.

2. Consider the Sturm-Liouville problem described by (8.75). Derive the solution (8.81) using the image method.
3. Show that a series (8.86) yields (8.81).

9

Green's Functions: Applications

9.1 Currents in Free Space

The free-space Green's function is used to evaluate electromagnetic wave scattering and radiation from currents in free space. This section investigates two examples using the free-space Green's function, namely, radiation from a sheet current and radiation from a shell current.

9.1.1 Radiation from Sheet Current

Consider electromagnetic radiation from a constant electric current sheet of infinite extent $\overline{J} = \hat{z}J\delta(x)$. The geometry of the radiating structure is shown in Fig. 9.1, where the medium wavenumber is $k = \omega\sqrt{\mu\epsilon}$. The field has no vari-

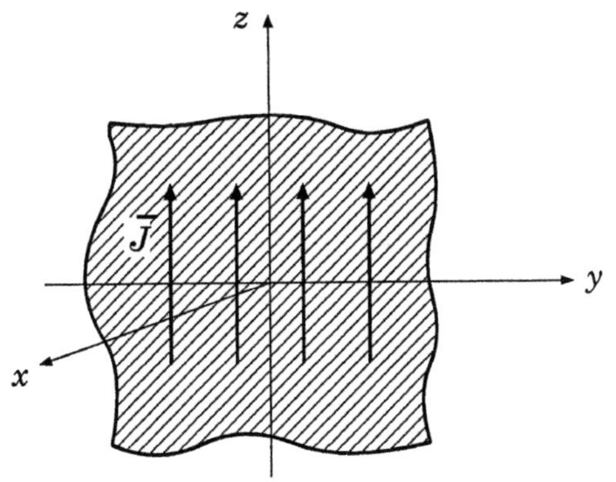

Fig. 9.1. Current sheet in free space.

ation with respect to y and z $\left(\frac{\partial}{\partial y} = \frac{\partial}{\partial z} = 0\right)$. The corresponding Helmholtz equation for $\overline{A} = \hat{z}A_z(x)$ with a delta source is

$$\left(\frac{d^2}{dx^2} + k^2\right)A_z(x) = -\mu J\delta(x) . \tag{9.1}$$

The solution is

$$A_z(x) = \frac{i\mu J}{2k}e^{ik|x|} . \tag{9.2}$$

Therefore

$$\overline{E} = \hat{z}E_z(x) = \hat{z}i\omega A_z(x)$$

$$= -\hat{z}\frac{J}{2}\sqrt{\frac{\mu}{\epsilon}}e^{ik|x|} . \tag{9.3}$$

It is also possible to derive $\overline{E}$ based on the boundary condition at $x = 0$, which requires

$$\overline{J} = \hat{x} \times \left(\overline{H}\big|_{x=0^+} - \overline{H}\big|_{x=0^-}\right) . \tag{9.4}$$

Letting

$$\overline{H}\big|_{x=0^+} = \hat{y}H_0 \tag{9.5}$$

$$\overline{H}\big|_{x=0^-} = -\hat{y}H_0 \tag{9.6}$$

results in

$$H_0 = \frac{J}{2} . \tag{9.7}$$

For the region $x > 0$, the current sheet produces a uniform plane wave that propagates in the $+x$-direction with the wavenumber k as

$$H_y(x) = \frac{J}{2}e^{ikx} \tag{9.8}$$

$$E_z(x) = -\frac{J}{2}\sqrt{\frac{\mu}{\epsilon}}e^{ikx} . \tag{9.9}$$

Similarly for the region $x < 0$

$$H_y(x) = -\frac{J}{2}e^{-ikx} \tag{9.10}$$

$$E_z(x) = -\frac{J}{2}\sqrt{\frac{\mu}{\epsilon}}e^{-ikx} . \tag{9.11}$$

Equations (9.9) and (9.11) are identical with (9.3).

9.1.2 Radiation from Shell Current

Consider radiation from an infinitely long current shell source located at $\rho = \rho'$ in Fig. 9.2. The shell current is given by $\overline{J} = \hat{z} I \frac{\delta(\rho - \rho')}{2\pi\rho}$. The derivation for the problem solution is available in [12]. The corresponding Helmholtz equation with a delta source using cylindrical coordinates (ρ, ϕ) is

$$\left[\frac{1}{\rho}\frac{\partial}{\partial \rho}\left(\rho \frac{\partial}{\partial \rho}\right) + \frac{1}{\rho^2}\frac{\partial^2}{\partial \phi^2} + k^2\right] A_z(\rho, \phi; \rho', \phi')$$

$$= -\mu \frac{I\delta(\rho - \rho')}{2\pi\rho} . \tag{9.12}$$

The following presents three different approaches to derive the solution $A_z(\rho, \phi; \rho', \phi')$.

Approach 1

Let us rewrite the governing equation for the two-dimensional free-space Green's function

$$\left[\frac{1}{\rho}\frac{\partial}{\partial \rho}\left(\rho \frac{\partial}{\partial \rho}\right) + \frac{1}{\rho^2}\frac{\partial^2}{\partial \phi^2} + k^2\right] g(\rho, \phi; \rho', \phi')$$

$$= -\frac{\delta(\rho - \rho')}{\rho}\delta(\phi - \phi') . \tag{9.13}$$

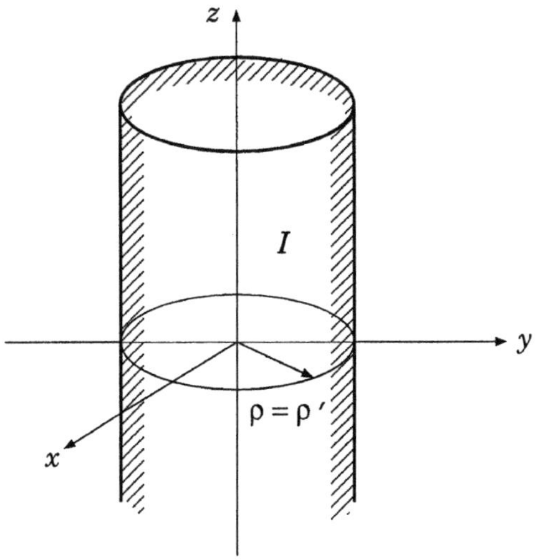

Fig. 9.2. Infinitely long current shell source located at $\rho = \rho'$.

Equation (9.12) is similar to (9.13), which describes the response to an infinitesimally thin line current at $\rho = \rho'$ and $\phi = \phi'$. The current shell is considered as a collection of infinitesimally thin line currents at $\rho = \rho'$ and $\phi = \phi'$. The response from an infinitesimally thin line current at $\rho = \rho'$ and $\phi = \phi'$ is given by the two-dimensional free-space Green's function $g(\rho, \phi; \rho', \phi')$. Therefore, the response to the current shell is obtained by

$$\begin{aligned} A_z(\rho, \phi; \rho', \phi') &= \int_0^{2\pi} g(\rho, \phi; \rho', \phi') \frac{\mu I}{2\pi}\, d\phi' \\ &= \int_0^{2\pi} \sum_{m=-\infty}^{\infty} g_m(\rho; \rho', \phi') e^{im\phi} \frac{\mu I}{2\pi}\, d\phi' \end{aligned} \tag{9.14}$$

where

$$g_m(\rho; \rho', \phi') = \frac{i}{4} \begin{cases} H_m^{(1)}(k\rho') J_m(k\rho) e^{-im\phi'} & \text{for } \rho < \rho' \\ J_m(k\rho') H_m^{(1)}(k\rho) e^{-im\phi'} & \text{for } \rho > \rho' \, . \end{cases} \tag{9.15}$$

Since

$$\int_0^{2\pi} e^{-im\phi'}\, d\phi' = 2\pi \delta_{m0} \tag{9.16}$$

$A_z(\rho; \rho')$ is written as

$$A_z(\rho; \rho') = \frac{i\mu I}{4} \begin{cases} H_0^{(1)}(k\rho') J_0(k\rho) & \text{for } \rho < \rho' \\ J_0(k\rho') H_0^{(1)}(k\rho) & \text{for } \rho > \rho' \, . \end{cases} \tag{9.17}$$

Hence

$$\overline{E} = \hat{z} E_z = \hat{z} i\omega A_z(\rho; \rho') \, . \tag{9.18}$$

Approach 2

Here, an approach based on the cylindrical wave propagation characteristics is used. Since $A_z(\rho; \rho')$ has no azimuthal variation in ϕ $\left(\frac{\partial}{\partial \phi} = 0 \right)$, (9.12) becomes

$$\frac{1}{\rho} \frac{d}{d\rho} \left[\rho \frac{dA_z(\rho; \rho')}{d\rho} \right] + k^2 A_z(\rho; \rho') = -\mu \frac{I \delta(\rho - \rho')}{2\pi\rho} \, . \tag{9.19}$$

When $\rho \neq \rho'$, (9.19) reduces to a Bessel's equation whose solutions are in the form of $J_0(k\rho)$, $N_0(k\rho)$, $H_0^{(1)}(k\rho)$, $H_0^{(2)}(k\rho)$, $\cdots$. The Bessel function of the first kind, $J_0(k\rho)$, is not singular at $\rho = 0$. The Hankel function of the first kind, $H_0^{(1)}(k\rho)$, represents the outgoing wave that vanishes at infinity. Therefore, $A_z(\rho;\rho')$ is chosen as

$$A_z(\rho;\rho') = \begin{cases} AJ_0(k\rho) & \text{for } \rho < \rho' \\ \\ BH_0^{(1)}(k\rho) & \text{for } \rho > \rho' \, . \end{cases} \tag{9.20}$$

Multiplying (9.19) by ρ and integrating from $\rho = \rho' - \Delta$ to $\rho = \rho' + \Delta$ yields

$$\rho \frac{dA_z(\rho;\rho')}{d\rho}\bigg|_{\rho'-\Delta}^{\rho'+\Delta} + \int_{\rho'-\Delta}^{\rho'+\Delta} k^2 A_z(\rho;\rho')\rho \, d\rho = -\frac{\mu I}{2\pi} \, . \tag{9.21}$$

Since $\dfrac{dA_z(\rho;\rho')}{d\rho}$ and $A_z(\rho;\rho')$ are discontinuous and continuous, respectively, at $\rho = \rho'$,

$$\rho \frac{dA_z(\rho;\rho')}{d\rho}\bigg|_{\rho'-\Delta}^{\rho'+\Delta} = \rho' \left[-AkJ_0'(k\rho') + BkH_0^{(1)'}(k\rho') \right] = -\frac{\mu I}{2\pi} \tag{9.22}$$

$$A_z(\rho;\rho')\bigg|_{\rho'-\Delta}^{\rho'+\Delta} = -AJ_0(k\rho') + BH_0^{(1)}(k\rho') = 0 \, . \tag{9.23}$$

Solving (9.22) and (9.23) for A and B, and substituting A and B into $A_z(\rho;\rho')$ gives

$$A_z(\rho;\rho') = \frac{i\mu I}{4} \begin{cases} H_0^{(1)}(k\rho')J_0(k\rho) & \text{for } \rho < \rho' \\ \\ J_0(k\rho')H_0^{(1)}(k\rho) & \text{for } \rho > \rho' \, . \end{cases} \tag{9.24}$$

Approach 3

In this approach, $A_z(\rho;\rho')$ is derived starting from a transform approach. It is expedient to express $A_z(\rho;\rho')$ based on the Hankel transform representation

$$A_z(\rho;\rho') = \int_0^\infty \tilde{A}_z(\zeta;\rho')J_n(\zeta\rho)\zeta \, d\zeta \tag{9.25}$$

$$\tilde{A}_z(\zeta;\rho') = \int_0^\infty A_z(\rho;\rho')J_n(\zeta\rho)\rho \, d\rho \, . \tag{9.26}$$

Since $A_z(\rho;\rho')$ has no azimuthal variation in ϕ, the parameter $n = 0$ is chosen in view of Bessel's equation. Hence

$$A_z(\rho;\rho') = \int_0^\infty \tilde{A}_z(\zeta;\rho')J_0(\zeta\rho)\zeta \, d\zeta \tag{9.27}$$

$$\frac{\delta(\rho - \rho')}{\rho} = \int_0^\infty J_0(\zeta\rho)J_0(\zeta\rho')\zeta \, d\zeta \, . \tag{9.28}$$

Substituting $A_z(\rho;\rho')$ and $\dfrac{\delta(\rho-\rho')}{\rho}$ into (9.19) and noting

$$\frac{1}{\rho}\frac{d}{d\rho}\left[\rho\frac{dJ_0(\zeta\rho)}{d\rho}\right] = -\zeta^2 J_0(\zeta\rho) \tag{9.29}$$

yields

$$\tilde{A}_z(\zeta;\rho') = \frac{\mu I}{2\pi}\frac{J_0(\zeta\rho')}{(\zeta^2-k^2)} . \tag{9.30}$$

Let us first evaluate $A_z(\rho;\rho')$ when $\rho < \rho'$. When $\tilde{A}_z(\zeta;\rho')$ and

$$J_0(\zeta\rho') = \frac{1}{2}\left[H_0^{(1)}(\zeta\rho') + H_0^{(2)}(\zeta\rho')\right] \tag{9.31}$$

are substituted into (9.27), $A_z(\rho;\rho')$ is written as

$$\begin{aligned} A_z(\rho;\rho') = &\frac{\mu I}{2\pi}\int_0^\infty \frac{J_0(\zeta\rho)H_0^{(1)}(\zeta\rho')}{2(\zeta^2-k^2)}\zeta\, d\zeta \\ &+\frac{\mu I}{2\pi}\int_0^\infty \frac{J_0(\zeta\rho)H_0^{(2)}(\zeta\rho')}{2(\zeta^2-k^2)}\zeta\, d\zeta . \end{aligned} \tag{9.32}$$

Since $H_0^{(1)}(-u) = -H_0^{(2)}(u)$ and $J_0(-u) = J_0(u)$,

$$A_z(\rho;\rho') = \frac{\mu I}{2\pi}\int_{-\infty}^\infty \frac{J_0(\zeta\rho)H_0^{(1)}(\zeta\rho')}{2(\zeta^2-k^2)}\zeta\, d\zeta . \tag{9.33}$$

Contour integration is performed in the complex ζ-plane, as shown in Fig. 8.7. Note that the integral evaluation along the path Γ_2 vanishes since $\rho < \rho'$. The result is

$$A_z(\rho;\rho') = \frac{i\mu I}{4}H_0^{(1)}(k\rho')J_0(k\rho) . \tag{9.34}$$

Similarly, when $\rho > \rho'$,

$$A_z(\rho;\rho') = \frac{i\mu I}{4}H_0^{(1)}(k\rho)J_0(k\rho') . \tag{9.35}$$

Note that the solution $A_z(\rho;\rho')$ is identical with (9.17).

9.2 Line Current in Rectangular Waveguide

This section investigates three cases dealing with radiation from line currents in a rectangular waveguide. The first is radiation in a parallel-plate waveguide, the second is radiation in a shorted parallel-plate waveguide, and the third is radiation in a rectangular waveguide.

9.2.1 Radiation in Parallel-Plate Waveguide

Let us consider electromagnetic radiation from an infinitely long electric line current $\overline{J}(x', y')$ within a conducting parallel-plate waveguide. The geometry of the radiating problem is shown in Fig. 9.3. Two different approaches are presented to analyze the problem.

Approach 1

The Helmholtz equation that governs radiation from $\overline{J}(x', y') = \hat{z}J\delta(x - x')\delta(y - y')$ is

$$\left(\nabla^2 + k^2\right)\overline{A}(x, y; x', y') = -\mu\overline{J}(x', y') . \tag{9.36}$$

Using rectangular coordinates (x, y), it can be rewritten as

$$\left(\frac{\partial^2}{\partial x^2} + \frac{\partial^2}{\partial y^2} + k^2\right)A_z(x, y; x', y') = -\mu J\delta(x - x')\delta(y - y') . \tag{9.37}$$

The boundary condition requires

$$E_z(x, y; x', y')\Big|_{x=0,a} = i\omega A_z(x, y; x', y')\Big|_{x=0,a} = 0 . \tag{9.38}$$

Let us choose eigenfunctions

$$\psi_m(x) = \sqrt{\frac{2}{a}}\sin\frac{m\pi}{a}x \tag{9.39}$$

and expand $A_z(x, y; x', y')$ in terms of $\psi_m(x)$ as

$$A_z(x, y; x', y') = \sum_{m=1}^{\infty} a_m(y; x', y')\psi_m(x) . \tag{9.40}$$

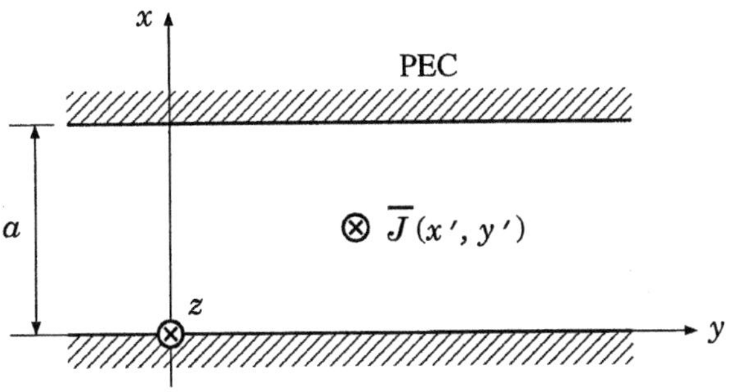

Fig. 9.3. Line current $\overline{J}$ in PEC parallel-plate waveguide.

The delta function $\delta(x-x')$ is also written as

$$\delta(x-x') = \sum_{m=1}^{\infty} \psi_m(x)\psi_m(x') \,. \tag{9.41}$$

Substituting $A_z(x,y;x',y')$ and $\delta(x-x')$ into (9.37) yields

$$\left[\frac{d^2}{dy^2} - \left(\frac{m\pi}{a}\right)^2 + k^2\right] a_m(y;x',y') = -\mu J \psi_m(x')\delta(y-y') \,. \tag{9.42}$$

The solution to (9.42) is the following one-dimensional free-space Green's function

$$a_m(y;x',y') = \frac{i\mu J \psi_m(x')}{2\zeta_m} \exp\left(i\zeta_m |y-y'|\right) \tag{9.43}$$

where $\zeta_m = \sqrt{k^2 - \left(\frac{m\pi}{a}\right)^2}$.

The vector potential and electric field inside the waveguide are

$$A_z(x,y;x',y') = \sum_{m=1}^{\infty} \frac{i\mu J}{2\zeta_m} \psi_m(x)\psi_m(x') \exp\left(i\zeta_m |y-y'|\right) \tag{9.44}$$

$$\overline{E} = \hat{z}E_z = \hat{z}i\omega A_z(x,y;x',y') \,. \tag{9.45}$$

Approach 2

It is also possible to obtain $A_z(x,y;x',y')$ using different expressions for the Green's function. The second approach uses the continuous mode representations for $A_z(x,y;x',y')$ and $\delta(y-y')$ based on the Fourier transform along the y-direction. Let

$$A_z(x,y;x',y') = \frac{1}{2\pi}\int_{-\infty}^{\infty} \tilde{A}_z e^{i\zeta y}\, d\zeta \tag{9.46}$$

$$\delta(y-y') = \frac{1}{2\pi}\int_{-\infty}^{\infty} e^{i\zeta(y-y')}\, d\zeta \,. \tag{9.47}$$

Substituting $A_z(x,y;x',y')$ and $\delta(y-y')$ into (9.37) gives

$$\left(\frac{d^2}{dx^2} + \kappa^2\right)\tilde{A}_z = -\mu J \delta(x-x') e^{-i\zeta y'} \tag{9.48}$$

where $\kappa = \sqrt{k^2 - \zeta^2}$. Since $\tilde{A}_z$ vanishes at $x=0$ and a, $\tilde{A}_z$ is given by the one-dimensional closed-space Green's function

$$\tilde{A}_z = -\mu J e^{-i\zeta y'} \begin{cases} \dfrac{\sin\kappa(x'-a)}{\kappa\sin\kappa a}\sin\kappa x & \text{for } 0 \le x < x' \\[2ex] \dfrac{\sin\kappa x'}{\kappa\sin\kappa a}\sin\kappa(x-a) & \text{for } x' < x \le a\,. \end{cases} \tag{9.49}$$

Let us rewrite $A_z(x,y;x',y')$ for $x - x' > 0$ as

$$A_z(x,y;x',y') = \int_{-\infty}^{\infty} \underbrace{-\frac{\mu J}{2\pi\kappa}\frac{\sin\kappa(x-a)\sin\kappa x'}{\sin\kappa a}e^{i\zeta(y-y')}}_{f(\zeta)}\,d\zeta\,. \tag{9.50}$$

It would be expedient to use residue calculus and contour integration to evaluate $A_z(x,y;x',y')$. The function $f(\zeta)$ has singularities when $\sin\kappa a = 0$. The equation $\sin\kappa a = 0$ contains simple poles at $\kappa = \kappa_m$ ($\zeta = \zeta_m$) where $\kappa_m = \dfrac{m\pi}{a}$ and $\zeta_m = \sqrt{k^2 - \kappa_m^2}$ for $m = 1, 2, \cdots$.

First consider when $y - y' > 0$. In view of Fig. 9.4, contour integration is performed to obtain

$$A_z(x,y;x',y') + \int_{\Gamma_2} f(\zeta)\,d\zeta = 2\pi i \sum_{m=1}^{\infty} \operatorname*{Res}_{\zeta=\zeta_m} f(\zeta) \tag{9.51}$$

Note that the function $f(\zeta)$ is bounded when $y - y' > 0$ and $x - x' > 0$. Since $\int_{\Gamma_2} f(\zeta)\,d\zeta \to 0$ as $R \to \infty$, $A_z(x,y;x',y')$ is given as follows:

$$\begin{aligned} A_z(x,y;x',y') &= 2\pi i \sum_{m=1}^{\infty} \operatorname*{Res}_{\zeta=\zeta_m} f(\zeta) \\ &= 2\pi i \sum_{m=1}^{\infty} \operatorname*{Res}_{\zeta=\zeta_m}\left[-\frac{\mu J}{2\pi\kappa}\frac{\sin\kappa(x-a)\sin\kappa x'}{\sin\kappa a}e^{i\zeta(y-y')}\right] \\ &= -i\mu J \sum_{m=1}^{\infty}\left[\frac{\sin\kappa(x-a)\sin\kappa x'}{\kappa\dfrac{d(\sin\kappa a)}{d\zeta}}e^{i\zeta(y-y')}\right]_{\zeta=\zeta_m} \\ &= -i\mu J \sum_{m=1}^{\infty}\left[\frac{\sin\kappa(x-a)\sin\kappa x'}{-a\zeta\cos\kappa a}e^{i\zeta(y-y')}\right]_{\zeta=\zeta_m} \\ &= \sum_{m=1}^{\infty}\frac{i\mu J}{\zeta_m a}\sin(\kappa_m x)\sin(\kappa_m x')\exp\left[i\zeta_m(y-y')\right]\,. \end{aligned} \tag{9.52}$$

Similarly, it is possible to show that $A_z(x,y;x',y')$ for $x - x' < 0$ reduces to (9.52) . Equation (9.52) agrees with (9.44) for $y - y' > 0$.

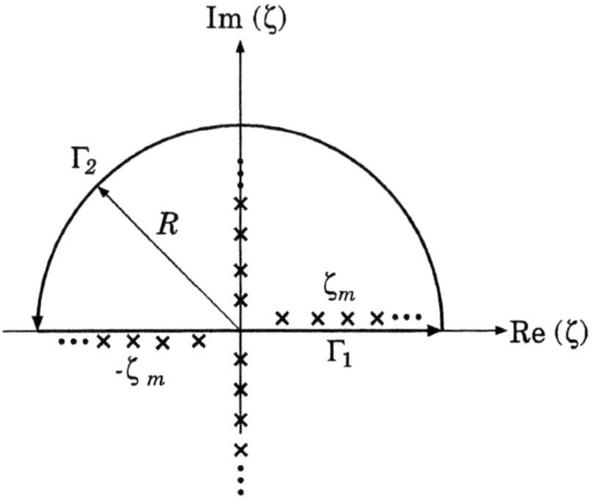

Fig. 9.4. Complex ζ-plane with simple poles.

For $y - y' < 0$, $A_z(x, y; x', y')$ is also shown to be

$$A_z(x, y; x', y') = \sum_{m=1}^{\infty} \frac{i\mu J}{\zeta_m a} \sin(\kappa_m x) \sin(\kappa_m x') \exp\left[-i\zeta_m (y - y')\right] \ . \quad (9.53)$$

9.2.2 Radiation in Shorted Parallel-Plate Waveguide

Let us consider radiation when a perfect conducting plane is placed at $y = 0$ in a parallel-plate waveguide. The problem geometry is shown in Fig. 9.5. The Helmholtz equation that governs radiation from $\overline{J}(x', y') = \hat{z} J \delta(x - x')\delta(y - y')$ is

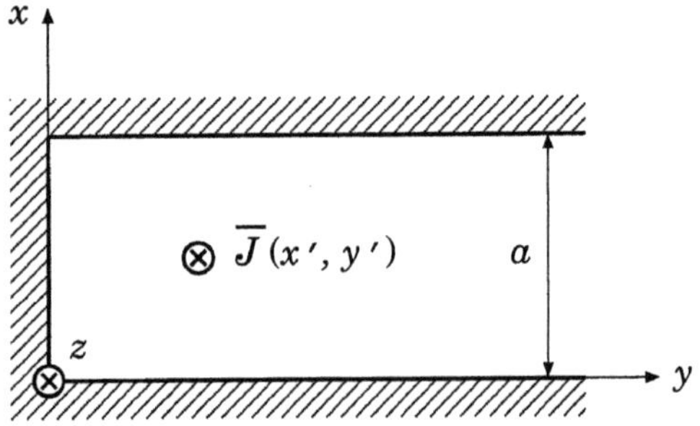

Fig. 9.5. Line current $\overline{J}$ in shorted parallel-plate waveguide.

$$\left(\nabla^2 + k^2\right) \overline{A}(x, y; x', y') = -\mu \overline{J}(x', y') . \tag{9.54}$$

The boundary conditions require

$$A_z(x, y; x', y')\Big|_{x=0,a} = A_z(x, y; x', y')\Big|_{y=0} = 0 . \tag{9.55}$$

Approach 1

Following the previous procedure, the solution can be written as

$$A_z(x, y; x', y') = \sum_{m=1}^{\infty} a_m(y; x', y')\psi_m(x) \tag{9.56}$$

where $\psi_m(x) = \sqrt{\frac{2}{a}} \sin \frac{m\pi}{a} x$. Then, $a_m(y; x', y')$ must satisfy $a_m(0; x', y') = 0$ and for $y \geq 0$

$$\left[\frac{d^2}{dy^2} - \left(\frac{m\pi}{a}\right)^2 + k^2\right] a_m(y; x', y') = -\mu J \psi_m(x')\delta(y - y') . \tag{9.57}$$

The solution to (9.57) is the one-dimensional half-space Green's function that satisfies the radiation condition at $y = \infty$

$$a_m(y; x', y') = \frac{\mu J \psi_m(x')}{\zeta_m} \begin{cases} \sin \zeta_m y' e^{i\zeta_m y} & \text{for } y' < y \\ e^{i\zeta_m y'} \sin \zeta_m y & \text{for } 0 \leq y < y' \end{cases} \tag{9.58}$$

where $\zeta_m = \sqrt{k^2 - \left(\frac{m\pi}{a}\right)^2}$. Hence, the vector potential and electric field are

$$A_z(x, y; x', y') = \sum_{m=1}^{\infty} \frac{\mu J}{\zeta_m} \psi_m(x)\psi_m(x') \cdot \begin{cases} \sin \zeta_m y' e^{i\zeta_m y} & \text{for } y' < y \\ e^{i\zeta_m y'} \sin \zeta_m y & \text{for } 0 \leq y < y' \end{cases} \tag{9.59}$$

$$\overline{E} = \hat{z} E_z = \hat{z} i\omega A_z(x, y; x', y') . \tag{9.60}$$

Approach 2

It is possible to transform the original problem in Fig. 9.5 into an equivalent one, as shown in Fig. 9.6, based on the image method. The total field for $y > 0$ is the sum of two contributions, the real current $\overline{J}(x', y')$ and the image current $-\overline{J}(x', -y')$. The total vector potential due to the real and image currents is

$$A_z(x,y;x',y') = \sum_{m=1}^{\infty} \frac{i\mu J}{2\zeta_m} \psi_m(x)\psi_m(x') \left(e^{i\zeta_m|y-y'|} - e^{i\zeta_m|y+y'|}\right) \quad (9.61)$$

which reduces to (9.59).

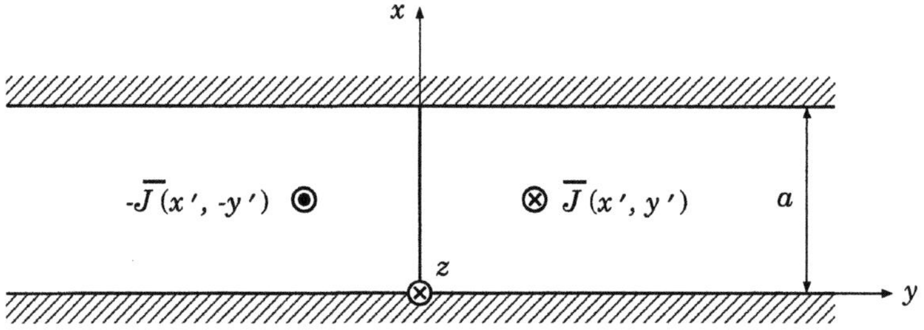

Fig. 9.6. Equivalent problem based on image method.

9.2.3 Radiation in Rectangular Waveguide

Let us consider radiation from a line current in a rectangular waveguide, as shown in Fig. 9.7. The Helmholtz equation

$$\left(\frac{\partial^2}{\partial x^2} + \frac{\partial^2}{\partial y^2} + k^2\right) A_z(x,y;x',y') = -\mu J \delta(x-x')\delta(y-y') \quad (9.62)$$

is subject to the boundary conditions

$$A_z(x,y;x',y')\Big|_{x=0,a} = A_z(x,y;x',y')\Big|_{y=0,b} = 0 \,. \quad (9.63)$$

Following the previous procedure, the solution is given by

$$A_z(x,y;x',y') = \sum_{m=1}^{\infty} a_m(y;x',y')\psi_m(x) \quad (9.64)$$

where $\psi_m(x) = \sqrt{\frac{2}{a}} \sin \frac{m\pi}{a} x$ and $a_m(y;x',y')$ satisfies

$$\left[\frac{d^2}{dy^2} - \left(\frac{m\pi}{a}\right)^2 + k^2\right] a_m(y;x',y') = -\mu J \psi_m(x')\delta(y-y') \,. \quad (9.65)$$

The solution $a_m(y;x',y')$ is known as the one-dimensional closed-space Green's function

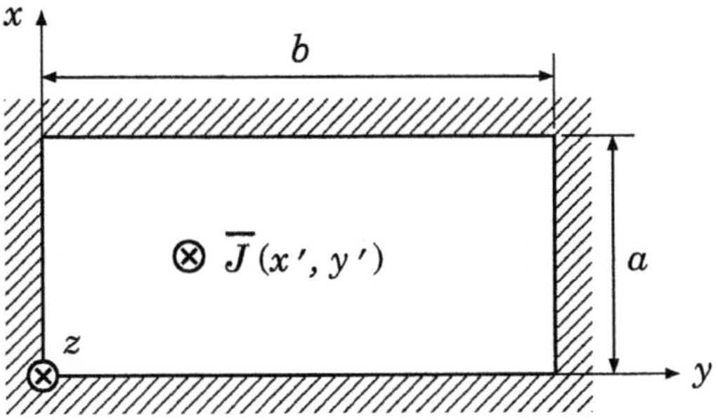

Fig. 9.7. Line current $\overline{J}$ in rectangular waveguide.

$$a_m(y; x', y') = -\frac{\mu J \psi_m(x')}{\zeta_m \sin \zeta_m b} \cdot \begin{cases} \sin \zeta_m(y' - b) \sin \zeta_m y & \text{for } 0 \le y < y' \\ \sin \zeta_m y' \sin \zeta_m (y - b) & \text{for } y' < y \le b \end{cases} \tag{9.66}$$

where $\zeta_m = \sqrt{k^2 - \left(\frac{m\pi}{a}\right)^2}$. Hence

$$A_z(x, y; x', y') = \sum_{m=1}^{\infty} -\frac{\mu J}{\zeta_m \sin \zeta_m b} \psi_m(x) \psi_m(x') \cdot \begin{cases} \sin \zeta_m(y' - b) \sin \zeta_m y & \text{for } 0 \le y < y' \\ \sin \zeta_m y' \sin \zeta_m (y - b) & \text{for } y' < y \le b\,. \end{cases} \tag{9.67}$$

The electric field is

$$\overline{E} = \hat{z} E_z = \hat{z} i\omega A_z(x, y; x', y')\,. \tag{9.68}$$

9.3 Line Current in Circular Waveguide

Consider radiation from an electric line current at (ρ', ϕ') inside a perfectly conducting circular waveguide. The geometry of the radiating problem is shown in Fig. 9.8. No field variation is assumed with respect to z $\left(\frac{\partial}{\partial z} = 0\right)$. The Helmholtz equation with a delta source of $\overline{J} = \hat{z}\frac{J}{\rho}\delta(\rho - \rho')\delta(\phi - \phi')$ is written as

$$\left(\nabla^2 + k^2\right) A_z(\rho, \phi; \rho', \phi') = -\frac{\mu J}{\rho}\delta(\rho - \rho')\delta(\phi - \phi') \tag{9.69}$$

where $k \left(= \omega\sqrt{\mu\epsilon}\right)$ is the wavenumber. Equation (9.69) is explicitly rewritten as

$$\left[\frac{1}{\rho}\frac{\partial}{\partial\rho}\left(\rho\frac{\partial}{\partial\rho}\right) + \frac{1}{\rho^2}\frac{\partial^2}{\partial\phi^2} + k^2\right] A_z(\rho, \phi; \rho', \phi')$$
$$= -\frac{\mu J}{\rho}\delta(\rho - \rho')\delta(\phi - \phi') \, . \tag{9.70}$$

This section illustrates two different approaches to obtain $A_z(\rho, \phi; \rho', \phi')$.

9.3.1 Approach 1

The solution for the problem is derived in [13] using eigenfunction expansions. Let us represent $\delta(\phi - \phi')$ and $A_z(\rho, \phi; \rho', \phi')$ in terms of eigenfunctions. Expanding $\delta(\phi - \phi')$ and $A_z(\rho, \phi; \rho', \phi')$ in Fourier series yields

$$\delta(\phi - \phi') = \sum_{m=-\infty}^{\infty} \frac{1}{2\pi} e^{im(\phi - \phi')} \tag{9.71}$$

$$A_z(\rho, \phi; \rho', \phi') = \sum_{m=-\infty}^{\infty} \mu J A_z^m(\rho; \rho') e^{im(\phi - \phi')} \, . \tag{9.72}$$

Substituting $\delta(\phi - \phi')$ and $A_z(\rho, \phi; \rho', \phi')$ into (9.70) gives

$$\left[\frac{1}{\rho}\frac{d}{d\rho}\left(\rho\frac{d}{d\rho}\right) - \frac{m^2}{\rho^2} + k^2\right] A_z^m(\rho; \rho') = -\frac{\delta(\rho - \rho')}{2\pi\rho} \, . \tag{9.73}$$

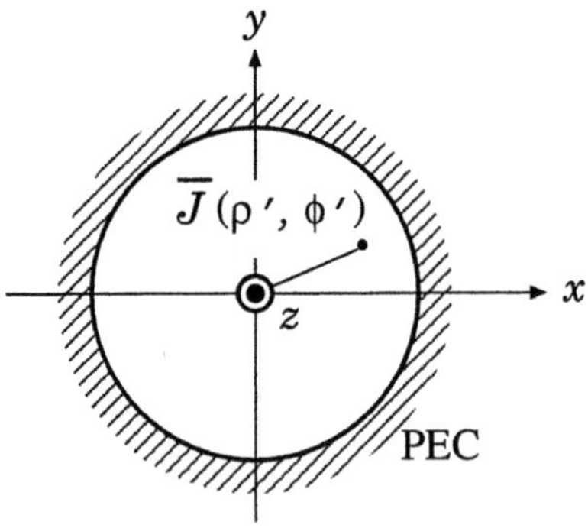

Fig. 9.8. Line current $\overline{J}$ in PEC circular waveguide.

When $\rho \neq \rho'$, (9.73) is a Bessel's equation.

The boundary condition requires that the tangential electric field ($E_z = i\omega A_z$) must vanish at $\rho = a$. It is assumed that $A_z^m(\rho;\rho')$ is given by

$$A_z^m(\rho;\rho') = \begin{cases} A_m J_m(k\rho) & \text{for } \rho < \rho' \\ B_m \Omega_m(k\rho) & \text{for } \rho > \rho' \end{cases} \tag{9.74}$$

where

$$\Omega_m(k\rho) = J_m(k\rho)N_m(ka) - N_m(k\rho)J_m(ka) \,. \tag{9.75}$$

The choice (9.74) automatically satisfies the boundary condition at $\rho = a$ since $\Omega_m(ka) = 0$. It is necessary to impose another boundary condition near $\rho = \rho'$. Multiplying (9.73) by ρ and integrating from $\rho = \rho' - \Delta$ to $\rho = \rho' + \Delta$ yields

$$\rho \frac{dA_z^m(\rho;\rho')}{d\rho}\bigg|_{\rho'-\Delta}^{\rho'+\Delta} + \int_{\rho'-\Delta}^{\rho'+\Delta} \left(-\frac{m^2}{\rho^2} + k^2\right) A_z^m(\rho;\rho')\rho\, d\rho = -\frac{1}{2\pi} \,. \tag{9.76}$$

Since $\dfrac{dA_z^m(\rho;\rho')}{d\rho}$ and $A_z^m(\rho;\rho')$ are discontinuous and continuous, respectively, at $\rho = \rho'$, the boundary conditions are

$$k\rho' \left[-A_m J_m'(k\rho') + B_m \Omega_m'(k\rho')\right] = -\frac{1}{2\pi} \tag{9.77}$$

$$A_m J_m(k\rho') = B_m \Omega_m(k\rho') \,. \tag{9.78}$$

Solving (9.77) and (9.78) for A_m and B_m gives

$$\begin{aligned} B_m &= \frac{J_m(k\rho')}{2\pi k\rho' \left[\Omega_m(k\rho')J_m'(k\rho') - J_m(k\rho')\Omega_m'(k\rho')\right]} \\ &= \frac{J_m(k\rho')}{2\pi k\rho' J_m(ka) \underbrace{\left[J_m(k\rho')N_m'(k\rho') - N_m(k\rho')J_m'(k\rho')\right]}_{\frac{2}{\pi k\rho'}}} \\ &= \frac{J_m(k\rho')}{4J_m(ka)} \end{aligned} \tag{9.79}$$

$$A_m = \frac{\Omega_m(k\rho')}{4J_m(ka)} \,. \tag{9.80}$$

Substituting A_m and B_m into (9.74) gives

$$A_z(\rho,\phi;\rho',\phi') = \frac{\mu J}{4} \sum_{m=-\infty}^{\infty} \begin{cases} \Omega_m(k\rho')\dfrac{J_m(k\rho)}{J_m(ka)}e^{im(\phi-\phi')} & \text{for} \quad \rho < \rho' \\ \Omega_m(k\rho)\dfrac{J_m(k\rho')}{J_m(ka)}e^{im(\phi-\phi')} & \text{for} \quad \rho > \rho' \,. \end{cases} \tag{9.81}$$

The electric field is, therefore

$$\overline{E} = \hat{z}E_z = \hat{z}i\omega A_z(\rho,\phi;\rho',\phi') \,. \tag{9.82}$$

9.3.2 Approach 2

A solution to (9.69) consists of the incident (A_z^i) and scattered (A_z^s) potentials, where the incident term is a response to a delta source in free space and the scattered term is due to the presence of a boundary at $\rho = a$. The incident potential is the two-dimensional free-space Green's function and the scattered potential is given in terms of Bessel functions of the first kind. Hence

$$A_z^i(\rho,\phi;\rho',\phi') = \frac{i\mu J}{4} \sum_{m=-\infty}^{\infty} \begin{cases} H_m^{(1)}(k\rho')J_m(k\rho)e^{im(\phi-\phi')} & \text{for} \quad \rho < \rho' \\ J_m(k\rho')H_m^{(1)}(k\rho)e^{im(\phi-\phi')} & \text{for} \quad \rho > \rho' \end{cases} \tag{9.83}$$

$$A_z^s(\rho,\phi;\rho',\phi') = \frac{i\mu J}{4} \sum_{m=-\infty}^{\infty} A_m J_m(k\rho)e^{im(\phi-\phi')} \,. \tag{9.84}$$

Therefore, the total potential $A_z(\rho,\phi;\rho',\phi')$ is

$$\begin{aligned} A_z&(\rho,\phi;\rho',\phi') \\ &= A_z^i(\rho,\phi;\rho',\phi') + A_z^s(\rho,\phi;\rho',\phi') \\ &= \frac{i\mu J}{4} \sum_{m=-\infty}^{\infty} \end{aligned}$$

$$\begin{cases} \left[H_m^{(1)}(k\rho')J_m(k\rho) + A_m J_m(k\rho)\right] e^{im(\phi-\phi')} & \text{for} \quad \rho < \rho' \\ \left[J_m(k\rho')H_m^{(1)}(k\rho) + A_m J_m(k\rho)\right] e^{im(\phi-\phi')} & \text{for} \quad \rho > \rho' \,. \end{cases} \tag{9.85}$$

To determine the unknown coefficient A_m, the boundary condition must be utilized. The boundary condition at a PEC waveguide surface requires $A_z(\rho,\phi;\rho',\phi')\Big|_{\rho=a} = 0$. Therefore

$$A_m = -\frac{J_m(k\rho')}{J_m(ka)} H_m^{(1)}(ka) \,. \tag{9.86}$$

Since

$$H_m^{(1)}(k\rho')J_m(k\rho) + A_m J_m(k\rho) = -i\frac{J_m(k\rho)\Omega_m(k\rho')}{J_m(ka)} \tag{9.87}$$

$$J_m(k\rho')H_m^{(1)}(k\rho) + A_m J_m(k\rho) = -i\frac{J_m(k\rho')\Omega_m(k\rho)}{J_m(ka)} \tag{9.88}$$

the total potential $A_z(\rho,\phi;\rho',\phi')$ reduces to

$$A_z(\rho,\phi;\rho',\phi') = \frac{\mu J}{4}\sum_{m=-\infty}^{\infty} \begin{cases} \Omega_m(k\rho')\dfrac{J_m(k\rho)}{J_m(ka)}e^{im(\phi-\phi')} & \text{for } \rho<\rho' \\[2ex] \Omega_m(k\rho)\dfrac{J_m(k\rho')}{J_m(ka)}e^{im(\phi-\phi')} & \text{for } \rho>\rho' \,. \end{cases} \tag{9.89}$$

The expression (9.89) agrees with (9.81).

9.4 Sheet Current in Parallel-Plate Waveguide

Let us consider electromagnetic radiation from an electric current sheet $\overline{J}(x,z)$ within a PEC parallel-plate waveguide. The geometry of the radiating problem is shown in Fig. 9.9. The Helmholtz equation that governs radiation from $\overline{J}(x,z) = \hat{z}J(z)\delta(x-x')$ is

$$\left(\nabla^2 + k^2\right) A_z(x,z;x') = -\mu J(z)\delta(x-x') \tag{9.90}$$

where $k\ (=\omega\sqrt{\mu\epsilon})$ is the wavenumber. There is no field variation in the y-direction $\left(\dfrac{\partial}{\partial y} = 0\right)$. Using rectangular coordinates (x,z), it can be rewritten as

$$\left(\frac{\partial^2}{\partial x^2} + \frac{\partial^2}{\partial z^2} + k^2\right) A_z(x,z;x') = -\mu J(z)\delta(x-x') \,. \tag{9.91}$$

The respective field components are given in terms of the magnetic vector

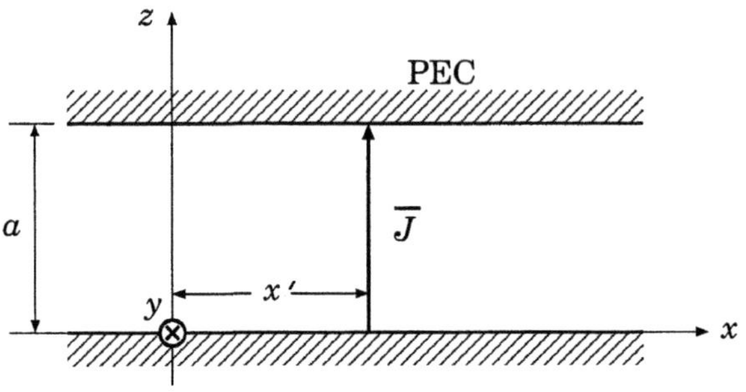

Fig. 9.9. Sheet current $\overline{J}$ in PEC parallel-plate waveguide.

potential as

$$E_x = \frac{i}{\omega\mu\epsilon}\frac{\partial^2 A_z}{\partial x \partial z} \tag{9.92}$$

$$E_y = 0 \tag{9.93}$$

$$E_z = \frac{i}{\omega\mu\epsilon}\left(\frac{\partial^2}{\partial z^2} + k^2\right) A_z \tag{9.94}$$

$$H_x = 0 \tag{9.95}$$

$$H_y = -\frac{1}{\mu}\frac{\partial A_z}{\partial x} \tag{9.96}$$

$$H_z = 0 \,. \tag{9.97}$$

Since $E_x(x, z; x')$ must vanish at $z = 0$ and a, $A_z(x, z; x')$ is represented in terms of the eigenfunction $\psi_m(z) = \sqrt{\frac{2}{a}} \cos\left(\frac{m\pi}{a} z\right)$ as

$$A_z(x, z; x') = \sum_{m=0}^{\infty} a_m(x; x')\psi_m(z) \,. \tag{9.98}$$

Substituting $A_z(x, z; x')$ into (9.91) yields

$$\sum_{m=0}^{\infty} \left[\frac{d^2}{dx^2} - \left(\frac{m\pi}{a}\right)^2 + k^2\right] a_m(x; x')\psi_m(z)$$

$$= -\mu J(z)\delta(x - x') \,. \tag{9.99}$$

Further simplification is possible by utilizing the orthogonality property of $\psi_m(z)$. Multiplying (9.99) by $\psi_n(z)$ and integrating over $(0,a)$ results in

$$\left[\frac{d^2}{dx^2} - \left(\frac{m\pi}{a}\right)^2 + k^2\right] \varepsilon_m a_m(x;x') = -\mu b_m \delta(x-x') \tag{9.100}$$

where $\varepsilon_0 = 2$, $\varepsilon_1 = \varepsilon_2 = \cdots = 1$, and

$$b_m = \int_0^a J(z)\psi_m(z)\,dz\,. \tag{9.101}$$

The solution to (9.100) is the one-dimensional free-space Green's function as

$$\varepsilon_m a_m(x;x') = \frac{i\mu b_m}{2\zeta_m} \exp\left(i\zeta_m|x-x'|\right) \tag{9.102}$$

where $\zeta_m = \sqrt{k^2 - \left(\frac{m\pi}{a}\right)^2}$. The magnetic vector potential inside the waveguide is thus

$$A_z(x,z;x') = \sum_{m=0}^{\infty} \frac{i\mu b_m}{2\zeta_m \varepsilon_m} \psi_m(z) \exp\left(i\zeta_m|x-x'|\right)\,. \tag{9.103}$$

Equation (9.103) is the formal solution to the radiation problem depicted in Fig. 9.9.

Let us also consider the special case of

$$J(z) = J_0 \cos\left(\frac{p\pi}{a}z\right) \tag{9.104}$$

where p is an integer. Substituting $J(z)$ into (9.101) gives

$$b_m = J_0 \sqrt{\frac{a}{2}}\delta_{mp}\varepsilon_m \tag{9.105}$$

where δ_{mp} is the Kronecker delta. The vector potential is

$$A_z(x,z;x') = \frac{i\mu}{2\zeta_p} J_0 \cos\left(\frac{p\pi}{a}z\right) \exp\left(i\zeta_p|x-x'|\right)\,. \tag{9.106}$$

Hence, for instance, the radiation fields for $x > x'$ are given by

$$E_x = \frac{iJ_0}{2\omega\epsilon}\frac{p\pi}{a} \sin\left(\frac{p\pi}{a}z\right) \exp\left[i\zeta_p(x-x')\right] \tag{9.107}$$

$$E_z = -\frac{J_0}{2\omega\epsilon}\zeta_p \cos\left(\frac{p\pi}{a}z\right) \exp\left[i\zeta_p(x-x')\right] \tag{9.108}$$

$$H_y = \frac{J_0}{2} \cos\left(\frac{p\pi}{a}z\right) \exp\left[i\zeta_p(x-x')\right]\,. \tag{9.109}$$

The radiated power delivery along the x-direction for the real ζ_p is

$$\begin{aligned} P_{rad} &= \int_S \frac{1}{2} \text{Re} \left(\overline{E} \times \overline{H}^* \right) \cdot d\overline{s} \\ &= \frac{1}{2} \text{Re} \int_0^a \left(-E_z H_y^* \right) dz \\ &= \frac{|J_0|^2 \zeta_p a \varepsilon_p}{16 \omega \epsilon} . \end{aligned} \quad (9.110)$$

The power supplied by the current source $\overline{J}$ is

$$\begin{aligned} P_i &= -\frac{1}{2} \int_V \overline{E} \cdot \overline{J}^* dv \\ &= -\frac{1}{2} \int_0^a E_z J_0^* \cos\left(\frac{p\pi}{a} z\right) dz \\ &= \frac{|J_0|^2 \zeta_p a \varepsilon_p}{8 \omega \epsilon} . \end{aligned} \quad (9.111)$$

The radiated power delivery along the $\pm x$-direction is $2P_{rad}$, and the power conservation can be checked through the relation $2P_{rad} = P_i$.

Problems for Chapter 9

1. A constant magnetic current sheet $\overline{M} = \hat{z} M \delta(x)$ is placed in free space. The geometry of the radiating problem is shown in Fig. 9.10. Evaluate its radiation field for $x > 0$ and $x < 0$.

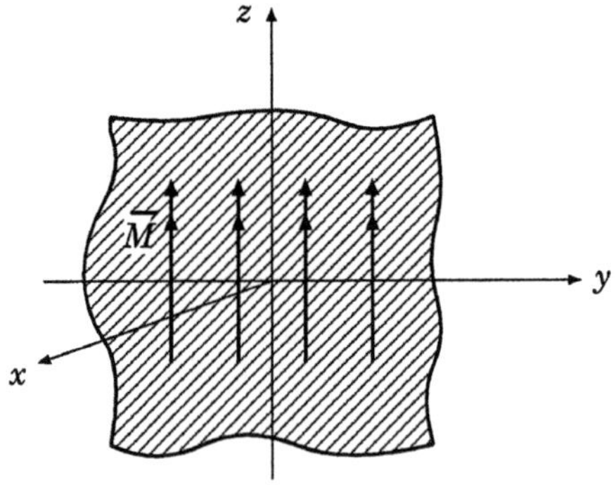

Fig. 9.10. Current sheet in free space.

2. Prove (9.35) using residue calculus.
3. Consider an electric line current $\overline{J}(x', y') = \hat{z}\delta(x - x')\delta(y - y')$ within a PMC parallel-plate waveguide, as shown in Fig. 9.11. Evaluate its radiation field.

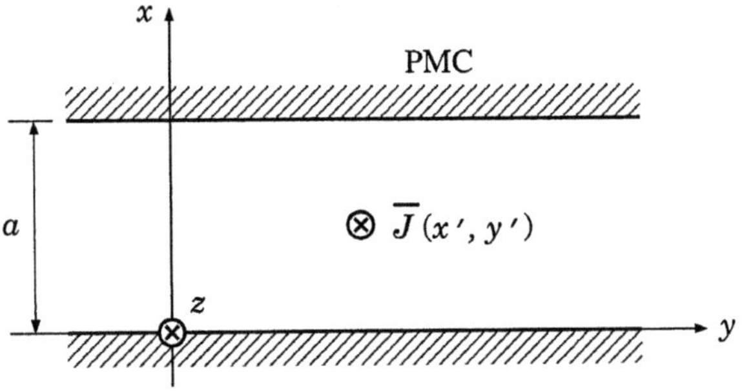

Fig. 9.11. Line current $\overline{J}$ in PMC parallel-plate waveguide

10

Antenna Radiation

10.1 Antenna Fundamentals

Antennas have been used as electromagnetic radiating devices in wireless communications, radars, and nondestructive testing. They are regarded as energy transducers that transform an electric current into an electromagnetic wave for the transmitting antenna and vice versa for the receiving antenna. The macroscopic relation between an electromagnetic wave and an electric current can be described in terms of Maxwell's equations. Thus, to understand the characteristics of the radiation from antennas, Maxwell's equations must be solved. Let us first consider the radiation from an antenna that is modeled as the current sources $\overline{J}$ and $\overline{M}$ in free space. This is a common approach to model antennas. The problem geometry is shown in Fig. 10.1. The radiation

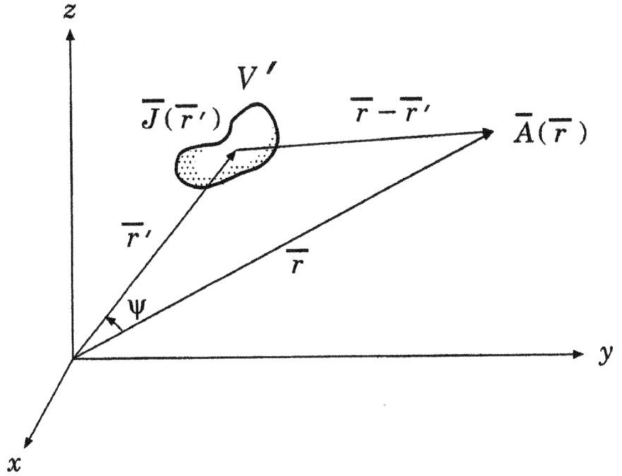

Fig. 10.1. Vector potential and current density in free space.

field is governed by the Helmholtz equations for the vector potentials $\overline{A}$ and $\overline{F}$ as

$$(\nabla^2 + k^2)\,\overline{A} = -\mu\overline{J} \tag{10.1}$$

$$(\nabla^2 + k^2)\,\overline{F} = -\epsilon\overline{M} \tag{10.2}$$

where $k \left(= \omega\sqrt{\mu\epsilon}\right)$ is the wavenumber of the medium. The free-space solutions in three dimensions are

$$\overline{A}(\overline{r}) = \frac{\mu}{4\pi}\int_{V'} \overline{J}(\overline{r}')\frac{e^{ik|\overline{r}-\overline{r}'|}}{|\,\overline{r}-\overline{r}'\,|}\,dv' \tag{10.3}$$

$$\overline{F}(\overline{r}) = \frac{\epsilon}{4\pi}\int_{V'} \overline{M}(\overline{r}')\frac{e^{ik|\overline{r}-\overline{r}'|}}{|\,\overline{r}-\overline{r}'\,|}\,dv' \tag{10.4}$$

where $\overline{r}$ and $\overline{r}'$ designate the observation and source points, respectively.

Let us consider radiation in a lossless medium where μ and ϵ are real. It is of practical interest to evaluate the radiation field in the far zone $kr \gg 1$. The field in the far zone is called the far field and decays as $\frac{1}{r}$. Let us consider the far-zone approximation using $r \gg r'$

$$\begin{aligned}|\,\overline{r}-\overline{r}'\,| &= \sqrt{(\overline{r}-\overline{r}')\cdot(\overline{r}-\overline{r}')}\\ &\approx \sqrt{\underbrace{\overline{r}\cdot\overline{r}}_{r^2} - \underbrace{2\overline{r}\cdot\overline{r}'}_{2rr'\cos\psi}}\\ &\approx r - r'\cos\psi\,.\end{aligned} \tag{10.5}$$

It is convenient to represent the far field using spherical coordinates (r,θ,ϕ). Substituting (10.5) into $\overline{A}(\overline{r})$ and $\overline{F}(\overline{r})$ gives

$$\begin{aligned}\overline{A}(\overline{r}) &\approx \frac{\mu e^{ikr}}{4\pi r}\underbrace{\int_{V'} \overline{J}(\overline{r}')e^{-ikr'\cos\psi}\,dv'}_{\text{function of }\theta\text{ and }\phi}\\ &\equiv \hat{r}A_r + \hat{\theta}A_\theta + \hat{\phi}A_\phi\end{aligned} \tag{10.6}$$

$$\begin{aligned}\overline{F}(\overline{r}) &\approx \frac{\epsilon e^{ikr}}{4\pi r}\underbrace{\int_{V'} \overline{M}(\overline{r}')e^{-ikr'\cos\psi}\,dv'}_{\text{function of }\theta\text{ and }\phi}\\ &\equiv \hat{r}F_r + \hat{\theta}F_\theta + \hat{\phi}F_\phi\,.\end{aligned} \tag{10.7}$$

Consider the magnetic field due to $\overline{J}(\overline{r}')$

$$\overline{H}_e = \frac{1}{\mu}\nabla \times \overline{A}$$

$$= \frac{1}{\mu}\left\{ \frac{\hat{r}}{r\sin\theta}\left[\frac{\partial(A_\phi \sin\theta)}{\partial\theta} - \frac{\partial A_\theta}{\partial\phi}\right] + \frac{\hat{\theta}}{r}\left[\frac{1}{\sin\theta}\frac{\partial A_r}{\partial\phi} - \frac{\partial(rA_\phi)}{\partial r}\right] + \frac{\hat{\phi}}{r}\left[\frac{\partial(rA_\theta)}{\partial r} - \frac{\partial A_r}{\partial\theta}\right]\right\} . \tag{10.8}$$

Since

$$\frac{\hat{\theta}}{r}\frac{\partial(rA_\phi)}{\partial r} \approx \hat{\theta}ikA_\phi \tag{10.9}$$

$$\frac{\hat{\phi}}{r}\frac{\partial(rA_\theta)}{\partial r} \approx \hat{\phi}ikA_\theta \tag{10.10}$$

the far field $\overline{H}_e$ can be obtained by collecting the $\frac{1}{r}$ terms from (10.8). The result is

$$\overline{H}_e = \frac{ik}{\mu}\left(-\hat{\theta}A_\phi + \hat{\phi}A_\theta\right) . \tag{10.11}$$

Note

$$\overline{E}_e = i\omega\overline{A} + \frac{i}{\omega\mu\epsilon}\nabla\left(\nabla\cdot\overline{A}\right) . \tag{10.12}$$

Since $\frac{i}{\omega\mu\epsilon}\nabla\left(\nabla\cdot\overline{A}\right) = -\hat{r}i\omega A_r$ in the far zone, the far field $\overline{E}_e$ is shown to be

$$\overline{E}_e = i\omega\left(\hat{\theta}A_\theta + \hat{\phi}A_\phi\right) . \tag{10.13}$$

Note that the far fields $\overline{E}_e$ and $\overline{H}_e$ satisfy the relation

$$\overline{H}_e = \frac{\hat{r}}{\eta}\times\overline{E}_e \tag{10.14}$$

where $\eta = \sqrt{\frac{\mu}{\epsilon}}$ is the intrinsic impedance of the medium. The far fields $\overline{E}_e$ and $\overline{H}_e$ only have $\hat{\theta}$ and $\hat{\phi}$ components in spherical coordinates, thereby constituting a TEM wave that propagates in the $\hat{r}$-direction. Similarly, the far fields $\overline{E}_m$ and $\overline{H}_m$ due to $\overline{M}(\overline{r}')$ are

$$\overline{E}_m = -\frac{ik}{\epsilon}\left(-\hat{\theta}F_\phi + \hat{\phi}F_\theta\right) \tag{10.15}$$

$$\overline{H}_m = i\omega\left(\hat{\theta}F_\theta + \hat{\phi}F_\phi\right)$$

$$= \frac{\hat{r}}{\eta} \times \overline{E}_m \,. \tag{10.16}$$

Hence, the total far fields can be obtained from

$$\overline{E} = \overline{E}_e + \overline{E}_m$$

$$= \underbrace{\hat{\theta}\, i\omega\left(A_\theta + \eta F_\phi\right)}_{E_\theta} + \underbrace{\hat{\phi}\, i\omega\left(A_\phi - \eta F_\theta\right)}_{E_\phi} \tag{10.17}$$

$$\overline{H} = \overline{H}_e + \overline{H}_m$$

$$= \underbrace{\hat{\theta}\, i\omega\left(F_\theta - \frac{A_\phi}{\eta}\right)}_{-\frac{E_\phi}{\eta}} + \underbrace{\hat{\phi}\, i\omega\left(F_\phi + \frac{A_\theta}{\eta}\right)}_{\frac{E_\theta}{\eta}} \,. \tag{10.18}$$

The total time-average radiated power through a surface S, which is a sphere of radius r, is

$$P_{rad} = \oint_S \frac{1}{2}\,\mathrm{Re}\,\left(\overline{E} \times \overline{H}^*\right) \cdot d\overline{s}$$

$$= \int_0^{2\pi}\int_0^{\pi} \frac{1}{2\eta}\left(|E_\theta|^2 + |E_\phi|^2\right) r^2 \sin\theta \, d\theta \, d\phi \,. \tag{10.19}$$

10.2 Wire Antennas

Wire antennas can be made from conducting wires. Simple wire antennas that are commonly used in radio communication systems include dipole, monopole, and loop antennas. This section investigates the radiation characteristics of dipole and loop antennas.

10.2.1 Dipole Antenna

Consider the problem of radiation from a center-fed thin, linear dipole antenna that carries an electric current density $\overline{J}(\overline{r}')$, as shown in Fig. 10.2. A reasonable assumption for the electric current density is given by

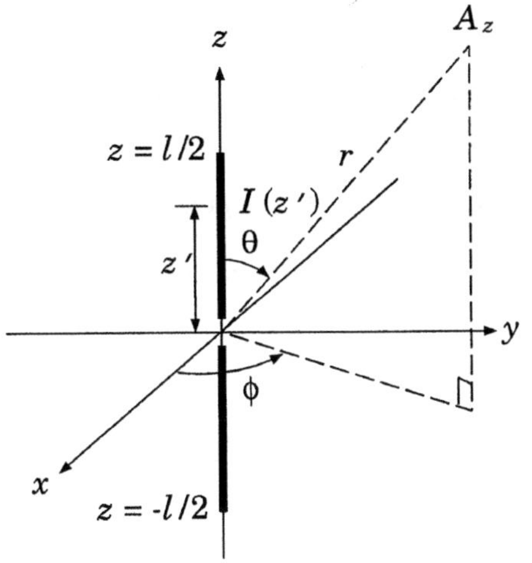

Fig. 10.2. Dipole antenna of length l in free space.

$$\overline{J}(\overline{r}') = \underbrace{\hat{z}\, I_0 \sin\left[k\left(\frac{l}{2} - |z'|\right)\right]}_{I(z')} \delta(x')\delta(y') \tag{10.20}$$

where $k \left(= \omega\sqrt{\mu\epsilon}\right)$ is the wavenumber. The current distribution $I(z')$ approximately assumes a sinusoidal variation that requires the condition $I\left(z' = \pm\frac{l}{2}\right) = 0$. Substituting $\overline{J}(\overline{r}')$ into

$$\overline{A}(\overline{r}) = \frac{\mu e^{ikr}}{4\pi r}\int_{V'} \overline{J}(\overline{r}') e^{-ikr'\cos\psi}\, dv' \tag{10.21}$$

yields

$$A_z = \mu\frac{e^{ikr} I_0}{4\pi r}\left\{\int_{-\frac{l}{2}}^{0} \sin\left[k\left(\frac{l}{2} + z'\right)\right] e^{ikz'\cos(\pi-\theta)} dz' + \int_0^{\frac{l}{2}} \sin\left[k\left(\frac{l}{2} - z'\right)\right] e^{-ikz'\cos\theta}\, dz'\right\}. \tag{10.22}$$

Integrating with the formula

$$\int e^{\alpha u}\sin(\beta u + \gamma)\, du = \frac{e^{\alpha u}\left[\alpha\sin(\beta u + \gamma) - \beta\cos(\beta u + \gamma)\right]}{\alpha^2 + \beta^2} \tag{10.23}$$

gives the result

$$A_z = \mu \frac{I_0 e^{ikr}}{2\pi kr} \left[\frac{\cos\left(\frac{kl}{2}\cos\theta\right) - \cos\left(\frac{kl}{2}\right)}{\sin^2\theta} \right] . \tag{10.24}$$

Since $\hat{z} = \hat{r}\cos\theta - \hat{\theta}\sin\theta$, the far field is given by

$$E_\theta = -i\omega A_z \sin\theta \tag{10.25}$$

$$H_\phi = \frac{E_\theta}{\eta} \tag{10.26}$$

where $\eta = \sqrt{\frac{\mu}{\epsilon}}$ is the intrinsic impedance of the medium. The total radiated power is given by

$$\begin{aligned} P_{rad} &= \int_0^{2\pi} \int_0^{\pi} \frac{1}{2\eta} |E_\theta|^2 r^2 \sin\theta \, d\theta \, d\phi \\ &= \frac{\eta I_0^2}{4\pi} \int_0^{\pi} \frac{\left[\cos\left(\frac{kl}{2}\cos\theta\right) - \cos\left(\frac{kl}{2}\right)\right]^2}{\sin\theta} d\theta . \end{aligned} \tag{10.27}$$

Integration must be performed to further investigate the radiated power characteristics.

Let us consider two special cases, namely, a half-wave dipole $\left(l = \frac{\lambda}{2}\right)$ and a small dipole. For the half-wave dipole antenna, the far-zone electric field is

$$E_\theta = -i\eta \frac{I_0 e^{ikr}}{2\pi r} \frac{\cos\left(\frac{\pi}{2}\cos\theta\right)}{\sin\theta} \tag{10.28}$$

where the maximum radiation occurs at $\theta = 90°$.

For the small dipole antenna ($kl \ll 1$), the far-zone electric field is

$$E_\theta = -i\eta \frac{I_0 e^{ikr}}{4\pi r} \left(\frac{kl}{2}\right)^2 \sin\theta . \tag{10.29}$$

10.2.2 Circular Loop Antenna

Consider the problem of an infinitesimally thin circular antenna of radius a that carries a constant current I_0. The current density using cylindrical coordinates is

$$\overline{J}(\overline{r}') = \hat{\phi}' I_0 \delta(\rho' - a)\delta(z') . \tag{10.30}$$

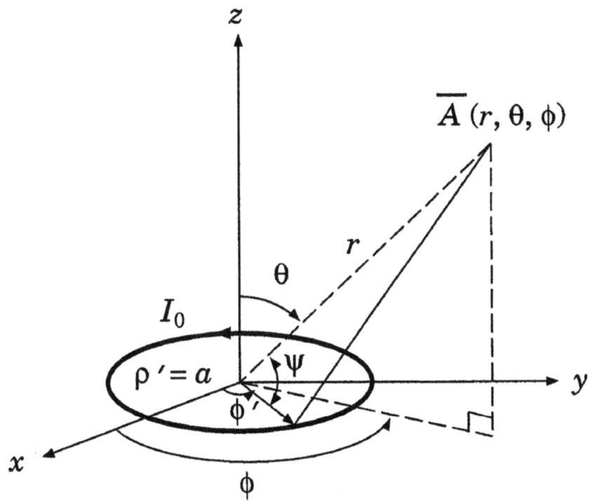

Fig. 10.3. Loop antenna of radius a in free space.

The geometry of the radiating problem is shown in Fig. 10.3. The magnetic vector potential $\overline{A}(\overline{r})$ is given by

$$\overline{A}(\overline{r}) = \frac{\mu e^{ikr}}{4\pi r} \int_{V'} \overline{J}(\overline{r}')e^{-ikr'\cos\psi}\,dv' \,. \tag{10.31}$$

The phase term $r'\cos\psi$ is shown to be

$$\begin{aligned} r'\cos\psi &= \overline{r'}\cdot\hat{r} \\ &= x'\sin\theta\cos\phi + y'\sin\theta\sin\phi \\ &= a\sin\theta\cos(\phi-\phi') \,. \end{aligned} \tag{10.32}$$

Based on the vector identity

$$\hat{\phi}' = \hat{r}\sin\theta\sin(\phi-\phi') + \hat{\theta}\cos\theta\sin(\phi-\phi') + \hat{\phi}\cos(\phi-\phi') \tag{10.33}$$

the $\hat{\phi}$ component of $\overline{A}$ is given by

$$A_\phi = \frac{\mu e^{ikr}}{4\pi r} \int_0^{2\pi} I_0\cos(\phi-\phi')e^{-ika\sin\theta\cos(\phi-\phi')}a\,d\phi' \,. \tag{10.34}$$

The generating function for the Bessel function gives

$$e^{i\xi\cos\psi} = J_0(\xi) + 2\sum_{n=1}^{\infty} i^n J_n(\xi)\cos n\psi \,. \tag{10.35}$$

Substituting the phase term

$$e^{-ika\sin\theta\cos(\phi-\phi')} = J_0(-ka\sin\theta) + 2\sum_{n=1}^{\infty} i^n J_n(-ka\sin\theta)\cos n(\phi-\phi') \qquad (10.36)$$

into (10.34) gives

$$A_\phi = -i\frac{a\mu I_0 e^{ikr}}{2r} J_1(ka\sin\theta)\ . \qquad (10.37)$$

Similarly, the $\hat{\theta}$ and $\hat{r}$ components of $\overline{A}$ are shown to be zero, $A_\theta = A_r = 0$. Therefore, the far-zone electric field is

$$\overline{E} = i\omega\overline{A} = \hat{\phi}\frac{a\omega\mu I_0 e^{ikr}}{2r} J_1(ka\sin\theta)\ . \qquad (10.38)$$

Let us consider an infinitesimally small antenna ($ka \ll 1$) where

$$J_1(ka\sin\theta) \approx \frac{ka\sin\theta}{2}\ . \qquad (10.39)$$

The far field is

$$\overline{E} = \hat{\phi}\frac{e^{ikr}}{4r}(ka)^2 I_0 \eta \sin\theta \qquad (10.40)$$

where $\eta = \sqrt{\frac{\mu}{\epsilon}}$ is the intrinsic impedance of the medium. The maximum radiation occurs at $\theta = 90°$ and the total radiated power is given by

$$P_{rad} = \int_0^{2\pi}\int_0^{\pi} \frac{1}{2\eta}|E_\phi|^2 r^2 \sin\theta\, d\theta\, d\phi = \frac{\pi}{12}\eta(ka)^4 I_0^2\ . \qquad (10.41)$$

10.3 Aperture Antenna

Apertures in a conducting plane are the basic radiating elements for aperture antennas. This section investigates the radiation from a two-dimensional aperture (slit) on a ground plane. Consider the problem of radiation from an open-ended parallel-plate waveguide with an infinite conducting flange, as shown in Fig. 10.4. A rigorous theory, based on a Fourier transform and mode matching, is presented in [14]. Assume that an incident TE wave $E_y^i(x,z)$

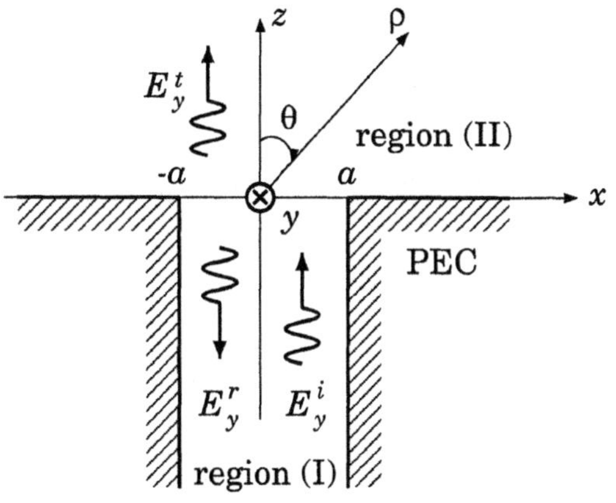

Fig. 10.4. Aperture antenna on ground plane.

propagates upward along a parallel-plate waveguide. The field has no variation with respect to y $\left(\frac{\partial}{\partial y}=0\right)$. Let the transmitted ($z \geq 0$) and reflected ($z \leq 0$) waves be denoted by $E_y^t(x,z)$ and $E_y^r(x,z)$, respectively. The fields $E_y^{i,r}(x,z)$ and $E_y^t(x,z)$ should then satisfy the corresponding Helmholtz equations

$$\left(\frac{\partial^2}{\partial x^2}+\frac{\partial^2}{\partial z^2}+k_0^2\right)E_y^{i,r}(x,z)=0 \tag{10.42}$$

$$\left(\frac{\partial^2}{\partial x^2}+\frac{\partial^2}{\partial z^2}+k^2\right)E_y^{t}(x,z)=0 \tag{10.43}$$

where the wavenumbers in regions (I) ($z \leq 0$) and (II) ($z \geq 0$) are k_0 $\left(=\omega\sqrt{\mu_0\epsilon_0}\right)$ and k $\left(=\omega\sqrt{\mu\epsilon}\right)$, respectively.

The incident and reflected electric fields in region (I) are

$$E_y^i(x,z)=\sin a_1(x+a)\exp(i\beta_1 z) \tag{10.44}$$

$$E_y^r(x,z)=\sum_{m=1}^{\infty}A_m\sin a_m(x+a)\exp(-i\beta_m z) \tag{10.45}$$

where $a_m=\dfrac{m\pi}{2a}$ and $\beta_m=\sqrt{k_0^2-a_m^2}$.

From Faraday's law, the magnetic field is

$$H_x^{i,r}(x,z)=\frac{i}{\omega\mu_0}\frac{\partial E_y^{i,r}(x,z)}{\partial z}\,. \tag{10.46}$$

Hence

$$H_x^i(x,z) = -\frac{\beta_1}{\omega\mu_0}\sin a_1(x+a)\exp(i\beta_1 z) \tag{10.47}$$

$$H_x^r(x,z) = \sum_{m=1}^{\infty}\frac{\beta_m}{\omega\mu_0}A_m \sin a_m(x+a)\exp(-i\beta_m z)\ . \tag{10.48}$$

The fields in region (II) are assumed to be

$$E_y^t(x,z) = \int_{-\infty}^{\infty}\tilde{E}^t(\zeta)e^{i\kappa z}e^{-i\zeta x}\,d\zeta \tag{10.49}$$

$$H_x^t(x,z) = -\int_{-\infty}^{\infty}\frac{\kappa}{\omega\mu}\tilde{E}^t(\zeta)e^{i\kappa z}e^{-i\zeta x}\,d\zeta \tag{10.50}$$

where $\kappa = \sqrt{k^2-\zeta^2}$.

To determine the unknown coefficients, A_m and $\tilde{E}^t(\zeta)$, the continuities of the tangential electric and magnetic fields must be enforced at $z=0$. The continuity of the tangential electric fields yields

$$E_y^t(x,0) = \begin{cases} E_y^i(x,0)+E_y^r(x,0) & \text{for } |x|<a \\ 0 & \text{otherwise}\ . \end{cases} \tag{10.51}$$

Applying the inverse Fourier transform $\left[\frac{1}{2\pi}\int_{-\infty}^{\infty}(\cdot)e^{i\zeta x}\,dx\right]$ to (10.51) yields

$$\begin{aligned}\tilde{E}^t(\zeta) &= \frac{1}{2\pi}\sum_{m=1}^{\infty}(A_m+\delta_{m1})\int_{-a}^{a}\sin a_m\,(x+a)\,e^{i\zeta x}\,dx \\ &= \frac{1}{2\pi}\sum_{m=1}^{\infty}(A_m+\delta_{m1})a_m a^2 F_m(\zeta a)\end{aligned} \tag{10.52}$$

where δ_{m1} is the Kronecker delta and

$$F_m(u) = \frac{\left[e^{iu}(-1)^m - e^{-iu}\right]}{u^2-\left(\frac{m\pi}{2}\right)^2}\ . \tag{10.53}$$

The continuity of the tangential magnetic field for the aperture $\big(|x|<a$ and $z=0\big)$ gives

$$H_x^i(x,0)+H_x^r(x,0) = H_x^t(x,0)\ . \tag{10.54}$$

The orthogonality of sinusoidal functions can be utilized to obtain simultaneous equations for the modal coefficient A_m. Hence, (10.54) is multiplied by $\sin a_p\,(x+a)$ and integrated over $|x|<a$ $\left[\int_{-a}^{a}(\cdot)\sin a_p\,(x+a)\,dx\right]$ to obtain

$$\frac{1}{\mu_0}(-\delta_{1p} + A_p)\beta_p a = -\frac{1}{2\pi\mu}\sum_{m=1}^{\infty}(\delta_{1m} + A_m)a_p a_m a^2 I(k) \qquad (10.55)$$

where

$$I(k) = \int_{-\infty}^{\infty} a^2 \kappa F_m(\zeta a) F_p(-\zeta a)\, d\zeta\ . \qquad (10.56)$$

An evaluation of $I(k)$ is available in Section 7.4.3. The simultaneous equations (10.55) can be solved for the modal coefficient A_m.

Let us obtain the far-zone transmitted field ($k\rho \gg 1$) based on the saddle-point method [9]. Substituting the relations

$$z = \rho\cos\theta \qquad (10.57)$$

$$x = \rho\sin\theta \qquad (10.58)$$

$$\zeta = k\sin\psi \qquad (10.59)$$

$$\kappa = \sqrt{k^2 - \zeta^2} = k\cos\psi \qquad (10.60)$$

into the transmitted field (10.49) gives

$$E_y^t(\rho,\theta) = \int \tilde{E}^t(\zeta)\exp\Big[ik\rho\underbrace{\cos(\psi+\theta)}_{f(\psi)}\Big]k\cos\psi\, d\psi\ . \qquad (10.61)$$

The condition $f'(\psi_0) = 0$ yields the saddle point $\psi_0 = -\theta$. Hence

$$E_y^t(\rho,\theta) \approx \tilde{E}^t(\zeta)k\cos\psi e^{ik\rho}\sqrt{\frac{2\pi}{ik\rho}}\Bigg|_{\psi=-\theta}$$

$$= \exp\left[i\left(k\rho - \frac{\pi}{4}\right)\right]\sqrt{\frac{k}{2\pi\rho}}\cos\theta$$

$$\cdot\sum_{m=1}^{\infty}(A_m + \delta_{m1})a_m a^2 F_m(-ka\sin\theta)\ . \qquad (10.62)$$

10.4 Groove-Backed Antenna

This section discusses the radiation from a line current that is placed inside a cavity. This type of antenna is called a cavity-backed antenna and has been used to detect underground targets. For simplicity, it is assumed that a cavity is a two-dimensional rectangular groove. This type of antenna is referred to as a groove-backed antenna. Consider a rectangular groove (width $2a$ and depth d) in a perfectly conducting plane, as shown in Fig. 10.5. A z-oriented electric line current is located inside a groove, thus radiating electromagnetic waves

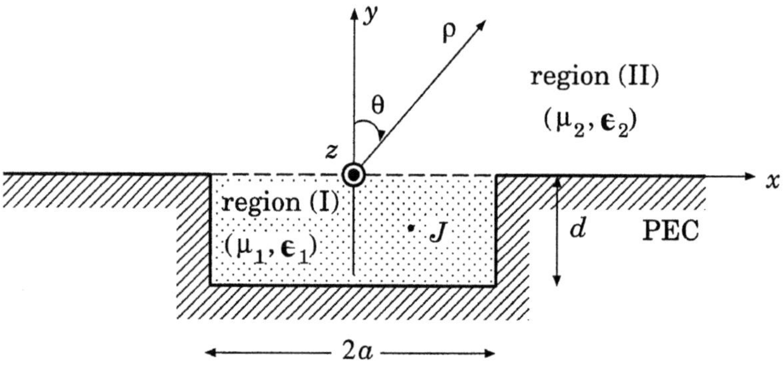

Fig. 10.5. Rectangular groove in perfectly conducting plane.

upward. Regions (I) and (II) denote the groove interior ($-a \leq x \leq a$ and $-d \leq y \leq 0$) and half space ($y \geq 0$) above a conducting plane, respectively. The magnetic vector potential in region (I) satisfies

$$\left(\nabla^2 + k_1^2\right) A_z\left(x, y; x', y'\right) = -\mu_1 J \delta(x - x')\delta(y - y') \tag{10.63}$$

where $\delta(\cdot)$ is the delta function and k_1 $(= \omega\sqrt{\mu_1 \epsilon_1})$ is the wavenumber in region (I). The relation

$$E_z = i\omega A_z \tag{10.64}$$

results in

$$\left(\nabla^2 + k_1^2\right) E_z^I\left(x, y; x', y'\right) = -i\omega\mu_1 J \delta(x - x')\delta(y - y') . \tag{10.65}$$

Note that the radiated electric field only has a z-component and there is no field variation in z $\left(\dfrac{\partial}{\partial z} = 0\right)$. The following are two different approaches to derive the solution to (10.65).

10.4.1 Approach 1

It is convenient to separate the original problem into two parts based on the superposition principle, as shown in Fig. 10.6. The electric field in region (I) consists of the primary and scattered components as

$$E_z^I\left(x, y; x', y'\right) = E_z^p\left(x, y; x', y'\right) + E_z^s\left(x, y; x', y'\right) \tag{10.66}$$

where the primary field $E_z^p\left(x, y; x', y'\right)$ is considered as the Green's function resulting from a line current placed in a PEC rectangular box. An analysis

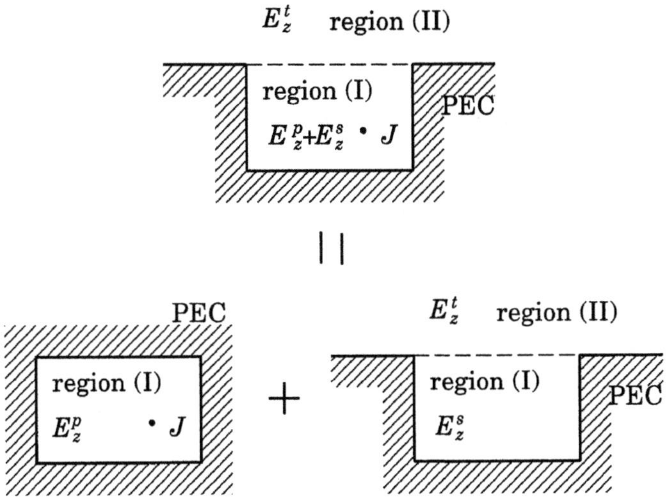

Fig. 10.6. Equivalent problem.

for the primary field $E_z^p(x, y; x', y')$ is available in Section 9.2.3. The primary electric field is represented in terms of the rectangular waveguide modes as

$$\begin{aligned} E_z^p(x, y; x', y') &\\ = -i\omega\mu_1 J \sum_{m=1}^{\infty} \frac{\psi_m(x)\psi_m(x')}{\xi_m \sin \xi_m d} &\\ \cdot \begin{cases} \sin \xi_m y' \sin \xi_m (y+d) & \text{for} \quad -d \le y < y' \\ \sin \xi_m (y'+d) \sin \xi_m y & \text{for} \quad y' < y \le 0 \end{cases} \end{aligned} \tag{10.67}$$

where $a_m = \dfrac{m\pi}{2a}$, $\xi_m = \sqrt{k_1^2 - a_m^2}$, and

$$\psi_m(x) = \frac{1}{\sqrt{a}} \sin a_m(x+a) . \tag{10.68}$$

The scattered field in region (I) must satisfy

$$\left(\nabla^2 + k_1^2\right) E_z^s(x, y; x', y') = 0 . \tag{10.69}$$

The solution $E_z^s(x, y; x', y')$ to the Helmholtz equation takes the form of a series

$$E_z^s(x, y; x', y')$$

$$= -i\omega\mu_1 J \sum_{m=1}^{\infty} \tilde{c}_m \frac{\psi_m(x)\psi_m(x')}{\xi_m \sin \xi_m d} \sin \xi_m y' \sin \xi_m (y+d)$$

$$\equiv \sum_{m=1}^{\infty} c_m \sin a_m(x+a) \sin \xi_m (y+d) \,. \tag{10.70}$$

The total transmitted field in region (II) satisfies

$$\left(\nabla^2 + k_2^2\right) E_z^t(x, y; x', y') = 0 \tag{10.71}$$

where k_2 $(= \omega\sqrt{\mu_2\epsilon_2})$ is the wavenumber in region (II), and the solution is chosen to be

$$E_z^t(x, y; x', y') = \frac{1}{2\pi} \int_{-\infty}^{\infty} \widetilde{E}_z^t(\zeta) e^{-i\zeta x + i\kappa y} \, d\zeta \tag{10.72}$$

where $\kappa = \sqrt{k_2^2 - \zeta^2}$.

Two unknown coefficients, c_m and $\widetilde{E}_z^t(\zeta)$, are introduced to represent the fields, thus requiring two boundary conditions. A set of simultaneous equations for the discrete modal coefficients c_m can be obtained by enforcing the boundary conditions at y=0. The boundary condition for the E_z continuity at $y = 0$ requires

$$E_z^t(x, 0; x', y') = \begin{cases} E_z^I(x, 0; x', y') & \text{for} \quad |x| < a \\ 0 & \text{otherwise} \,. \end{cases} \tag{10.73}$$

Taking the inverse Fourier transform of $E_z^t(x, 0; x', y')$ gives

$$\int_{-\infty}^{\infty} E_z^t(x, 0; x', y') e^{i\zeta x} \, dx = \int_{-a}^{a} E_z^s(x, 0; x', y') e^{i\zeta x} \, dx \,. \tag{10.74}$$

Substituting (10.72) into (10.74) yields

$$\widetilde{E}_z^t(\zeta) = \sum_{m=1}^{\infty} c_m \sin(\xi_m d) a_m a^2 F_m(\zeta a) \tag{10.75}$$

where

$$F_m(u) = \frac{\left[e^{iu}(-1)^m - e^{-iu}\right]}{u^2 - \left(\frac{m\pi}{2}\right)^2} \,. \tag{10.76}$$

Since $H_x = \dfrac{1}{i\omega\mu}\dfrac{\partial E_z}{\partial y}$, the H_x continuity at $y = 0$ is given by

$$\begin{aligned} H_x^t(x,0;x',y') &= H_x^I(x,0;x',y') \\ &= H_x^s(x,0;x',y') + H_x^p(x,0;x',y') \,. \end{aligned} \tag{10.77}$$

Equation (10.77) can be rewritten as

$$\begin{aligned} \int_{-\infty}^{\infty} \frac{i\kappa}{2\pi\mu_2} \widetilde{E}_z^t(\zeta) e^{-i\zeta x}\, d\zeta &= \sum_{m=1}^{\infty} c_m \frac{\xi_m}{\mu_1} \sin a_m(x+a) \cos \xi_m d \\ &- i\omega J \sum_{m=1}^{\infty} \frac{\sin a_m(x+a) \sin a_m(x'+a)}{a \sin \xi_m d} \sin \xi_m(y'+d) \,. \end{aligned} \tag{10.78}$$

Substituting (10.75) into (10.78), multiplying (10.78) by $\sin a_n(x+a)$, and integrating with respect to x from $-a$ to a yields a set of simultaneous equations for the discrete modal coefficient c_m

$$\begin{aligned} &\frac{ia_n}{2\pi\mu_2} \sum_{m=1}^{\infty} c_m a_m \sin(\xi_m d) a^2 I(k) \\ &= \frac{c_n \xi_n a}{\mu_1} \cos \xi_n d - \frac{i\omega J}{\sin \xi_n d} \sin a_n(x'+a) \sin \xi_n(y'+d) \end{aligned} \tag{10.79}$$

where

$$I(k) = \int_{-\infty}^{\infty} a^2 F_m(\zeta a) F_n(-\zeta a) \kappa \, d\zeta \,. \tag{10.80}$$

An evaluation of $I(k)$ is available in Section 7.4.3. Computations must be performed to solve (10.79) for the discrete modal coefficient c_m.

Once c_m is determined, it is possible to evaluate the radiation field $E_z^t(x,y;x',y')$ using (10.72). Let us derive the far-zone transmitted field based on the saddle-point method [9]. Let

$$y = \rho \cos\theta \tag{10.81}$$

$$x = \rho \sin\theta \tag{10.82}$$

$$\zeta = k_2 \sin\psi \tag{10.83}$$

$$\kappa = \sqrt{k_2^2 - \zeta^2} = k_2 \cos\psi \,. \tag{10.84}$$

Consequently, $E_z^t(x,y;x',y')$ is given by

$$\begin{aligned} E_z^t(\rho,\theta) &= \frac{1}{2\pi} \int \tilde{E}_z^t(\zeta) e^{ik_2\rho\cos(\psi+\theta)} k_2 \cos\psi \, d\psi \\ &\approx \frac{1}{2\pi} \tilde{E}_z^t(\zeta) k_2 \cos\psi e^{ik_2\rho} \sqrt{\frac{2\pi}{ik_2\rho}} \Bigg|_{\psi=-\theta} \,. \end{aligned} \tag{10.85}$$

10.4.2 Approach 2

It is also possible to solve the antenna problem in Fig. 10.5 without recourse to the superposition principle and primary field $E_z^p(x, y; x', y')$, as shown in Fig. 10.6. The governing equation in region (I) is

$$\left(\nabla^2 + k_1^2\right) E_z^I(x, y; x', y') = -i\omega\mu_1 J\delta(x - x')\delta(y - y') . \tag{10.86}$$

Let us represent $E_z^I(x, y; x', y')$ in terms of the eigenfunction $\psi_m(x)$ as

$$E_z^I(x, y; x', y') = \sum_{m=1}^{\infty} i\omega\mu_1 J\psi_m(x')\psi_m(x)Y(y; y') \tag{10.87}$$

where $\psi_m(x) = \sqrt{\frac{1}{a}} \sin \frac{m\pi}{2a}(x + a)$. Then, $Y(y; y')$ satisfies

$$\left[\frac{d^2}{dy^2} - \left(\frac{m\pi}{2a}\right)^2 + k_1^2\right] Y(y; y') = -\delta(y - y') . \tag{10.88}$$

A Green's function is chosen to satisfy the boundary condition at $y = -d$ as

$$Y(y; y') = \begin{cases} A \sin \xi_m(y + d) & \text{for} \quad -d \le y < y' \\ B \sin \xi_m(y + \gamma) & \text{for} \quad y' < y \le 0 \end{cases} \tag{10.89}$$

where $\xi_m = \sqrt{k_1^2 - \left(\frac{m\pi}{2a}\right)^2}$ and the unknown coefficients are A, B, and γ.

Let us determine the unknown coefficients using the boundary conditions at $y = y'$. Since $\frac{dY(y; y')}{dy}$ is discontinuous at $y = y'$, the condition

$$\left.\frac{dY(y; y')}{dy}\right|_{y'-\Delta}^{y'+\Delta} = -1 \tag{10.90}$$

yields

$$\xi_m B \cos \xi_m(y' + \gamma) - \xi_m A \cos \xi_m(y' + d) = -1 . \tag{10.91}$$

Also the condition

$$Y(y; y')\Big|_{y'-\Delta}^{y'+\Delta} = 0 \tag{10.92}$$

produces

$$A \sin \xi_m(y' + d) = B \sin \xi_m(y' + \gamma) . \tag{10.93}$$

Solving (10.91) and (10.93) for A and B yields

$$A = -\frac{\sin \xi_m (y' + \gamma)}{\xi_m \sin \xi_m (d - \gamma)} \tag{10.94}$$

$$B = -\frac{\sin \xi_m (y' + d)}{\xi_m \sin \xi_m (d - \gamma)} \,. \tag{10.95}$$

Hence, $E_z^I(x, y; x', y')$ is given as

$$\begin{aligned} &E_z^I(x, y; x', y') \\ &= -i\omega\mu_1 J \sum_{m=1}^{\infty} \frac{\psi_m(x)\,\psi_m(x')}{\xi_m \sin \xi_m (d - \gamma)} \\ &\cdot \begin{cases} \sin \xi_m (y' + \gamma) \sin \xi_m (y + d) & \text{for} \quad -d \le y < y' \\ \sin \xi_m (y' + d) \sin \xi_m (y + \gamma) & \text{for} \quad y' < y \le 0 \,. \end{cases} \end{aligned} \tag{10.96}$$

This is another representation for $E_z^I(x, y; x', y')$ in terms of an unknown γ. It is also possible to show that $E_z^p(x, y; x', y') + E_z^s(x, y; x', y')$ yields (10.96) by substituting

$$\tilde{c}_m = \frac{\sin \xi_m \gamma \sin \xi_m (y' + d)}{\sin \xi_m (d - \gamma) \sin \xi_m y'} \tag{10.97}$$

into (10.70).

10.5 Slit Array Antenna

Antennas with a single radiating element are inherently narrow-band, low-gain, and wide-beamwidth systems. Antennas with a wide band, high gain, and narrow beamwidth are often required in radio communications. To meet these antenna requirements, array antennas consisting of a multiple number of radiating elements have been utilized. This section investigates an array antenna that consists of a multiple-slit parallel-plate waveguide. The slit array antenna considered in this section is relatively easy to analyze and provides a rigorous solution based on the mode matching approach.

10.5.1 Field Analysis

A detailed analysis of the radiation from a multiple-slit parallel-plate waveguide has already been performed in [15] using the technique of a Fourier transform and mode matching. The notations used in this section are somewhat similar to those in [15]. The antenna geometry is shown in Fig. 10.7.

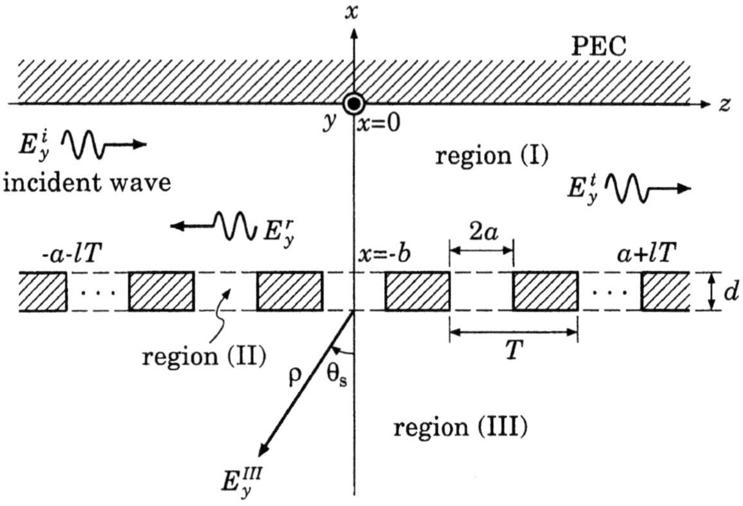

Fig. 10.7. Slit parallel-plate waveguide.

Consider TE wave radiation from an N number of slits of width $2a$ and depth d in a parallel-plate waveguide. Regions (I), (II), and (III) denote the parallel-plate interior ($k_1 = \omega\sqrt{\mu_1\epsilon_1}$), N ($= L_1 + L_2 + 1$) slits ($k_2 = \omega\sqrt{\mu_2\epsilon_2}$), and lower half-space $\left(k_3 = \omega\sqrt{\mu_3\epsilon_3} = \frac{2\pi}{\lambda}\right)$, respectively. Assume that an incident wave $E_y^i(x,z)$ excites region (I). Throughout regions (I), (II), and (III), the electric field has a y-component. There is no field variation with respect to y $\left(\frac{\partial}{\partial y} = 0\right)$. The scattered fields in regions (I), (II), and (III) must satisfy the Helmholtz equations

$$\left(\nabla^2 + k_1^2\right) E_y^I(x,z) = 0 \tag{10.98}$$

$$\left(\nabla^2 + k_2^2\right) E_y^{II}(x,z) = 0 \tag{10.99}$$

$$\left(\nabla^2 + k_3^2\right) E_y^{III}(x,z) = 0\,. \tag{10.100}$$

The total electric field in region (I) has the following incident and scattered components

$$E_y^i(x,z) = \exp(ik_{zs}z)\sin k_{xs}(x+b) \tag{10.101}$$

$$E_y^I(x,z) = \frac{i}{\pi}\int_{-\infty}^{\infty} \tilde{E}_y^I(\zeta)\sin(\kappa_1 x)e^{-i\zeta z}\,d\zeta \tag{10.102}$$

where $0 < s < \frac{k_1 b}{\pi}$ (s: integer), $k_{xs} = \frac{s\pi}{b}$, $k_{zs} = \sqrt{k_1^2 - k_{xs}^2}$, and $\kappa_1 = \sqrt{k_1^2 - \zeta^2}$. The scattered field $E_y^I(x,z)$ is expressed in a Fourier transform representation that satisfies the boundary condition $E_y^I(0,z) = 0$.

In region (II) ($lT - a \le z \le lT + a$, $-d - b \le x \le -b$, and $l = -L_1, \cdots, L_2$), the electric field is represented in terms of the discrete mode summation

$$E_y^{II}(x,z) = \sum_{m=1}^{\infty} \sin a_m(z + a - lT) \cdot \left[b_m^l \cos \xi_m(x+b) + c_m^l \sin \xi_m(x+b)\right] \qquad (10.103)$$

where $a_m = \frac{m\pi}{2a}$ and $\xi_m = \sqrt{k_2^2 - a_m^2}$. Note that the field $E_y^{II}(x,z)$ is chosen to satisfy the boundary condition $E_y^{II}(x, lT \pm a) = 0$.

The total field in region (III) is

$$E_y^{III}(x,z) = \frac{1}{2\pi} \int_{-\infty}^{\infty} \tilde{E}_y^{III}(\zeta) e^{-i\zeta z - i\kappa_3(x+b+d)}\, d\zeta \qquad (10.104)$$

where $\kappa_3 = \sqrt{k_3^2 - \zeta^2}$. The field $E_y^{III}(x,z)$ in region (III) satisfies the radiation condition $E_y^{III}(-\infty, z) \to 0$.

Four unknown modal coefficients–$\tilde{E}_y^I(\zeta)$, b_m^l, c_m^l, and $\tilde{E}_y^{III}(\zeta)$–are used to represent the scattered fields, thereby requiring four boundary conditions for the continuities of the tangential fields–$E_y(-b,z)$, $H_z(-b,z)$, $E_y(-b-d,z)$, and $H_z(-b-d,z)$. The continuity of the tangential electric field at $x = -b$ is given by

$$E_y^I(-b,z) = \begin{cases} E_y^{II}(-b,z) & \text{for } |z - lT| < a \\ 0 & \text{otherwise}. \end{cases} \qquad (10.105)$$

Applying the inverse Fourier transform $\frac{1}{2\pi}\int_{-\infty}^{\infty} (\cdot) e^{i\zeta z}\, dz$ to (10.105) gives

$$\tilde{E}_y^I(\zeta) = \frac{i}{2\sin(\kappa_1 b)} \sum_{l=-L_1}^{L_2} \sum_{m=1}^{\infty} b_m^l a_m e^{i\zeta lT} a^2 F_m(a\zeta) \qquad (10.106)$$

where

$$F_m(u) = \frac{\left[e^{iu}(-1)^m - e^{-iu}\right]}{u^2 - \left(\frac{m\pi}{2}\right)^2}. \qquad (10.107)$$

Since $H_z = -\frac{i}{\omega\mu}\frac{\partial E_y}{\partial x}$, the continuity of the tangential magnetic field for the aperture of slits ($rT - a < z < rT + a, x = -b$, and $r = -L_1, \cdots, L_2$)

$$H_z^i(-b,z)+H_z^I(-b,z)=H_z^{II}(-b,z) \tag{10.108}$$

can be rewritten as

$$\frac{k_{xs}}{\mu_1}\exp(ik_{zs}z)+\frac{i}{\pi\mu_1}\int_{-\infty}^{\infty}\tilde{E}_y^I(\zeta)\kappa_1\cos(\kappa_1 b)e^{-i\zeta z}\,d\zeta$$
$$=\frac{1}{\mu_2}\sum_{m=1}^{\infty}\xi_m c_m^r\sin a_m(z+a-rT)\,. \tag{10.109}$$

It is useful to employ the orthogonality property of $\sin a_m(z+a-rT)$ for further simplification. Multiplying (10.109) by $\sin a_n(z+a-rT)$ and integrating with respect to z for $(rT-a<z<rT+a)$ gives

$$\frac{k_{xs}}{\mu_1}a_n\exp(ik_{zs}rT)a^2F_n(k_{zs}a)$$
$$-\frac{1}{2\pi\mu_1}\sum_{l=-L_1}^{L_2}\sum_{m=1}^{\infty}b_m^l a_m a_n a^2 I_1=\frac{c_n^r\xi_n a}{\mu_2} \tag{10.110}$$

where

$$I_1=\int_{-\infty}^{\infty}\cot(\kappa_1 b)a^2F_m(\zeta a)F_n(-\zeta a)\kappa_1 e^{i(l-r)\zeta T}\,d\zeta\,. \tag{10.111}$$

It is necessary to transform the integral I_1 into a fast convergent series based on residue calculus [15]. The result is

$$I_1=\frac{2\pi}{aa_m^2}\sqrt{k_1^2-a_m^2}\cot\left(\sqrt{k_1^2-a_m^2}\,b\right)\delta_{mn}\delta_{lr}$$
$$-i\sum_{v=1}^{\infty}\frac{2\pi\kappa_1^2 A}{b(\zeta^2-a_m^2)(\zeta^2-a_n^2)\zeta a^2}\Bigg|_{\zeta=\sqrt{k_1^2-(v\pi/b)^2}} \tag{10.112}$$

where δ_{mn} is the Kronecker delta and

$$A=[(-1)^{m+n}+1]e^{i\zeta|l-r|T}$$
$$-(-1)^m e^{i\zeta|(l-r)T+2a|}-(-1)^n e^{i\zeta|(l-r)T-2a|}\,. \tag{10.113}$$

Similarly, the boundary conditions for the tangential field continuities must be enforced at $x=-b-d$. Applying the inverse Fourier transform to the tangential electric field continuity at $x=-b-d$

$$E_y^{III}(-b-d,z)=\begin{cases}E_y^{II}(-b-d,z) & \text{for } |z-lT|<a\\ 0 & \text{otherwise}\end{cases} \tag{10.114}$$

gives

$$\tilde{E}_y^{III}(\zeta) = \sum_{l=-L_1}^{L_2} \sum_{m=1}^{\infty}$$

$$\left[b_m^l \cos(\xi_m d) - c_m^l \sin(\xi_m d)\right] a_m a^2 e^{i\zeta lT} F_m(\zeta a) \,. \quad (10.115)$$

The continuity of the tangential magnetic field for the aperture of slits ($rT - a < z < rT + a, x = -b - d$, and $r = -L_1, \cdots, L_2$) requires

$$H_z^{III}(-b-d, z) = H_z^{II}(-b-d, z) \,. \quad (10.116)$$

Multiplying (10.116) by $\sin a_n(z + a - rT)$ and integrating with respect to z for $(rT - a < z < rT + a)$ finally gives

$$\sum_{l=-L_1}^{L_2} \sum_{m=1}^{\infty} \frac{a_m a_n}{2\pi\mu_3} \left[b_m^l \cos(\xi_m d) - c_m^l \sin(\xi_m d)\right] a^2 I_2$$

$$= \frac{i}{\mu_2} a\xi_n \left[b_n^r \sin(\xi_n d) + c_n^r \cos(\xi_n d)\right] \quad (10.117)$$

where

$$I_2 = \int_{-\infty}^{\infty} a^2 F_m(\zeta a) F_n(-\zeta a) \kappa_3 e^{i(l-r)\zeta T} \, d\zeta \,. \quad (10.118)$$

It is possible to evaluate I_2 using a numerical quadrature scheme. The set of simultaneous equations (10.110) and (10.117) must be solved for the unknown modal coefficients, b_m^l and c_m^l, to further investigate the scattering and radiation characteristics.

10.5.2 Reflection, Transmission, and Far Field

The antenna radiation pattern and reflection coefficient are important parameters to describe the characteristics of an antenna. This subsection presents the reflection coefficient and far-zone radiation. Let us evaluate the total scattered fields at $z = \pm\infty$ using residue calculus. The scattered fields considered at $z = \pm\infty$ are the transmitted (E_y^t) and reflected (E_y^r) components of the scattered wave in region (I), respectively. Substituting $\tilde{E}_y^I(\zeta)$ into $E_y^I(x, z)$ yields

$$E_y^I(x, z) = \int_{-\infty}^{\infty} -\frac{1}{2\pi \sin(\kappa_1 b)}$$

$$\cdot \sum_{l=-L_1}^{L_2} \sum_{m=1}^{\infty} b_m^l a_m e^{i\zeta lT} a^2 F_m(a\zeta) \sin(\kappa_1 x) e^{-i\zeta z} \, d\zeta$$

$$\equiv \int_{-\infty}^{\infty} f(\zeta) \, d\zeta \,. \quad (10.119)$$

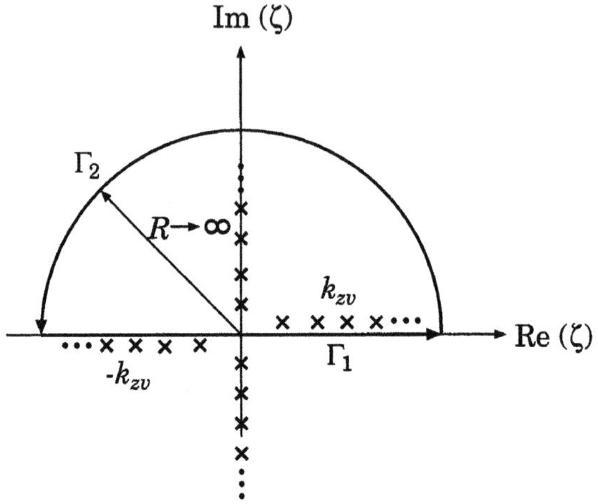

Fig. 10.8. Infinite number of simple poles in complex ζ-plane.

Note that the simple poles associated with the condition $\sin(\kappa_1 b) = 0$ yield $\zeta = k_{zv} = \sqrt{k_1^2 - k_{xv}^2}$ and $k_{xv} = \dfrac{v\pi}{b}$ where $v = 1, 2, 3, \cdots$. Consider when $z = -\infty$. Performing contour integration, as shown in Fig. 10.8, gives

$$\int_{\Gamma_1} f(\zeta)\, d\zeta + \int_{\Gamma_2} f(\zeta)\, d\zeta = 2\pi i \sum_v \operatorname*{Res}_{\zeta = k_{zv}} f(\zeta) \,. \tag{10.120}$$

Since

$$\int_{\Gamma_1} f(\zeta)\, d\zeta = E_y^I(x, -\infty) \tag{10.121}$$

$$\int_{\Gamma_2} f(\zeta)\, d\zeta = 0 \tag{10.122}$$

the scattered field at $z = -\infty$ becomes

$$\begin{aligned} E_y^I(x, -\infty) &= 2\pi i \sum_v \operatorname*{Res}_{\zeta = k_{zv}} f(\zeta) \\ &= \sum_v K_v^- \sin k_{xv}(x+b) \exp(-ik_{zv} z) \end{aligned} \tag{10.123}$$

where $0 < v < \dfrac{k_1 b}{\pi}$ and

$$K_v^- = i \sum_{l=-L_1}^{L_2} \sum_{m=1}^{\infty} b_m^l a_m k_{xv} \exp(+ik_{zv} lT) a^2 \frac{F_m(+k_{zv} a)}{k_{zv} b} \,. \tag{10.124}$$

Similarly, the transmitted field at $z = +\infty$ is given by

$$E_y^I(x, +\infty) = \sum_v K_v^+ \sin k_{xv}(x+b) \exp(+ik_{zv}z) \tag{10.125}$$

where

$$K_v^+ = i \sum_{l=-L_1}^{L_2} \sum_{m=1}^{\infty} b_m^l a_m k_{xv} \exp(-ik_{zv}lT) a^2 \frac{F_m(-k_{zv}a)}{k_{zv}b} \,. \tag{10.126}$$

Note that the series representation $E_y^I(x, \pm\infty)$ is fast convergent and numerically efficient.

Let us evaluate the far-zone scattered field in region (III) at $k_3\rho \gg 1$ using the saddle-point method [9]. Substituting the relations

$$-(x+b+d) = \rho\cos\theta_s \tag{10.127}$$

$$-z = \rho\sin\theta_s \tag{10.128}$$

$$\zeta = k_3 \sin\psi \tag{10.129}$$

into (10.104) gives

$$\begin{aligned} E_y^{III}(\rho, \theta_s) &= \frac{1}{2\pi} \int \tilde{E}_y^{III}(\zeta) e^{ik_3\rho\cos(\psi-\theta_s)} k_3 \cos\psi \, d\psi \\ &\approx \sqrt{\frac{k_3}{2\pi\rho}} \exp\left(ik_3\rho - i\frac{\pi}{4}\right) \cos\theta_s \tilde{E}_y^{III}(k_3\sin\theta_s) \,. \end{aligned} \tag{10.130}$$

10.6 Slotted Coaxial Line Antenna

A slotted coaxial line is one of the basic radiating structures for microwave antennas. The characteristics of the radiation from a multiple-slotted coaxial line have been considered in [16] based on mode matching and a Fourier transform. This section investigates the radiation from a coaxial line that is circumferentially slotted, as shown in Fig. 10.9, using mode matching and a Fourier transform.

10.6.1 Field Analysis

Let us assume that a TEM wave is incident from below a coaxial line. The inner and outer radii of the coaxial line are a and b, respectively, and the thickness of the outer conductor is $(c-b)$. The width of the slot is d and the permittivity of the dielectric material in the coaxial line is ϵ_1. Regions (I), (II), and (III) denote the coaxial line interior ($a \le \rho \le b$), slotted aperture ($b \le \rho \le c$ and $0 \le z \le d$), and surrounding medium ($\rho \ge c$), respectively, and their respective wavenumbers are $k_j = \omega\sqrt{\mu\epsilon_j}$ for $j = 1, 2$, and 3.

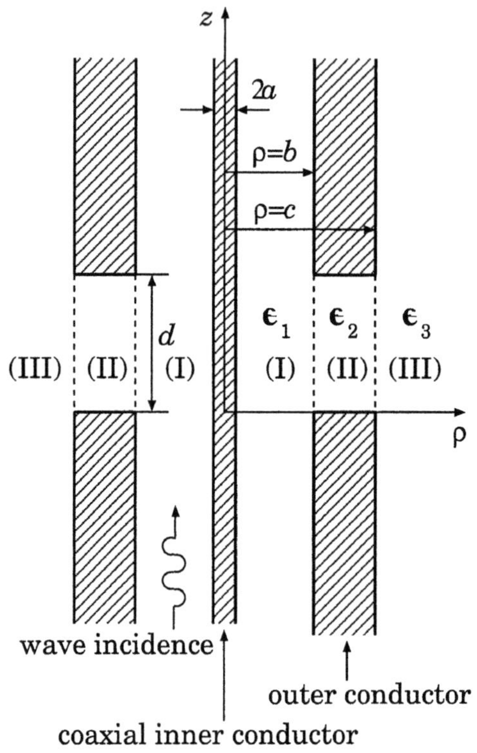

Fig. 10.9. Circumferentially slotted coaxial line.

The incident TEM wave is given by

$$E_\rho^i(\rho, z) = \frac{e^{ik_1 z}}{\rho} \tag{10.131}$$

$$H_\phi^i(\rho, z) = \frac{e^{ik_1 z}}{\eta_1 \rho} \tag{10.132}$$

where $\eta_1 = \sqrt{\frac{\mu}{\epsilon_1}}$. Since the incident wave has no variation in the azimuthal ϕ-direction, the scattered wave has no dependence on ϕ such that $\left(\frac{\partial}{\partial \phi}\right) = 0$. The scattered wave can be represented in terms of the vector potentials A_z and F_z (or equivalently E_z and H_z). Table 10.1 indicates that the scattered wave is a TM wave, where the magnetic vector potential A_z (or E_z) is nonzero while $F_z = 0$.

Table 10.1. Field representations using cylindrical coordinates

Fields	TM wave	TE wave
E_ρ	$\frac{i}{\omega\mu\epsilon}\frac{\partial^2 A_z}{\partial\rho\,\partial z}$	0
E_ϕ	0	$\frac{1}{\epsilon}\frac{\partial F_z}{\partial\rho}$
E_z	$\frac{i}{\omega\mu\epsilon}\left(\frac{\partial^2}{\partial z^2}+k^2\right)A_z$	0
H_ρ	0	$\frac{i}{\omega\mu\epsilon}\frac{\partial^2 F_z}{\partial\rho\,\partial z}$
H_ϕ	$-\frac{1}{\mu}\frac{\partial A_z}{\partial\rho}$	0
H_z	0	$\frac{i}{\omega\mu\epsilon}\left(\frac{\partial^2}{\partial z^2}+k^2\right)F_z$

The fields in region (I) consist of incident and scattered waves. The scattered magnetic vector potential $A_z^I(\rho,z)$ should satisfy the Helmholtz equation

$$\left(\nabla^2+k_1^2\right)A_z^I(\rho,z)=0\,. \tag{10.133}$$

Following the field representation in [16], the magnetic vector potential $A_z^I(\rho,z)$ can be represented by a Fourier transform as

$$A_z^I(\rho,z)=\frac{1}{2\pi}\int_{-\infty}^{\infty}\tilde{A}^I(\zeta)R(\kappa\rho)e^{-i\zeta z}\,d\zeta\,. \tag{10.134}$$

Substituting $A_z^I(\rho,z)$ into the Helmholtz equation shows that $R(\kappa\rho)$ should be the zeroth order Bessel function and Hankel function and $\kappa=\sqrt{k_1^2-\zeta^2}$. Let us represent the scattered electric field in region (I)

$$E_z^I(\rho,z)=\frac{1}{2\pi}\int_{-\infty}^{\infty}\underbrace{\frac{i}{\omega\mu\epsilon_1}\left(-\zeta^2+k_1^2\right)\tilde{A}^I(\zeta)}_{\tilde{E}^I(\zeta)}\,R(\kappa\rho)e^{-i\zeta z}\,d\zeta\,. \tag{10.135}$$

Due to the boundary condition $E_z^I(a,z)=0$, it is necessary to choose

$$R(\kappa\rho)=J_0(\kappa\rho)N_0(\kappa a)-N_0(\kappa\rho)J_0(\kappa a) \tag{10.136}$$

where $J_n(\cdot)$ is the nth order Bessel function of the first kind and $N_n(\cdot)$ is the nth order Bessel function of the second kind. Similarly, the magnetic field in region (I) is

$$H_\phi^I(\rho,z) = \frac{i\omega\epsilon_1}{2\pi}\int_{-\infty}^{\infty}\frac{\tilde{E}^I(\zeta)}{\kappa}R'(\kappa\rho)e^{-i\zeta z}\,d\zeta \tag{10.137}$$

where

$$\begin{aligned} R'(\kappa\rho) &= \frac{\partial R(\kappa\rho)}{\partial(\kappa\rho)} \\ &= -J_1(\kappa\rho)N_0(\kappa a) + N_1(\kappa\rho)J_0(\kappa a)\,. \end{aligned} \tag{10.138}$$

Similarly, the fields in region (III) can be written as

$$E_z^{III}(\rho,z) = \frac{1}{2\pi}\int_{-\infty}^{\infty}\tilde{E}^{III}(\zeta)H_0^{(1)}(\chi\rho)e^{-i\zeta z}\,d\zeta \tag{10.139}$$

$$H_\phi^{III}(\rho,z) = -\frac{i\omega\epsilon_3}{2\pi}\int_{-\infty}^{\infty}\frac{\tilde{E}^{III}(\zeta)}{\chi}H_1^{(1)}(\chi\rho)e^{-i\zeta z}d\zeta \tag{10.140}$$

where $\chi = \sqrt{k_3^2-\zeta^2}$ and $H_n^{(1)}(\cdot)$ is the nth order Hankel function of the first kind.

The fields in region (II) are

$$E_z^{II}(\rho,z) = \sum_{m=0}^{\infty} S(\kappa_m\rho)\cos(a_m z) \tag{10.141}$$

$$H_\phi^{II}(\rho,z) = \sum_{m=0}^{\infty}\frac{i\omega\epsilon_2}{\kappa_m}S'(\kappa_m\rho)\cos(a_m z) \tag{10.142}$$

where $a_m = \dfrac{m\pi}{d}$, $\kappa_m = \sqrt{k_2^2-a_m^2}$, and

$$S(\kappa_m\rho) = p_m\left[\frac{H_0^{(1)}(\kappa_m\rho)}{H_1^{(1)}(\kappa_m b)}\right] + q_m\left[\frac{H_0^{(2)}(\kappa_m\rho)}{H_1^{(2)}(\kappa_m c)}\right] \tag{10.143}$$

$$S'(\kappa_m\rho) = -p_m\left[\frac{H_1^{(1)}(\kappa_m\rho)}{H_1^{(1)}(\kappa_m b)}\right] - q_m\left[\frac{H_1^{(2)}(\kappa_m\rho)}{H_1^{(2)}(\kappa_m c)}\right]\,. \tag{10.144}$$

Four unknown coefficients–$\tilde{E}^I(\zeta)$, $\tilde{E}^{III}(\zeta)$, p_m, and q_m–are introduced to represent the fields in regions (I), (II), and (III). Four boundary conditions are also necessary to determine the four unknown coefficients. Let us constitute a set of simultaneous equations for the modal coefficients p_m and q_m. The continuity of the tangential electric field at $\rho = b$ is written as

$$E_z^I(b,z) = \begin{cases} E_z^{II}(b,z) & \text{for } 0<z<d \\ 0 & \text{otherwise}\,. \end{cases} \tag{10.145}$$

Applying an inverse Fourier transform to (10.145) gives

$$\tilde{E}^I(\zeta) = \sum_{m=0}^{\infty} \frac{S(\kappa_m b)}{R(\kappa b)} \underbrace{\int_0^d \cos(a_m z) e^{i\zeta z}\, dz}_{G_m(\zeta)} \tag{10.146}$$

where

$$G_m(\zeta) = \frac{-i\zeta[(-1)^m e^{i\zeta d} - 1]}{\zeta^2 - a_m^2} \,. \tag{10.147}$$

The continuity of the tangential magnetic field at $\rho = b$ for $0 < z < d$ is

$$H_\phi^i(b, z) + H_\phi^I(b, z) = H_\phi^{II}(b, z) \,. \tag{10.148}$$

Multiplying (10.148) by $\cos\left(\frac{s\pi}{d} z\right)$ and integrating with respect to z from 0 to d yields

$$\sum_{m=0}^{\infty} \left[\frac{H_0^{(1)}(\kappa_m b)}{H_1^{(1)}(\kappa_m b)} I_1 + \frac{\epsilon_2}{\epsilon_1} \frac{d}{2} \frac{\varepsilon_m}{\kappa_m} \delta_{ms} \right] p_m$$
$$+ \sum_{m=0}^{\infty} \left[\frac{H_0^{(2)}(\kappa_m b)}{H_1^{(2)}(\kappa_m c)} I_1 + \frac{\epsilon_2}{\epsilon_1} \frac{d}{2} \frac{\varepsilon_m}{\kappa_m} \frac{H_1^{(2)}(\kappa_m b)}{H_1^{(2)}(\kappa_m c)} \delta_{ms} \right] q_m$$
$$= \frac{i}{k_1 b} G_s(k_1) \tag{10.149}$$

where δ_{ms} is the Kronecker delta, $\varepsilon_0 = 2$, $\varepsilon_1 = \varepsilon_2 = \cdots = 1$, and

$$I_1 = \frac{1}{2\pi} \int_{-\infty}^{\infty} \frac{R'(\kappa b)}{\kappa R(\kappa b)} G_s(-\zeta) G_m(\zeta)\, d\zeta \tag{10.150}$$

$$R(\kappa b) = J_0(\kappa b) N_0(\kappa a) - N_0(\kappa b) J_0(\kappa a) \tag{10.151}$$

$$R'(\kappa b) = \left. \frac{\partial R(\kappa \rho)}{\partial (\kappa \rho)} \right|_{\rho = b}$$
$$= -J_1(\kappa b) N_0(\kappa a) + N_1(\kappa b) J_0(\kappa a) \,. \tag{10.152}$$

It is convenient to transform I_1 into a convergent series based on residue calculus. The derivation is shown in the next subsection.

Similarly, the continuity of the tangential electric field between regions (II) and (III) is

$$E_z^{III}(c, z) = \begin{cases} E_z^{II}(c, z) & \text{for } 0 < z < d \\ 0 & \text{otherwise} \,. \end{cases} \tag{10.153}$$

Applying an inverse Fourier transform to (10.153) gives

$$\tilde{E}^{III}(\zeta) = \sum_{m=0}^{\infty} \frac{S(\kappa_m c)}{H_0^{(1)}(\chi c)} G_m(\zeta) \, . \tag{10.154}$$

The continuity of the tangential magnetic field at $\rho = c$ for $0 < z < d$ gives

$$H_\phi^{II}(c, z) = H_\phi^{III}(c, z) \, . \tag{10.155}$$

Multiplying (10.155) by $\cos\left(\frac{s\pi}{d} z\right)$ and integrating with respect to z from 0 to d gives

$$\begin{aligned} &\sum_{m=0}^{\infty} \left[\frac{H_0^{(1)}(\kappa_m c)}{H_1^{(1)}(\kappa_m b)} I_2 - \frac{\epsilon_2}{\epsilon_3} \frac{d}{2} \frac{\varepsilon_m}{\kappa_m} \frac{H_1^{(1)}(\kappa_m c)}{H_1^{(1)}(\kappa_m b)} \delta_{ms} \right] p_m \\ &+ \sum_{m=0}^{\infty} \left[\frac{H_0^{(2)}(\kappa_m c)}{H_1^{(2)}(\kappa_m c)} I_2 - \frac{\epsilon_2}{\epsilon_3} \frac{d}{2} \frac{\varepsilon_m}{\kappa_m} \delta_{ms} \right] q_m = 0 \end{aligned} \tag{10.156}$$

where

$$I_2 = \frac{1}{2\pi} \int_{-\infty}^{\infty} \frac{H_1^{(1)}(\chi c)}{\chi H_0^{(1)}(\chi c)} G_m(\zeta) G_s(-\zeta) \, d\zeta \, . \tag{10.157}$$

It is possible to solve the simultaneous equations (10.149) and (10.156) for the discrete modal coefficients p_m and q_m.

10.6.2 Reflection, Transmission, and Far Field

The scattered field in region (I) is

$$\begin{aligned} H_\phi^I(\rho, z) &= \frac{i\omega\epsilon_1}{2\pi} \int_{-\infty}^{\infty} \frac{\tilde{E}^I(\zeta)}{\kappa} R'(\kappa\rho) e^{-i\zeta z} \, d\zeta \\ &= \frac{i\omega\epsilon_1}{2\pi} \int_{-\infty}^{\infty} \sum_{m=0}^{\infty} \frac{S(\kappa_m b)}{\kappa R(\kappa b)} G_m(\zeta) R'(\kappa\rho) e^{-i\zeta z} \, d\zeta \\ &\equiv \int_{-\infty}^{\infty} f(\zeta) \, d\zeta \, . \end{aligned} \tag{10.158}$$

Let us evaluate $H_\phi^I(\rho, -\infty)$ in the complex ζ-plane. Assume that only a TEM mode propagates in region (I), while all higher modes are evanescent. Consider poles at $\zeta = \pm k_1$. When $\zeta \to k_1$ ($\kappa \to 0$), note that $R(\kappa b) \to \frac{2}{\pi} \ln\left(\frac{a}{b}\right)$ and $R'(\kappa\rho) \to -\frac{2}{\pi\kappa\rho}$. Performing contour integration yields the reflected field

$$H_\phi^I(\rho,-\infty) = 2\pi i \operatorname*{Res}_{\zeta=k_1} f(\zeta)$$

$$= \sum_{m=0}^{\infty} \frac{G_m(k_1)}{2\ln\left(\dfrac{b}{a}\right)} S(\kappa_m b) \frac{e^{-ik_1 z}}{\eta_1 \rho} . \tag{10.159}$$

Similarly, the transmitted field is given by

$$H_\phi^I(\rho,+\infty) = \sum_{m=0}^{\infty} \frac{G_m(-k_1)}{2\ln\left(\dfrac{b}{a}\right)} S(\kappa_m b) \frac{e^{ik_1 z}}{\eta_1 \rho} . \tag{10.160}$$

To evaluate the far field in region (III) at observation point (r,θ), the saddle-point method can be used. Assume that $z = r\sin\theta$, $\rho = r\cos\theta$, $\zeta = k_3 \sin\psi$, and $\chi = k_3\cos\psi$. Since

$$\lim_{\chi\rho\to\infty} H_1^{(1)}(\chi\rho) \sim \sqrt{\frac{2}{\pi\chi\rho}} \exp\left[i\left(\chi\rho - \frac{3\pi}{4}\right)\right] \tag{10.161}$$

the far field in region (III) is

$$H_\phi^{III}(r,\theta) = -\frac{i\omega\epsilon_3}{2\pi} \int_{-\infty}^{\infty} \frac{1}{\chi}\sqrt{\frac{2}{\pi\chi\rho}} \exp\left[i\left(\chi\rho - \frac{3\pi}{4}\right)\right]$$

$$\cdot \sum_{m=0}^{\infty} \frac{S(\kappa_m c)}{H_0^{(1)}(\chi c)} G_m(\zeta) e^{-i\zeta z}\, d\zeta$$

$$\approx \sum_{m=0}^{\infty} \frac{iG_m(-k_3\sin\theta)}{\pi\cos\theta\; H_0^{(1)}(k_3 c\cos\theta)} \frac{e^{ik_3 r}}{\eta_3\; r} S(\kappa_m c) \tag{10.162}$$

where $\eta_3 = \sqrt{\dfrac{\mu}{\epsilon_3}}$ is the intrinsic impedance in region (III).

10.6.3 Residue Calculus for I_1

Consider

$$I_1 = \int_{-\infty}^{\infty} \frac{1}{2\pi} \frac{R'(\kappa b)}{\kappa R(\kappa b)}$$

$$\cdot \frac{\zeta^2\left[(-1)^{m+s} + 1 - (-1)^m e^{i\zeta d} - (-1)^s e^{-i\zeta d}\right]}{(\zeta^2 - a_m^2)(\zeta^2 - a_s^2)}\, d\zeta$$

$$
= \int_{-\infty}^{\infty} \frac{1}{2\pi} \frac{R'(\kappa b)}{\kappa R(\kappa b)}
$$

$$
\cdot \frac{\zeta^2 \left[(-1)^{m+s} + 1 - (-1)^m e^{i\zeta d} - (-1)^s e^{i\zeta d}\right]}{(\zeta^2 - a_m^2)(\zeta^2 - a_s^2)} \, d\zeta
$$

$$
\equiv \int_{-\infty}^{\infty} f(\zeta) \, d\zeta \, . \tag{10.163}
$$

Let us evaluate I_1 using contour integration in the complex ζ-plane, as shown in Fig. 10.10. Note that the term $R(\kappa b)$ yields an infinite number of simple poles at $\zeta = \zeta_v$ $(v = 1, 2, 3, \cdots)$. Two simple poles exist at $\zeta = \pm a_m$ when $m = s$ and two more simple poles exist at $\zeta = \pm k_1$. Performing contour integration in the upper half-plane along Γ, which consists of Γ_1, Γ_2, Γ_3, and Γ_4, gives

$$
\oint f(\zeta) \, d\zeta = \int_{\Gamma} f(\zeta) \, d\zeta
$$

$$
= 2\pi i \left[\operatorname*{Res}_{\zeta = k_1} f(\zeta) + \sum_{v=1}^{\infty} \operatorname*{Res}_{\zeta = \zeta_v} f(\zeta) \right] . \tag{10.164}
$$

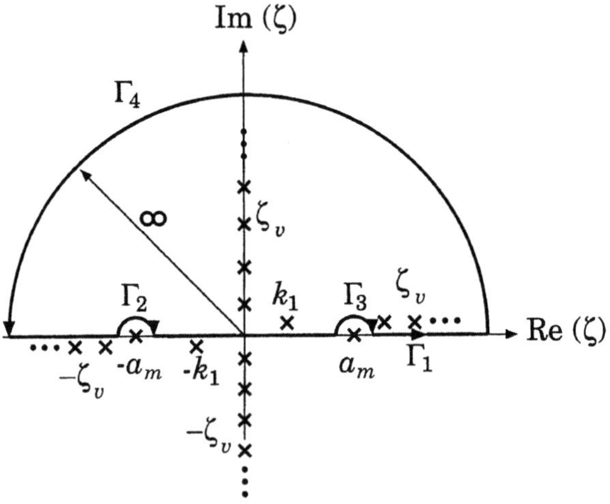

Fig. 10.10. Simple poles in complex ζ-plane.

$$\int_{\Gamma_1} f(\zeta)\,d\zeta = I_1 \tag{10.165}$$

$$\int_{\Gamma_2} f(\zeta)\,d\zeta = -\pi i \operatorname*{Res}_{\zeta=-a_m} f(\zeta) \equiv I_2 \tag{10.166}$$

$$\int_{\Gamma_3} f(\zeta)\,d\zeta = -\pi i \operatorname*{Res}_{\zeta=a_m} f(\zeta) \equiv I_3 \tag{10.167}$$

$$\int_{\Gamma_4} f(\zeta)\,d\zeta = 0\,. \tag{10.168}$$

Hence

$$I_1 = 2\pi i \left[\operatorname*{Res}_{\zeta=k_1} f(\zeta) + \sum_{v=1}^{\infty} \operatorname*{Res}_{\zeta=\zeta_v} f(\zeta) \right] - (I_2 + I_3)\,. \tag{10.169}$$

- Consider the contribution from $\zeta = -a_m$ for $m \neq 0$

$$\begin{aligned} I_2 &= -\pi i \operatorname*{Res}_{\zeta=-a_m} \left\{ \frac{R'(\kappa b)}{2\pi\kappa R(\kappa b)} \frac{2\zeta^2[1-(-1)^m e^{i\zeta d}]}{(\zeta - a_m)^2(\zeta + a_m)^2} \right\} \\ &= -\frac{d}{4} \frac{R'(\xi_m b)}{\xi_m R(\xi_m b)} \delta_{ms} \end{aligned} \tag{10.170}$$

  where $\xi_m = \sqrt{k_1^2 - a_m^2}$.
- Similarly, I_3 for $m \neq 0$ is

$$I_3 = -\frac{d}{4} \frac{R'(\xi_m b)}{\xi_m R(\xi_m b)} \delta_{ms}\,. \tag{10.171}$$

- When $m = s = 0$, $f(\zeta)$ contains a simple pole at $\zeta = 0$. Hence

$$I_2 + I_3 = -d \frac{R'(\xi_m b)}{\xi_m R(\xi_m b)} \delta_{ms}\,. \tag{10.172}$$

- When $\zeta = k_1$ $(\kappa \to 0)$

$$R(\kappa b) = \frac{2}{\pi} \ln\left(\frac{a}{b}\right) \tag{10.173}$$

$$R'(\kappa b) = -\frac{2}{\pi\kappa b}\,. \tag{10.174}$$

  Therefore

$$\begin{aligned} &2\pi i \operatorname*{Res}_{\zeta=k_1} f(\zeta) \\ &= -\frac{ik_1\left[1 + (-1)^{m+s} - (-1)^m e^{ik_1 d} - (-1)^s e^{ik_1 d}\right]}{2b \ln\left(\dfrac{b}{a}\right) \xi_m^2 \xi_s^2}\,. \end{aligned} \tag{10.175}$$

- Consider when $\zeta = \zeta_v$

$$\frac{\partial[R(\kappa b)]}{\partial\zeta} = \frac{b\zeta}{\kappa}\Big[J_1(\kappa b)N_0(\kappa a) - N_1(\kappa b)J_0(\kappa a)\Big]$$

$$+\frac{a\zeta}{\kappa}\Big[N_1(\kappa a)J_0(\kappa b) - J_1(\kappa a)N_0(\kappa b)\Big]. \quad (10.176)$$

Since

$$R(\kappa b) = J_0(\kappa b)N_0(\kappa a) - N_0(\kappa b)J_0(\kappa a) = 0 \quad (10.177)$$

$$N_1(\kappa a)J_0(\kappa b) - J_1(\kappa a)N_0(\kappa b)$$

$$= \frac{J_0(\kappa b)}{J_0(\kappa a)}\Big[N_1(\kappa a)J_0(\kappa a) - J_1(\kappa a)N_0(\kappa a)\Big]$$

$$= \frac{J_0(\kappa b)}{J_0(\kappa a)}\left(\frac{-2}{\pi\kappa a}\right) \quad (10.178)$$

$$R'(\kappa b) = N_1(\kappa b)J_0(\kappa a) - J_1(\kappa b)N_0(\kappa a)$$

$$= \frac{J_0(\kappa a)}{J_0(\kappa b)}\left(\frac{-2}{\pi\kappa b}\right) \quad (10.179)$$

the result is

$$\frac{\partial[R(\kappa b)]}{\partial\zeta} = \frac{2\zeta}{\pi\kappa^2}\frac{J_0(\kappa a)}{J_0(\kappa b)}\left[1 - \frac{J_0^2(\kappa b)}{J_0^2(\kappa a)}\right]. \quad (10.180)$$

Therefore

$$2\pi i \sum_{v=1}^{\infty} \operatorname*{Res}_{\zeta=\zeta_v} f(\zeta) =$$

$$-\sum_{v=1}^{\infty} \frac{i\zeta\left[1 + (-1)^{m+s} - (-1)^m e^{i\zeta d} - (-1)^s e^{i\zeta d}\right]}{b\left[1 - \frac{J_0^2(\kappa b)}{J_0^2(\kappa a)}\right](\zeta^2 - a_m^2)(\zeta^2 - a_s^2)}\Bigg|_{\zeta=\zeta_v}. \quad (10.181)$$

Collecting the results from (10.170), (10.171), (10.172), (10.175), and (10.181) finally gives

$$
\begin{aligned}
I_1 &= \frac{\varepsilon_m d}{2} \frac{R'(\xi_m b)}{\xi_m R(\xi_m b)} \delta_{ms} \\
&- \frac{ik_1 \left[1 + (-1)^{m+s} - (-1)^m e^{ik_1 d} - (-1)^s e^{ik_1 d}\right]}{2b \ln\left(\dfrac{b}{a}\right) \xi_m^2 \xi_s^2} \\
&- \sum_{v=1}^{\infty} \frac{i\zeta[1 + (-1)^{m+s} - (-1)^m e^{i\zeta d} - (-1)^s e^{i\zeta d}]}{b\left[1 - \dfrac{J_0^2(\kappa b)}{J_0^2(\kappa a)}\right](\zeta^2 - a_m^2)(\zeta^2 - a_s^2)}\Bigg|_{\zeta=\zeta_v}
\end{aligned}
\tag{10.182}
$$

where $\varepsilon_0 = 2$ and $\varepsilon_1 = \varepsilon_2 = \cdots = 1$.

10.7 Flanged Coaxial Line Antenna

Electromagnetic radiation from an open-ended coaxial line into a half space has been studied in terms of the material permittivity characterization. The material permittivity can be estimated from the reflection coefficient of an open-ended coaxial line that radiates into a dielectric material. Thus, it is of utmost importance to accurately predict the reflection coefficient of an open-ended coaxial line. A theoretical investigation of a coaxial line radiating into parallel plates has already been performed based on a Hankel transform and mode matching [17]. This section analyzes the electromagnetic radiation from a flanged coaxial line into a half space.

10.7.1 Field Representations

The problem geometry is shown in Fig. 10.11. The inner and outer radii of the coaxial line are a and b, respectively, while the permittivity of the dielectric material between $\rho = a$ and b is ϵ_1. Regions (I) and (II) denote the coaxial line interior ($a \le \rho \le b$ and $z \le 0$) and half space ($z \ge 0$), respectively, and their respective wavenumbers are $k_j = \omega\sqrt{\mu\epsilon_j}$ for $j = 1$ and 2.

Let us assume a TEM wave is incident from below a flanged coaxial line as

$$
H_\phi^i(\rho, z) = \frac{e^{ik_1 z}}{\rho} \tag{10.183}
$$

$$
E_\rho^i(\rho, z) = \eta_1 \frac{e^{ik_1 z}}{\rho} \tag{10.184}
$$

where $\eta_1 = \sqrt{\dfrac{\mu}{\epsilon_1}}$. Since the incident wave has no variation in the azimuthal ϕ-direction, the scattered wave has no dependence on ϕ. Table 10.1 in Section

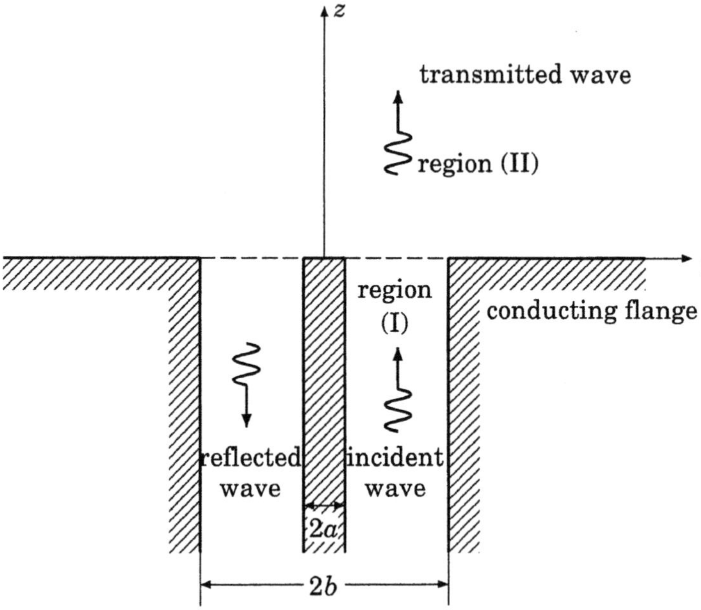

Fig. 10.11. Flanged coaxial line.

10.6 indicates that the scattered wave is a TM wave, where the magnetic vector potential A_z (or E_z) is nonzero.

The fields in region (I) consist of the incident and reflected waves. The reflected wave consists of a TEM wave and higher order TM waves. The vector potential associated with the reflected TM wave in region (I) is governed by the Helmholtz equation

$$\left(\nabla^2 + k_1^2\right) A_z^r(\rho, z) = 0 \tag{10.185}$$

which can be rewritten as

$$\left[\frac{1}{\rho}\frac{\partial}{\partial \rho}\left(\rho\frac{\partial}{\partial \rho}\right) + \frac{\partial^2}{\partial z^2} + k_1^2\right] A_z^r(\rho, z) = 0\ . \tag{10.186}$$

The scattering domain in region (I) is closed ($a \leq \rho \leq b$) in the radial direction and independent of the azimuthal direction. Hence, a linear combination of zeroth order Bessel functions of the first and second kinds is chosen as the radial modes. The solution to (10.186) is, therefore

$$A_z^r(\rho, z) = \sum_{n=1}^{\infty} a_n R_0(\xi_n \rho) \exp(-ik_z z) \tag{10.187}$$

where a_n is an unknown modal coefficient, $k_z = \sqrt{k_1^2 - \xi_n^2}$, and

$$R_0(\xi_n \rho) = J_0(\xi_n \rho) N_0(\xi_n b) - N_0(\xi_n \rho) J_0(\xi_n b) \, . \tag{10.188}$$

The boundary condition $E_z(\rho, z)\Big|_{\rho=b} = 0$ is automatically satisfied with $R_0(\xi_n b) = 0$. The boundary condition $E_z(\rho, z)\Big|_{\rho=a} = 0$ requires that an eigenvalue ξ_n satisfy the characteristic equation

$$R_0(\xi_n a) \equiv J_0(\xi_n a) N_0(\xi_n b) - N_0(\xi_n a) J_0(\xi_n b) = 0 \, . \tag{10.189}$$

Substituting the vector potential $A_z^r(\rho, z)$ into the field expressions in Table 10.1 in Section 10.6 produces the respective field components. The reflected TEM and TM ρ and ϕ components are

$$H_\phi^r(\rho, z) = c_0 \frac{\exp(-ik_1 z)}{\rho} + \sum_{n=1}^{\infty} c_n R_1(\xi_n \rho) \exp(-ik_z z) \tag{10.190}$$

$$\begin{aligned} E_\rho^r(\rho, z) &= -\frac{i}{\omega \epsilon_1} \frac{\partial H_\phi^r(\rho, z)}{\partial z} \\ &= -c_0 \eta_1 \frac{\exp(-ik_1 z)}{\rho} \\ &\quad - \frac{1}{\omega \epsilon_1} \sum_{n=1}^{\infty} c_n k_z R_1(\xi_n \rho) \exp(-ik_z z) \end{aligned} \tag{10.191}$$

where the modal coefficient $c_n = \dfrac{\xi_n}{\mu} a_n$ and

$$R_1(\xi_n \rho) = J_1(\xi_n \rho) N_0(\xi_n b) - N_1(\xi_n \rho) J_0(\xi_n b) \, . \tag{10.192}$$

The transmitted vector potential in region (II) is governed by the Helmholtz equation

$$\left[\frac{1}{\rho} \frac{\partial}{\partial \rho} \left(\rho \frac{\partial}{\partial \rho} \right) + \frac{\partial^2}{\partial z^2} + k_2^2 \right] A_z^t(\rho, z) = 0 \, . \tag{10.193}$$

Since the scattering domain in region (II) is open ($0 \le \rho < \infty$) in the radial direction, it is useful to employ the Hankel transform representation

$$\tilde{f}(\zeta) = \int_0^\infty f(\rho) J_0(\zeta \rho) \rho \, d\rho \tag{10.194}$$

$$f(\rho) = \int_0^\infty \tilde{f}(\zeta) J_0(\zeta \rho) \zeta \, d\zeta \, . \tag{10.195}$$

Let the transmitted potential be

$$A_z^t(\rho, z) = \int_0^\infty \tilde{A}_z^t(\zeta, z) J_0(\zeta\rho)\zeta \, d\zeta \, . \tag{10.196}$$

Substituting (10.196) into (10.193) and noting

$$\frac{1}{\rho}\frac{d}{d\rho}\left[\rho \frac{dJ_0(\zeta\rho)}{d\rho}\right] = -\zeta^2 J_0(\zeta\rho) \tag{10.197}$$

gives

$$\frac{d^2 \tilde{A}_z^t(\zeta, z)}{dz^2} + \kappa^2 \tilde{A}_z^t(\zeta, z) = 0 \tag{10.198}$$

where $\kappa = \sqrt{k_2^2 - \zeta^2}$. The solution

$$\tilde{A}_z^t(\zeta, z) = \tilde{A}_z^t(\zeta) e^{i\kappa z} \tag{10.199}$$

is chosen to satisfy the radiation condition $A_z^t(\rho, \infty) = 0$. Hence, the transmitted vector potential in region (II) is represented in terms of the continuous modes as

$$A_z^t(\rho, z) = \int_0^\infty \tilde{A}_z^t(\zeta) e^{i\kappa z} J_0(\zeta\rho)\zeta \, d\zeta \, . \tag{10.200}$$

Similarly, substituting the vector potential $A_z^t(\rho, z)$ into the field expressions in Table 10.1 in Section 10.6 produces the following transmitted field components in region (II)

$$H_\phi^t(\rho, z) = \int_0^\infty \tilde{H}(\zeta) e^{i\kappa z} J_1(\zeta\rho)\zeta \, d\zeta \tag{10.201}$$

$$\begin{aligned} E_\rho^t(\rho, z) &= -\frac{i}{\omega\epsilon_2}\frac{\partial H_\phi^t(\rho, z)}{\partial z} \\ &= \frac{1}{\omega\epsilon_2}\int_0^\infty \tilde{H}(\zeta)\kappa e^{i\kappa z} J_1(\zeta\rho)\zeta \, d\zeta \end{aligned} \tag{10.202}$$

where $\tilde{H}(\zeta) = \dfrac{\zeta}{\mu}\tilde{A}_z^t(\zeta)$.

10.7.2 Boundary Conditions

To determine the unknown coefficients c_n and $\tilde{H}(\zeta)$, two boundary conditions are necessary. The continuity of the tangential electric field at $z = 0$ requires

$$E_\rho^t(\rho, 0) = \begin{cases} E_\rho^i(\rho, 0) + E_\rho^r(\rho, 0) & \text{for } \ a < \rho < b \\ 0 & \text{otherwise} \end{cases} \tag{10.203}$$

which can be rewritten as

$$\frac{1}{\omega\epsilon_2}\int_0^{\infty}\tilde{H}(\zeta)\kappa J_1(\zeta\rho)\zeta\,d\zeta = \begin{cases}\eta_1\dfrac{1-c_0}{\rho}-\dfrac{1}{\omega\epsilon_1}\displaystyle\sum_{n=1}^{\infty}c_n k_z R_1(\xi_n\rho) & \text{for } a<\rho<b \\ 0 & \text{otherwise}\,.\end{cases} \tag{10.204}$$

Applying the Hankel transform $\int_0^{\infty}(\cdot)J_1(\zeta\rho)\rho\,d\rho$ to (10.204) yields

$$\begin{aligned}\frac{\tilde{H}(\zeta)\kappa}{\omega\epsilon_2} &= \int_a^b\left[\eta_1\frac{1-c_0}{\rho}-\frac{1}{\omega\epsilon_1}\sum_{n=1}^{\infty}c_n k_z R_1(\xi_n\rho)\right]J_1(\zeta\rho)\rho\,d\rho \\ &= \eta_1(1-c_0)\underbrace{\int_a^b J_1(\zeta\rho)\,d\rho}_{I_0} \\ &\quad -\frac{1}{\omega\epsilon_1}\sum_{n=1}^{\infty}c_n k_z\underbrace{\int_a^b R_1(\xi_n\rho)J_1(\zeta\rho)\rho\,d\rho}_{I_n}\,.\end{aligned} \tag{10.205}$$

For arbitrary Bessel functions Z_p and B_p [11, page 634],

$$\int Z_1(u)\,du = -Z_0(u) \tag{10.206}$$

$$\int uZ_p(\alpha u)B_p(\beta u)\,du = \frac{\beta uZ_p(\alpha u)B_{p-1}(\beta u)-\alpha uZ_{p-1}(\alpha u)B_p(\beta u)}{\alpha^2-\beta^2}\,. \tag{10.207}$$

Hence

$$I_0 = \int_a^b J_1(\zeta\rho)\,d\rho = \frac{J_0(\zeta a)-J_0(\zeta b)}{\zeta} \tag{10.208}$$

$$
\begin{aligned}
I_n &= \int_a^b \rho J_1(\zeta\rho) R_1(\xi_n\rho)\, d\rho \\
&= \left. \frac{\xi_n \rho J_1(\zeta\rho) R_0(\xi_n\rho) - \zeta\rho J_0(\zeta\rho) R_1(\xi_n\rho)}{\zeta^2 - \xi_n^2} \right|_a^b \\
&= \Big[\xi_n b J_1(\zeta b) R_0(\xi_n b) - \zeta b J_0(\zeta b) R_1(\xi_n b) \\
&\quad - \xi_n a J_1(\zeta a) R_0(\xi_n a) + \zeta a J_0(\zeta a) R_1(\xi_n a)\Big] \left(\zeta^2 - \xi_n^2\right)^{-1} . \qquad (10.209)
\end{aligned}
$$

Further simplification of I_n is possible with the relations

$$
R_0(\xi_n b) = 0 \qquad (10.210)
$$

$$
R_0(\xi_n a) = 0 \qquad (10.211)
$$

$$
\begin{aligned}
R_1(\xi_n b) &= J_1(\xi_n b) N_0(\xi_n b) - N_1(\xi_n b) J_0(\xi_n b) \\
&= \frac{2}{\pi \xi_n b} \qquad (10.212)
\end{aligned}
$$

$$
\begin{aligned}
R_1(\xi_n a) &= J_1(\xi_n a) N_0(\xi_n b) - N_1(\xi_n a) J_0(\xi_n b) \\
&= J_1(\xi_n a) \left[\frac{N_0(\xi_n a) J_0(\xi_n b)}{J_0(\xi_n a)}\right] - N_1(\xi_n a) J_0(\xi_n b) \\
&= \frac{J_0(\xi_n b)}{J_0(\xi_n a)} \left[J_1(\xi_n a) N_0(\xi_n a) - N_1(\xi_n a) J_0(\xi_n a)\right] \\
&= \frac{J_0(\xi_n b)}{J_0(\xi_n a)} \frac{2}{\pi \xi_n a} . \qquad (10.213)
\end{aligned}
$$

Hence

$$
I_n = 2\zeta \frac{-J_0(\zeta b) J_0(\xi_n a) + J_0(\zeta a) J_0(\xi_n b)}{\pi \xi_n (\zeta^2 - \xi_n^2) J_0(\xi_n a)} . \qquad (10.214)
$$

Therefore, (10.205) is rewritten as

$$
\tilde{H}(\zeta) = \frac{\epsilon_2}{\epsilon_1} \frac{1}{\kappa} \left[(1 - c_0) k_1 I_0 - \sum_{n=1}^{\infty} c_n k_z I_n\right] . \qquad (10.215)
$$

The continuity of the tangential magnetic field at $z = 0$ for $a < \rho < b$ requires

$$
H_\phi^t(\rho, 0) = H_\phi^i(\rho, 0) + H_\phi^r(\rho, 0) \qquad (10.216)
$$

which can be rewritten as

$$\int_0^\infty \tilde{H}(\zeta) J_1(\zeta\rho)\zeta\, d\zeta = \frac{(1+c_0)}{\rho} + \sum_{n=1}^{\infty} c_n R_1(\xi_n \rho)\,. \tag{10.217}$$

Substituting $\tilde{H}(\zeta)$ into (10.217) gives

$$\int_0^\infty \frac{\epsilon_2}{\epsilon_1}\frac{1}{\kappa}\left[(1-c_0)k_1 I_0 - \sum_{n=1}^{\infty} c_n k_z I_n\right] J_1(\zeta\rho)\zeta\, d\zeta$$

$$= \frac{(1+c_0)}{\rho} + \sum_{n=1}^{\infty} c_n R_1(\xi_n \rho)\,. \tag{10.218}$$

To simplify (10.218), it is convenient to use the orthogonality

$$\int_a^b R_1(\xi_n\rho) R_1(\xi_p\rho)\rho\, d\rho = \begin{cases} \dfrac{2}{\pi^2\xi_p^2}\left[1 - \dfrac{J_0^2(\xi_p b)}{J_0^2(\xi_p a)}\right] & \text{for } n = p \\ 0 & \text{for } n \neq p\,. \end{cases} \tag{10.219}$$

Multiplying (10.218) by $\rho R_1(\xi_p \rho)$ and integrating from a to b gives

$$\int_0^\infty \frac{\epsilon_2}{\epsilon_1}\frac{1}{\kappa}\left[(1-c_0)k_1 I_0 - \sum_{n=1}^{\infty} c_n k_z I_n\right] I_p \zeta\, d\zeta$$

$$= c_p \frac{2}{\pi^2 \xi_p^2}\left[1 - \frac{J_0^2(\xi_p b)}{J_0^2(\xi_p a)}\right] \tag{10.220}$$

where

$$I_p = 2\zeta \frac{-J_0(\zeta b)J_0(\xi_p a) + J_0(\zeta a)J_0(\xi_p b)}{\pi \xi_p(\zeta^2 - \xi_p^2)J_0(\xi_p a)}\,. \tag{10.221}$$

Rewriting (10.220) in a compact form gives

$$(1-c_0)I_{0p} - \sum_{n=1}^{\infty} c_n I_{np} = c_p \frac{2}{\pi^2\xi_p^2}\left[1 - \frac{J_0^2(\xi_p b)}{J_0^2(\xi_p a)}\right] \tag{10.222}$$

where

$$I_{0p} = \frac{\epsilon_2}{\epsilon_1} k_1 \int_0^\infty \frac{I_0 I_p}{\kappa}\zeta\, d\zeta \tag{10.223}$$

$$I_{np} = \frac{\epsilon_2}{\epsilon_1} k_z \int_0^\infty \frac{I_n I_p}{\kappa}\zeta\, d\zeta\,. \tag{10.224}$$

While the unknown modal coefficients are $c_0, c_1, c_2, \cdots, c_n$, (10.222) produces an n number of equations ($p = 1, 2, 3, \cdots, n$). Therefore, one additional equation is needed. Integrating (10.218) from a to b with respect to ρ gives

$$\int_0^\infty \frac{\epsilon_2}{\epsilon_1}\frac{1}{\kappa}\left[(1-c_0)k_1 I_0 - \sum_{n=1}^{\infty} c_n k_z I_n\right] I_0 \zeta \, d\zeta$$

$$= (1+c_0)\ln\frac{b}{a} . \tag{10.225}$$

Equations (10.222) and (10.225) constitute a set of simultaneous equations for the modal coefficients, $c_0, c_1, c_2, \cdots, c_n$.

The radiated field in region (II) is

$$H_\phi^t(\rho, z) = \int_0^\infty \tilde{H}(\zeta) e^{i\kappa z} J_1(\zeta\rho)\zeta \, d\zeta . \tag{10.226}$$

Since

$$J_1(\zeta\rho) = \frac{1}{2}\left[H_1^{(1)}(\zeta\rho) + H_1^{(2)}(\zeta\rho)\right] \tag{10.227}$$

$H_\phi^t(\rho, z)$ is written as

$$H_\phi^t(\rho, z) = \int_0^\infty \tilde{H}(\zeta)\frac{e^{i\kappa z}}{2}\left[H_1^{(1)}(\zeta\rho) + H_1^{(2)}(\zeta\rho)\right]\zeta \, d\zeta . \tag{10.228}$$

$H_1^{(1)}(-\zeta\rho) = H_1^{(2)}(\zeta\rho)$ and $\tilde{H}(\zeta)$ is an odd function of ζ, so

$$H_\phi^t(\rho, z) = \int_{-\infty}^\infty \tilde{H}(\zeta)\frac{e^{i\kappa z}}{2} H_1^{(1)}(\zeta\rho)\zeta \, d\zeta . \tag{10.229}$$

To evaluate the far field in region (II) at observation point (r, θ) using spherical coordinates, the saddle-point method can be used. Assume that $z = r\cos\theta$, $\rho = r\sin\theta$, $\zeta = k_2 \sin\psi$, and $\kappa = k_2\cos\psi$. Since

$$\lim_{\zeta\rho\to\infty} H_1^{(1)}(\zeta\rho) \sim \sqrt{\frac{2}{\pi\zeta\rho}}\exp\left[i\left(\zeta\rho - \frac{3\pi}{4}\right)\right] \tag{10.230}$$

the far field in region (II) is

$$H_\phi^t(r, \theta) = \int_{-\infty}^\infty \tilde{H}(\zeta)\sqrt{\frac{1}{2\pi\zeta\rho}}\exp\left[ik_2 r\cos(\psi-\theta) - i\frac{3\pi}{4}\right]\zeta \, d\zeta$$

$$\approx -\tilde{H}(k_2\sin\theta)k_2\cos\theta\frac{\exp(ik_2 r)}{r} . \tag{10.231}$$

Problems for Chapter 10

1. Derive the far-zone expressions (10.11) and (10.13).
2. The magnetic Hertzian dipole has an infinitesimally small magnetic current $\overline{M} = \hat{z}\delta(x - x')\delta(y - y')\delta(z - z')$ in free space. Evaluate its radiation field in the far zone.
3. Consider the integral representation (10.111). Using residue calculus and contour integration, show that (10.111) is given by (10.112).
4. Consider TM wave radiation from an N number of slits of width $2a$ and depth d in a parallel-plate waveguide. The problem geometry is shown in Fig. 10.12. Assume that an incident wave $H_y^i(x, z)$ excites region (I). Throughout regions (I), (II), and (III), the magnetic field has a y-component. There is no field variation with respect to y. Derive the radiation field $H_y^s(x, z)$.

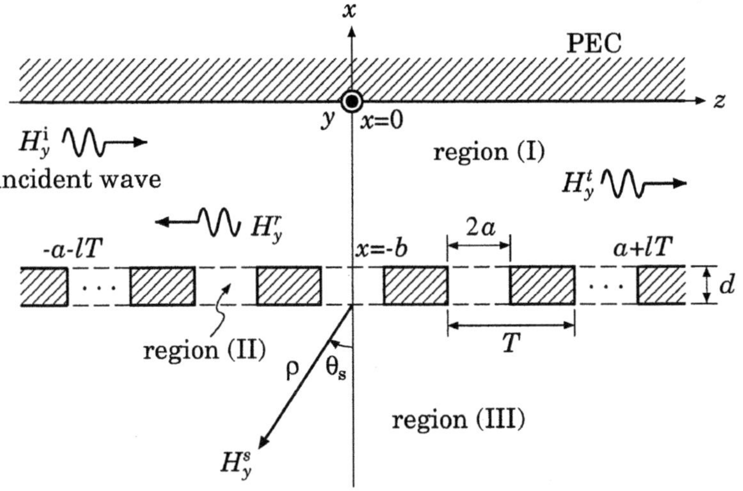

Fig. 10.12. Slit parallel-plate waveguide

11

Radiation Above Half Space

Antenna radiation above the earth's surface is an important subject in radio wave communications. The inhomogeneous earth terrain can be approximately modeled as a homogeneous half space. The boundary-value problem of radiation from a dipole antenna above a dielectric half space was first investigated by Arnold Sommerfeld and it thus became known as the Sommerfeld dipole problem. Electromagnetic wave radiation above a dielectric half space has been studied in [2, pp. 447-485] and [18]. This chapter investigates radiation from line currents and dipole sources in a half space, using a Fourier transform in the spectral domain. Some relevant discussion is available in [18].

11.1 Electric Line Source

Consider electromagnetic radiation from a horizontal electric line source $\overline{J}=\hat{y}\delta(x)\delta(z-z')$ that is located at $(0,z')$ in a half space. The geometry of the radiation problem is shown in Fig. 11.1. The magnetic vector potentials in the half space satisfy the Helmholtz equations

$$\left(\nabla^2+k_0^2\right)A_y^0(x,z)=-\mu_0\delta(x)\delta(z-z') \qquad \text{for } z\geq 0 \tag{11.1}$$

$$\left(\nabla^2+k_1^2\right)A_y^1(x,z)=0 \qquad \text{for } z\leq 0 \tag{11.2}$$

where k_0 $(=\omega\sqrt{\mu_0\epsilon_0})$ and k_1 $(=\omega\sqrt{\mu_1\epsilon_1})$ are the wavenumbers for $z>0$ and $z<0$, respectively. No field variation is assumed with respect to y, such as $\left(\dfrac{\partial}{\partial y}=0\right)$.

Since the problem domain in the x-axis is unbounded, it is expedient to express the vector potentials and delta function in terms of the one-dimensional inverse Fourier transform

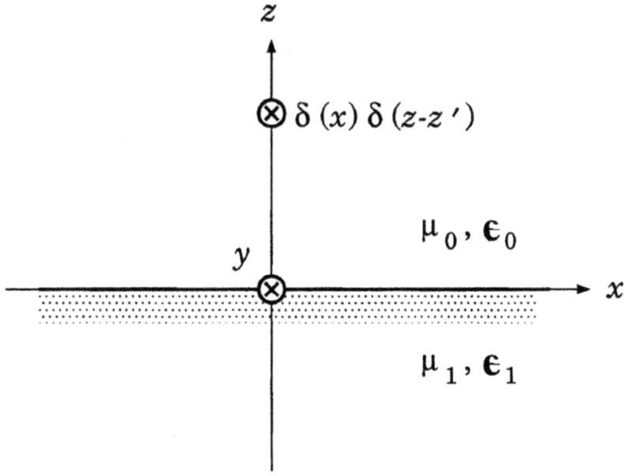

Fig. 11.1. Horizontal electric line source in half space.

$$A_y^{0,1}(x,z) = \frac{1}{2\pi}\int_{-\infty}^{\infty} \tilde{A}_y^{0,1}(\zeta,z)e^{i\zeta x}\,d\zeta \tag{11.3}$$

$$\delta(x) = \frac{1}{2\pi}\int_{-\infty}^{\infty} e^{i\zeta x}\,d\zeta \,. \tag{11.4}$$

Substituting $A_y^{0,1}(x,z)$ and $\delta(x)$ into (11.1) and (11.2) gives, respectively,

$$\left(\frac{d^2}{dz^2}+\kappa_0^2\right)\tilde{A}_y^0(\zeta,z) = -\mu_0\delta(z-z') \tag{11.5}$$

$$\left(\frac{d^2}{dz^2}+\kappa_1^2\right)\tilde{A}_y^1(\zeta,z) = 0 \tag{11.6}$$

where $\kappa_0 = \sqrt{k_0^2-\zeta^2}$ and $\kappa_1 = \sqrt{k_1^2-\zeta^2}$. The solution to (11.5) consists of the incident and scattered potentials. The incident potential is a response to the delta source in free space, while the scattered potential is due to reflection from the boundary at $z=0$. A graphical representation of the incident $\left[\tilde{A}_y^i(\zeta,z)\right]$ and scattered $\left[\tilde{A}_y^s(\zeta,z)\right]$ potentials is shown in Fig. 11.2. The total potential $\tilde{A}_y^0(\zeta,z)$ is, therefore, the sum of the two components

$$\tilde{A}_y^0(\zeta,z) = \tilde{A}_y^i(\zeta,z)+\tilde{A}_y^s(\zeta,z)$$

$$= \frac{i\mu_0}{2\kappa_0}\exp\left(i\kappa_0|z-z'|\right) + B\exp(i\kappa_0 z)\,. \tag{11.7}$$

Note that the choice $\exp(i\kappa_0 z)$ for the scattered potential satisfies the radiation condition, which requires that the field vanish at $z=\infty$. The solution to (11.6) is a wave that propagates in the $-z$-direction

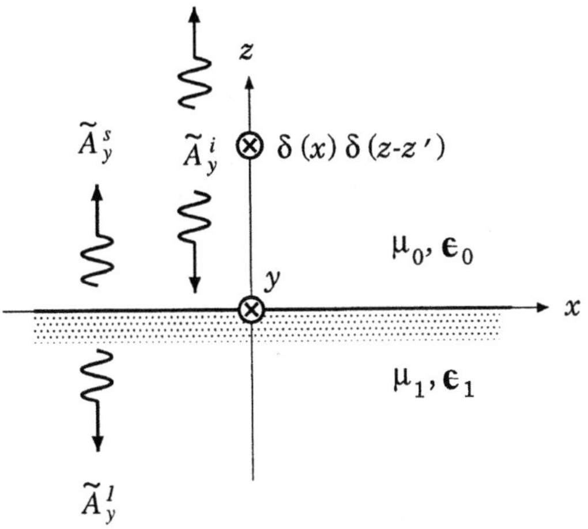

Fig. 11.2. Incident and scattered potentials.

$$\tilde{A}_y^1(\zeta, z) = C\exp(-i\kappa_1 z) \tag{11.8}$$

where the choice $\exp(-i\kappa_1 z)$ satisfies the radiation condition at $z = -\infty$.

To determine the unknown coefficients B and C, the boundary conditions for the tangential field continuities must be enforced at $z = 0$. Two boundary conditions, H_x and E_y continuities, are required. The field components can be evaluated from the magnetic vector potential $\overline{A}$. The nonzero field components are seen to be

$$E_y = i\omega A_y \tag{11.9}$$

$$H_x = -\frac{1}{\mu}\frac{\partial A_y}{\partial z} \tag{11.10}$$

$$H_z = \frac{1}{\mu}\frac{\partial A_y}{\partial x}\ . \tag{11.11}$$

The H_x continuity at $z = 0$

$$-\frac{1}{\mu_0}\frac{\partial A_y^0(x,z)}{\partial z}\bigg|_{z=0} = -\frac{1}{\mu_1}\frac{\partial A_y^1(x,z)}{\partial z}\bigg|_{z=0} \tag{11.12}$$

requires

$$\frac{1}{\mu_0}\frac{\partial \tilde{A}_y^0(\zeta,z)}{\partial z}\bigg|_{z=0} = \frac{1}{\mu_1}\frac{\partial \tilde{A}_y^1(\zeta,z)}{\partial z}\bigg|_{z=0}\ . \tag{11.13}$$

The E_y continuity condition at $z = 0$

$$i\omega A_y^0(x,0) = i\omega A_y^1(x,0) \tag{11.14}$$

gives

$$\tilde{A}_y^0(\zeta,0) = \tilde{A}_y^1(\zeta,0) . \tag{11.15}$$

Solving (11.13) and (11.15) for B and C yields

$$B = \left(\frac{\mu_1\kappa_0 - \mu_0\kappa_1}{\mu_1\kappa_0 + \mu_0\kappa_1}\right)\frac{i\mu_0\exp(i\kappa_0 z')}{2\kappa_0} \tag{11.16}$$

$$C = \left(\frac{2\mu_1\kappa_0}{\mu_1\kappa_0 + \mu_0\kappa_1}\right)\frac{i\mu_0\exp(i\kappa_0 z')}{2\kappa_0} . \tag{11.17}$$

Substituting B and (11.7) into (11.3) finally gives

$$A_y^0(x,z)$$
$$= \frac{i}{4\pi}\int_{-\infty}^{\infty}\frac{\exp(i\zeta x)}{\kappa_0}\mu_0$$
$$\cdot\left\{\exp(i\kappa_0|z-z'|) + \left(\frac{\mu_1\kappa_0 - \mu_0\kappa_1}{\mu_1\kappa_0 + \mu_0\kappa_1}\right)\exp\left[i\kappa_0(z+z')\right]\right\}d\zeta . \tag{11.18}$$

Considering the identity

$$\frac{i}{4}H_0^{(1)}(k_0\rho_1) = \frac{i}{4\pi}\int_{-\infty}^{\infty}\frac{e^{i\zeta x}}{\kappa_0}\exp(i\kappa_0|z-z'|)\,d\zeta \tag{11.19}$$

where $\rho_1 = \sqrt{x^2 + (z-z')^2}$, $A_y^0(x,z)$ can be rewritten as

$$A_y^0(x,z)$$
$$= \frac{i\mu_0}{4}H_0^{(1)}(k_0\rho_1)$$
$$+\frac{i\mu_0}{4\pi}\int_{-\infty}^{\infty}\left(\frac{\mu_1\kappa_0 - \mu_0\kappa_1}{\mu_1\kappa_0 + \mu_0\kappa_1}\right)\frac{\exp\left[i\zeta x + i\kappa_0(z+z')\right]}{\kappa_0}d\zeta . \tag{11.20}$$

Similarly, substituting C into (11.3) gives

$$A_y^1(x,z) = \frac{i\mu_1\mu_0}{2\pi}\int_{-\infty}^{\infty}\frac{\exp\left[i\left(\zeta x - \kappa_1 z + \kappa_0 z'\right)\right]}{(\mu_1\kappa_0 + \mu_0\kappa_1)}d\zeta . \tag{11.21}$$

Equation (11.20) consists of two contributions, namely, the direct term $\frac{i}{4}H_0^{(1)}(k_0\rho_1)$ and the scattered term containing the integral. The direct term accounts for the primary wave radiating from the electric line current in the absence of a boundary, while the scattering term is due to the reflected wave from the boundary at $z = 0$. The integrals in (11.20) and (11.21) must be evaluated to further investigate the characteristics of half-space radiation from the line current.

11.2 Vertical Electric Dipole

Consider electromagnetic radiation from a vertical electric dipole at $(0, 0, z')$ as

$$\overline{J} = \hat{z}\delta(\overline{r} - \overline{r}') = \hat{z}\delta(x)\delta(y)\delta(z - z') \ . \tag{11.22}$$

The geometry of the radiation problem is shown in Fig. 11.3. The magnetic vector potentials in a half space satisfy the Helmholtz equations

$$\left(\nabla^2 + k_0^2\right) A_z^0(x, y, z) = -\mu_0\delta(x)\delta(y)\delta(z - z') \quad \text{for} \ \ z \geq 0 \tag{11.23}$$

$$\left(\nabla^2 + k_1^2\right) A_z^1(x, y, z) = 0 \quad \text{for} \ \ z \leq 0 \tag{11.24}$$

where k_0 $(= \omega\sqrt{\mu_0\epsilon_0})$ and k_1 $(= \omega\sqrt{\mu_1\epsilon_1})$ are the wavenumbers for $z > 0$ and $z < 0$, respectively.

Since the x-y plane is unbounded, it is expedient to express the magnetic vector potentials in terms of the two-dimensional inverse Fourier transform

$$A_z^{0,1}(x, y, z) = \frac{1}{(2\pi)^2} \int_{-\infty}^{\infty} \int_{-\infty}^{\infty} \tilde{A}_z^{0,1}(\zeta, \eta, z) e^{i(\zeta x + \eta y)} \, d\zeta \, d\eta \ . \tag{11.25}$$

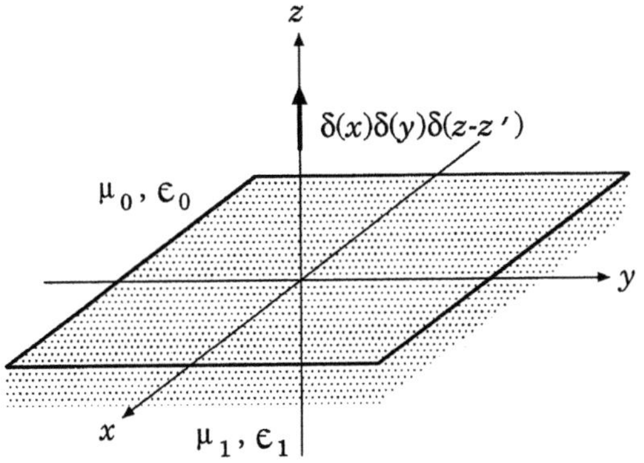

Fig. 11.3. Vertical electric point source in half space.

Substituting $A_z^{0,1}(x,y,z)$ into (11.23) and (11.24) gives, respectively,

$$\left(\frac{d^2}{dz^2}+\kappa_0^2\right)\tilde{A}_z^0(\zeta,\eta,z)=-\mu_0\delta(z-z') \tag{11.26}$$

$$\left(\frac{d^2}{dz^2}+\kappa_1^2\right)\tilde{A}_z^1(\zeta,\eta,z)=0 \tag{11.27}$$

where $\kappa_0=\sqrt{k_0^2-\zeta^2-\eta^2}$ and $\kappa_1=\sqrt{k_1^2-\zeta^2-\eta^2}$. The solution to (11.26) consists of the incident and scattered potentials

$$\tilde{A}_z^0(\zeta,\eta,z)=\tilde{A}_z^i(\zeta,\eta,z)+\tilde{A}_z^s(\zeta,\eta,z)$$

$$=\frac{i\mu_0}{2\kappa_0}\exp(i\kappa_0|z-z'|)+B\exp(i\kappa_0 z)\,. \tag{11.28}$$

The solution to (11.27) is assumed to be

$$\tilde{A}_z^1(\zeta,\eta,z)=C\exp(-i\kappa_1 z)\,. \tag{11.29}$$

To determine the unknown coefficients B and C, the boundary conditions for the tangential fields E_x, E_y, H_x, and H_y continuities must be enforced at $z=0$. The tangential field components can be evaluated by substituting $A_z^{0,1}(x,y,z)$ into Table 3.1 in Section 3.1. The H_x and H_y continuities

$$\left.\frac{1}{\mu_0}\frac{\partial A_z^0(x,y,z)}{\partial y}\right|_{z=0}=\left.\frac{1}{\mu_1}\frac{\partial A_z^1(x,y,z)}{\partial y}\right|_{z=0} \tag{11.30}$$

$$\left.\frac{1}{\mu_0}\frac{\partial A_z^0(x,y,z)}{\partial x}\right|_{z=0}=\left.\frac{1}{\mu_1}\frac{\partial A_z^1(x,y,z)}{\partial x}\right|_{z=0} \tag{11.31}$$

require

$$\frac{1}{\mu_0}\tilde{A}_z^0(\zeta,\eta,0)=\frac{1}{\mu_1}\tilde{A}_z^1(\zeta,\eta,0)\,. \tag{11.32}$$

Similarly, the E_x and E_y continuities

$$\left.\frac{1}{\mu_0\epsilon_0}\frac{\partial^2 A_z^0(x,y,z)}{\partial x\,\partial z}\right|_{z=0}=\left.\frac{1}{\mu_1\epsilon_1}\frac{\partial^2 A_z^1(x,y,z)}{\partial x\,\partial z}\right|_{z=0} \tag{11.33}$$

$$\left.\frac{1}{\mu_0\epsilon_0}\frac{\partial^2 A_z^0(x,y,z)}{\partial y\,\partial z}\right|_{z=0}=\left.\frac{1}{\mu_1\epsilon_1}\frac{\partial^2 A_z^1(x,y,z)}{\partial y\,\partial z}\right|_{z=0} \tag{11.34}$$

yield

$$\left.\frac{1}{\mu_0\epsilon_0}\frac{\partial\tilde{A}_z^0(\zeta,\eta,z)}{\partial z}\right|_{z=0}=\left.\frac{1}{\mu_1\epsilon_1}\frac{\partial\tilde{A}_z^1(\zeta,\eta,z)}{\partial z}\right|_{z=0}. \tag{11.35}$$

Solving (11.32) and (11.35) for B and C gives

$$B = \frac{i\mu_0}{2\kappa_0}\left(\frac{\epsilon_1\kappa_0 - \epsilon_0\kappa_1}{\epsilon_1\kappa_0 + \epsilon_0\kappa_1}\right)\exp(i\kappa_0 z') \tag{11.36}$$

$$C = i\mu_1\left(\frac{\epsilon_1}{\epsilon_1\kappa_0 + \epsilon_0\kappa_1}\right)\exp(i\kappa_0 z') . \tag{11.37}$$

Substituting B into $A_z^0(x,y,z)$ yields

$$A_z^0(x,y,z) = \frac{1}{(2\pi)^2}\int_{-\infty}^{\infty}\int_{-\infty}^{\infty}\frac{ie^{i(\zeta x+\eta y)}}{2\kappa_0}\mu_0 \cdot\left\{\exp(i\kappa_0|z-z'|) + \left(\frac{\epsilon_1\kappa_0 - \epsilon_0\kappa_1}{\epsilon_1\kappa_0 + \epsilon_0\kappa_1}\right)\exp\left[i\kappa_0(z+z')\right]\right\} d\zeta\, d\eta . \tag{11.38}$$

The identity for a three-dimensional free-space Green's function shows

$$\frac{1}{(2\pi)^2}\int_{-\infty}^{\infty}\int_{-\infty}^{\infty}\frac{i}{2\kappa_0}\exp(i\kappa_0|z-z'|)e^{i\zeta x+i\eta y}\, d\zeta\, d\eta = \frac{\exp(ik_0 r_1)}{4\pi r_1} \tag{11.39}$$

where $r_1 = \sqrt{x^2+y^2+(z-z')^2}$. Hence, $A_z^0(x,y,z)$ is rewritten as

$$A_z^0(x,y,z) = \mu_0\frac{\exp(ik_0 r_1)}{4\pi r_1} + \frac{i\mu_0}{8\pi^2}\int_{-\infty}^{\infty}\int_{-\infty}^{\infty}\left(\frac{\epsilon_1\kappa_0 - \epsilon_0\kappa_1}{\epsilon_1\kappa_0 + \epsilon_0\kappa_1}\right) \cdot\frac{\exp\left[i(\zeta x+\eta y) + i\kappa_0(z+z')\right]}{\kappa_0} d\zeta\, d\eta . \tag{11.40}$$

Similarly, substituting C into $A_z^1(x,y,z)$ gives

$$A_z^1(x,y,z) = \frac{i\epsilon_1}{4\pi^2}\int_{-\infty}^{\infty}\int_{-\infty}^{\infty} \mu_1\frac{\exp\left[i\left(\zeta x+\eta y - \kappa_1 z + \kappa_0 z'\right)\right]}{(\epsilon_1\kappa_0 + \epsilon_0\kappa_1)} d\zeta\, d\eta . \tag{11.41}$$

Let us introduce the polar coordinates (ξ, ψ) and (ρ, ϕ)

$$\zeta = \xi \cos\psi \tag{11.42}$$
$$\eta = \xi \sin\psi \tag{11.43}$$
$$x = \rho \cos\phi \tag{11.44}$$
$$y = \rho \sin\phi \tag{11.45}$$

to further simplify $A_z^0(x, y, z)$ and $A_z^1(x, y, z)$. First, consider integration with respect to ψ as

$$\begin{aligned}\int_0^{2\pi} e^{i(\zeta x + \eta y)}\, d\psi &= \int_0^{2\pi} \exp\left[i\rho\xi\cos(\psi - \phi)\right] d\psi \\ &= \int_0^{2\pi} \sum_{n=-\infty}^{\infty} i^n J_n(\xi\rho) e^{in(\psi-\phi)}\, d\psi \\ &= 2\pi J_0(\rho\xi)\ .\end{aligned} \tag{11.46}$$

Hence, the final polar forms for the potentials for $z \geq 0$ and $z \leq 0$ are, respectively,

$$\begin{aligned}A_z^0(\rho, z) &= \mu_0 \frac{\exp(ik_0 r_1)}{4\pi r_1} \\ &+ \frac{i\mu_0}{4\pi} \int_0^\infty \left(\frac{\epsilon_1\kappa_0 - \epsilon_0\kappa_1}{\epsilon_1\kappa_0 + \epsilon_0\kappa_1}\right) \\ &\cdot \frac{\exp\left[i\kappa_0(z + z')\right]}{\kappa_0} J_0(\rho\xi)\xi\, d\xi\end{aligned} \tag{11.47}$$

$$A_z^1(\rho, z) = \frac{i\epsilon_1}{2\pi} \int_0^\infty \mu_1 \frac{\exp\left[i(-\kappa_1 z + \kappa_0 z')\right]}{(\epsilon_1\kappa_0 + \epsilon_0\kappa_1)} J_0(\rho\xi)\xi\, d\xi \tag{11.48}$$

where

$$\rho = \sqrt{x^2 + y^2} \tag{11.49}$$
$$\kappa_0 = \sqrt{k_0^2 - \xi^2} \tag{11.50}$$
$$\kappa_1 = \sqrt{k_1^2 - \xi^2}\ . \tag{11.51}$$

11.3 Horizontal Electric Dipole

The geometry of the radiation problem is shown in Fig. 11.4. Consider electromagnetic radiation from a horizontal electric dipole located at $(0, 0, z')$ as

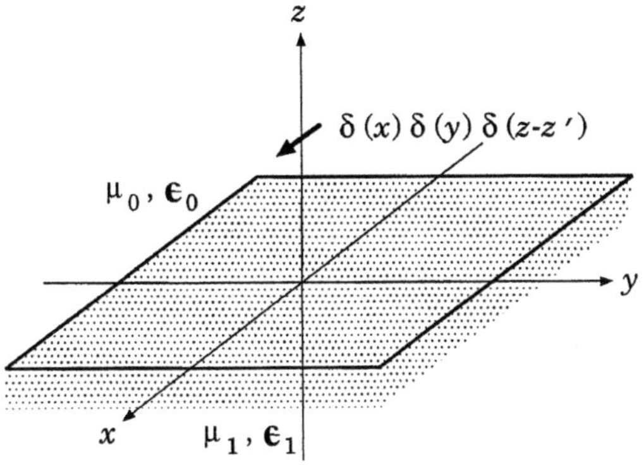

Fig. 11.4. Horizontal electric point source in half space.

$$\overline{J} = \hat{x}\delta(\overline{r} - \overline{r}') = \hat{x}\delta(x)\delta(y)\delta(z - z') \ . \tag{11.52}$$

The magnetic vector potentials in the half space satisfy the Helmholtz equations

$$\left(\nabla^2 + k_0^2\right)\overline{A}^0(x, y, z) = -\hat{x}\mu_0\delta(x)\delta(y)\delta(z - z') \quad \text{for } z \geq 0 \tag{11.53}$$

$$\left(\nabla^2 + k_1^2\right)\overline{A}^1(x, y, z) = 0 \quad \text{for } z \leq 0 \tag{11.54}$$

where k_0 $(= \omega\sqrt{\mu_0\epsilon_0})$ and k_1 $(= \omega\sqrt{\mu_1\epsilon_1})$ are the wavenumbers for $z > 0$ and $z < 0$, respectively. The problem of horizontal dipole radiation considered here is more complicated than the vertical case in the previous section since the radiation field is no longer azimuthally uniform. Consequently, the magnetic vector potential component A_x alone cannot satisfy the boundary condition at $z = 0$, thereby requiring an additional component. Let us use A_x and A_z to match the boundary conditions. The magnetic vector potentials A_x and A_z satisfy the following Helmholtz equations

$$\left(\nabla^2 + k_0^2\right) A_x^0(x, y, z) = -\mu_0\delta(x)\delta(y)\delta(z - z') \quad \text{for } z \geq 0 \tag{11.55}$$

$$\left(\nabla^2 + k_0^2\right) A_z^0(x, y, z) = 0 \quad \text{for } z \geq 0 \tag{11.56}$$

$$\left(\nabla^2 + k_1^2\right) A_x^1(x, y, z) = 0 \quad \text{for } z \leq 0 \tag{11.57}$$

$$\left(\nabla^2 + k_1^2\right) A_z^1(x, y, z) = 0 \quad \text{for } z \leq 0 \ . \tag{11.58}$$

Substituting

$$A_x^{0,1}(x,y,z) = \frac{1}{(2\pi)^2}\int_{-\infty}^{\infty}\int_{-\infty}^{\infty} \tilde{A}_x^{0,1}(\zeta,\eta,z)e^{i(\zeta x+\eta y)}\, d\zeta\, d\eta \quad (11.59)$$

$$A_z^{0,1}(x,y,z) = \frac{1}{(2\pi)^2}\int_{-\infty}^{\infty}\int_{-\infty}^{\infty} \tilde{A}_z^{0,1}(\zeta,\eta,z)e^{i(\zeta x+\eta y)}\, d\zeta\, d\eta. \quad (11.60)$$

into (11.55) through (11.58) gives, respectively,

$$\left(\frac{d^2}{dz^2} + \kappa_0^2\right)\tilde{A}_x^0(\zeta,\eta,z) = -\mu_0\delta(z-z') \quad (11.61)$$

$$\left(\frac{d^2}{dz^2} + \kappa_0^2\right)\tilde{A}_z^0(\zeta,\eta,z) = 0 \quad (11.62)$$

$$\left(\frac{d^2}{dz^2} + \kappa_1^2\right)\tilde{A}_x^1(\zeta,\eta,z) = 0 \quad (11.63)$$

$$\left(\frac{d^2}{dz^2} + \kappa_1^2\right)\tilde{A}_z^1(\zeta,\eta,z) = 0 \quad (11.64)$$

where $\kappa_0 = \sqrt{k_0^2 - \zeta^2 - \eta^2}$ and $\kappa_1 = \sqrt{k_1^2 - \zeta^2 - \eta^2}$. The solutions to (11.61) through (11.64) take the form of

$$\tilde{A}_x^0(\zeta,\eta,z) = \frac{i\mu_0}{2\kappa_0}\exp\left(i\kappa_0|z-z'|\right) + P\exp(i\kappa_0 z) \quad (11.65)$$

$$\tilde{A}_z^0(\zeta,\eta,z) = Q\exp(i\kappa_0 z) \quad (11.66)$$

$$\tilde{A}_x^1(\zeta,\eta,z) = R\exp(-i\kappa_1 z) \quad (11.67)$$

$$\tilde{A}_z^1(\zeta,\eta,z) = S\exp(-i\kappa_1 z)\,. \quad (11.68)$$

To determine the unknown coefficients–P, Q, R, and S, the boundary conditions for the tangential field (E_x, E_y, H_x, and H_y) continuities must be enforced at $z = 0$. The relations

$$\overline{E} = i\omega\overline{A} + \frac{i}{\omega\mu\epsilon}\nabla\left(\nabla\cdot\overline{A}\right) \quad (11.69)$$

$$\overline{H} = \frac{1}{\mu}\nabla\times\overline{A} \quad (11.70)$$

result in the field components

$$E_x = i\omega A_x + \frac{i}{\omega\mu\epsilon}\frac{\partial}{\partial x}\left(\frac{\partial A_x}{\partial x} + \frac{\partial A_z}{\partial z}\right) \tag{11.71}$$

$$E_y = \frac{i}{\omega\mu\epsilon}\frac{\partial}{\partial y}\left(\frac{\partial A_x}{\partial x} + \frac{\partial A_z}{\partial z}\right) \tag{11.72}$$

$$H_x = \frac{1}{\mu}\frac{\partial A_z}{\partial y} \tag{11.73}$$

$$H_y = -\frac{1}{\mu}\left(\frac{\partial A_z}{\partial x} - \frac{\partial A_x}{\partial z}\right) . \tag{11.74}$$

The H_x continuity at $z = 0$

$$\frac{1}{\mu_0}\tilde{A}_z^0(\zeta,\eta,0) = \frac{1}{\mu_1}\tilde{A}_z^1(\zeta,\eta,0) \tag{11.75}$$

yields

$$\frac{Q}{\mu_0} = \frac{S}{\mu_1} . \tag{11.76}$$

The E_y continuity at $z = 0$ is written as

$$\frac{1}{\mu_0\epsilon_0}\left[i\zeta\tilde{A}_x^0(\zeta,\eta,0) + \left.\frac{\partial\tilde{A}_z^0(\zeta,\eta,z)}{\partial z}\right|_{z=0}\right] =$$

$$\frac{1}{\mu_1\epsilon_1}\left[i\zeta\tilde{A}_x^1(\zeta,\eta,0) + \left.\frac{\partial\tilde{A}_z^1(\zeta,\eta,z)}{\partial z}\right|_{z=0}\right] . \tag{11.77}$$

Therefore

$$\frac{1}{\mu_0\epsilon_0}\left\{\zeta\left[\frac{i\mu_0}{2\kappa_0}\exp(i\kappa_0 z') + P\right] + \kappa_0 Q\right\} = \frac{1}{\mu_1\epsilon_1}\left(\zeta R - \kappa_1 S\right) . \tag{11.78}$$

Considering (11.77), the E_x continuity at $z = 0$ yields

$$\tilde{A}_x^0(\zeta,\eta,0) = \tilde{A}_x^1(\zeta,\eta,0) . \tag{11.79}$$

Therefore

$$\frac{i\mu_0}{2\kappa_0}\exp(i\kappa_0 z') + P = R . \tag{11.80}$$

Considering (11.75), the H_y continuity at $z = 0$ gives

$$\frac{1}{\mu_0}\left.\frac{\partial\tilde{A}_x^0(\zeta,\eta,z)}{\partial z}\right|_{z=0} = \frac{1}{\mu_1}\left.\frac{\partial\tilde{A}_x^1(\zeta,\eta,z)}{\partial z}\right|_{z=0} . \tag{11.81}$$

Therefore

$$\frac{\kappa_0}{\mu_0}\left[-\frac{i\mu_0}{2\kappa_0}\exp(i\kappa_0 z') + P\right] = -\frac{\kappa_1}{\mu_1}R\,. \tag{11.82}$$

From (11.80) and (11.82)

$$P = \frac{i\mu_0}{2\kappa_0}\left(\frac{\kappa_0\mu_1 - \kappa_1\mu_0}{\kappa_0\mu_1 + \kappa_1\mu_0}\right)\exp(i\kappa_0 z') \tag{11.83}$$

$$R = i\mu_0\left(\frac{\mu_1}{\kappa_0\mu_1 + \kappa_1\mu_0}\right)\exp(i\kappa_0 z')\,. \tag{11.84}$$

Substituting P and R into (11.76) and (11.78) yields

$$Q = i\mu_0\zeta\frac{(\mu_0\epsilon_0 - \mu_1\epsilon_1)}{(\kappa_0\epsilon_1 + \kappa_1\epsilon_0)(\kappa_0\mu_1 + \kappa_1\mu_0)}\exp(i\kappa_0 z') \tag{11.85}$$

$$S = \frac{\mu_1}{\mu_0}Q\,. \tag{11.86}$$

Substituting P, Q, R, and S into (11.65) through (11.68) and, subsequently, into (11.59) and (11.60) yields

$$A_x^0(x,y,z) = \frac{1}{(2\pi)^2}\int_{-\infty}^{\infty}\int_{-\infty}^{\infty}\frac{i\mu_0}{2\kappa_0}\Bigg[\exp(i\kappa_0|z-z'|) + \left(\frac{\kappa_0\mu_1 - \kappa_1\mu_0}{\kappa_0\mu_1 + \kappa_1\mu_0}\right)\exp(i\kappa_0 z' + i\kappa_0 z)\Bigg] \cdot e^{i(\zeta x + \eta y)}\,d\zeta\,d\eta \tag{11.87}$$

$$A_x^1(x,y,z) = \frac{1}{(2\pi)^2}\int_{-\infty}^{\infty}\int_{-\infty}^{\infty} i\mu_0\left(\frac{\mu_1}{\kappa_0\mu_1 + \kappa_1\mu_0}\right) \cdot\exp(i\kappa_0 z' - i\kappa_1 z)e^{i(\zeta x + \eta y)}\,d\zeta\,d\eta \tag{11.88}$$

$$A_z^0(x,y,z) = \frac{1}{(2\pi)^2}\int_{-\infty}^{\infty}\int_{-\infty}^{\infty} i\zeta\mu_0 \cdot\frac{(\mu_0\epsilon_0 - \mu_1\epsilon_1)\exp\left[i\kappa_0(z+z')\right]}{(\kappa_0\epsilon_1 + \kappa_1\epsilon_0)(\kappa_0\mu_1 + \kappa_1\mu_0)}e^{i(\zeta x + \eta y)}\,d\zeta\,d\eta \tag{11.89}$$

$$A_z^1(x,y,z) = \frac{1}{(2\pi)^2}\int_{-\infty}^{\infty}\int_{-\infty}^{\infty} i\zeta\mu_1 \cdot\frac{(\mu_0\epsilon_0 - \mu_1\epsilon_1)\exp\left(i\kappa_0 z' - i\kappa_1 z\right)}{(\kappa_0\epsilon_1 + \kappa_1\epsilon_0)(\kappa_0\mu_1 + \kappa_1\mu_0)}e^{i(\zeta x + \eta y)}\,d\zeta\,d\eta\,. \tag{11.90}$$

Let us transform (11.87) through (11.90) into single integrals, using the polar coordinates $\zeta = \xi \cos\psi$, $\eta = \xi \sin\psi$, $x = \rho\cos\phi$, and $y = \rho \sin\phi$. Hence

$$A_x^0(\rho,\phi,z) = \frac{1}{(2\pi)^2}\int_0^\infty \frac{i\mu_0}{2\kappa_0}\left[\exp(i\kappa_0|z-z'|) + \left(\frac{\kappa_0\mu_1-\kappa_1\mu_0}{\kappa_0\mu_1+\kappa_1\mu_0}\right)\exp(i\kappa_0 z' + i\kappa_0 z)\right] \cdot \underbrace{\int_0^{2\pi} e^{i\xi\rho\cos(\psi-\phi)}d\psi}_{2\pi J_0(\xi\rho)}\,\xi\, d\xi$$

$$= \frac{1}{2\pi}\int_0^\infty \frac{i\mu_0}{2\kappa_0}\left[\exp(i\kappa_0|z-z'|) + \left(\frac{\kappa_0\mu_1-\kappa_1\mu_0}{\kappa_0\mu_1+\kappa_1\mu_0}\right)\exp(i\kappa_0 z' + i\kappa_0 z)\right] J_0(\xi\rho)\xi\, d\xi \quad (11.91)$$

where $\kappa_0 = \sqrt{k_0^2 - \xi^2}$ and $\kappa_1 = \sqrt{k_1^2-\xi^2}$. Similarly

$$A_x^1(\rho,\phi,z) = \frac{1}{2\pi}\int_0^\infty \left(\frac{i\mu_0\mu_1}{\kappa_0\mu_1+\kappa_1\mu_0}\right) \cdot\exp(i\kappa_0 z' - i\kappa_1 z)J_0(\xi\rho)\xi\, d\xi\,. \quad (11.92)$$

Consider (11.89)

$$A_z^0(\rho,\phi,z) = \frac{1}{(2\pi)^2}\int_0^\infty i\mu_0 \frac{(\mu_0\epsilon_0-\mu_1\epsilon_1)}{(\kappa_0\epsilon_1+\kappa_1\epsilon_0)(\kappa_0\mu_1+\kappa_1\mu_0)} \cdot\exp\left[i\kappa_0(z+z')\right]\underbrace{\int_0^{2\pi}\cos\psi e^{i\xi\rho\cos(\psi-\phi)}d\psi}_{i2\pi J_1(\xi\rho)\cos\phi}\,\xi^2\, d\xi$$

$$= -\frac{\cos\phi}{2\pi}\int_0^\infty \frac{\mu_0(\mu_0\epsilon_0-\mu_1\epsilon_1)}{(\kappa_0\epsilon_1+\kappa_1\epsilon_0)(\kappa_0\mu_1+\kappa_1\mu_0)} \cdot\exp\left[i\kappa_0(z+z')\right]J_1(\xi\rho)\xi^2\, d\xi\,. \quad (11.93)$$

Similarly, $A_z^1(\rho,\phi,z)$ can be rewritten as

$$A_z^1(\rho,\phi,z) = -\frac{\cos\phi}{2\pi}\int_0^\infty \frac{\mu_1(\mu_0\epsilon_0-\mu_1\epsilon_1)}{(\kappa_0\epsilon_1+\kappa_1\epsilon_0)(\kappa_0\mu_1+\kappa_1\mu_0)} \cdot\exp(i\kappa_0 z' - i\kappa_1 z)J_1(\xi\rho)\xi^2\, d\xi\,. \quad (11.94)$$

The expressions (11.91) through (11.94) are the final forms for the vector potentials using cylindrical coordinates (ρ, ϕ, z). Computation of (11.91) through (11.94) is required to further investigate the characteristics of half-space radiation.

Problems for Chapter 11

1. Consider electromagnetic radiation from a horizontal magnetic line source $\overline{M}=\hat{y}\delta(x)\delta(z-z')$ that is located at $(0, z')$ in a half space. The geometry of the radiation problem is shown in Fig. 11.5. Evaluate the radiation field.

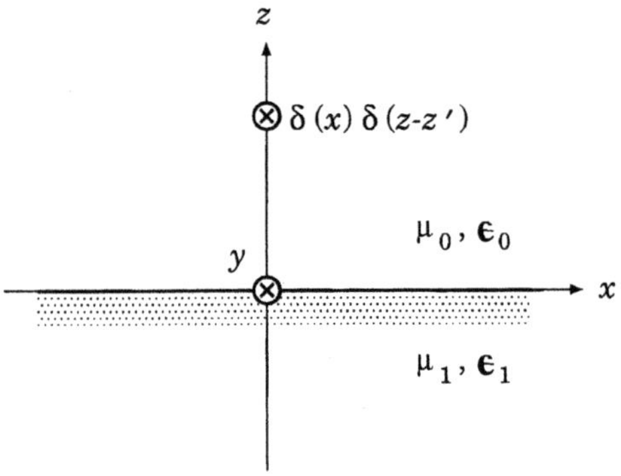

Fig. 11.5. Horizontal magnetic line source in half space.

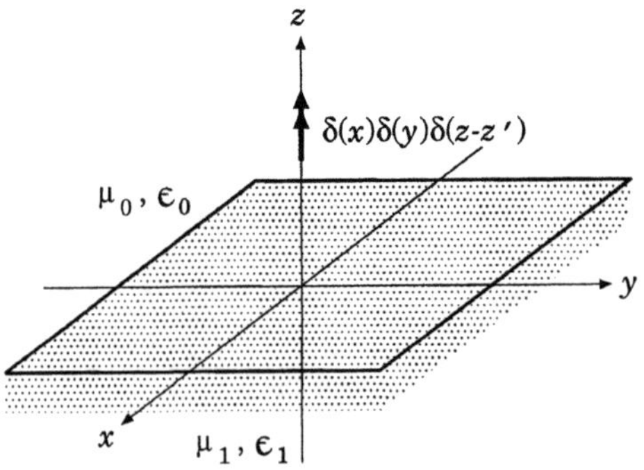

Fig. 11.6. Vertical magnetic point source in half space

2. Consider electromagnetic radiation from a vertical magnetic dipole at $(0, 0, z')$ as $\overline{M} = \hat{z}\delta(\overline{r} - \overline{r}') = \hat{z}\delta(x)\delta(y)\delta(z - z')$. The geometry of the radiation problem is shown in Fig. 11.6. Evaluate the radiation field.

References

1. S. Kozaki, "A new expression for the scattering of a Gaussian beam by a conducting cylinder", *IEEE Trans. Antennas Propagat.*, vol. 30, no. 5, pp. 881-887, Sept. 1982.
2. A. Ishimaru, *Electromagnetic Wave Propagation, Radiation, and Scattering.* Englewood Cliffs, NJ: Prentice-Hall, 1991.
3. Y. H. Cho and H. J. Eom, "Fourier transform analysis of a ridge waveguide and a rectangular coaxial line", *Radio Sci.*, vol. 36, no. 4, pp. 533-538, July-Aug. 2001.
4. H. J. Eom and Y. H. Cho, "Analysis of multiple groove guide", *Electron. Lett.*, vol. 35, no. 20, pp. 1749-1751, Sept. 1999.
5. S. Ramo, J. R. Whinnery, and T. Van Duzer, *Fields and Waves in Communication Electronics.* Third Edition, New York: John Wiley & Sons, 1993, pp. 723-727.
6. J. Helszajn, *Principles of Microwave Ferrite Engineering.* London: Wiley-Interscience, 1969, pp. 136-139.
7. R. F. Harrington, *Time-Harmonic Electromagnetic Fields.* New York: McGraw-Hill Book Company, 1961, pp. 106-110.
8. S. H. Kang, H. J. Eom, and T. J. Park, "TM-scattering from a slit in a thick conducting screen: revisited", *IEEE Trans. Microwave Theory Tech.*, vol. 41, no. 5, pp. 895-899, May 1993.
9. J. A. Kong, *Electromagnetic Wave Theory.* Second Edition, New York: John Wiley & Sons, 1990, pp. 307-310.
10. J. H. Lee and H. J. Eom, "Electrostatic potential through a circular aperture in a thick conducting plane", *IEEE Trans. Microwave Theory Tech.*, vol. 44, no. 2, pp. 341-343, Feb. 1996.
11. I. S. Gradshteyn and I. M. Ryzhik, *Table of Integrals, Series, and Products.* New York: Academic Press, 1965.
12. D. G. Dudley, *Mathematical Foundations for Electromagnetic Theory.* Piscataway, NJ: IEEE Press, 1994, pp. 166-168.
13. C. A. Balanis, *Advanced Engineering Electromagnetics.* New York: John Wiley & Sons, 1989, pp. 897-901.

14. T. J. Park and H. J. Eom, "Analytic solution for TE-mode radiation from a flanged parallel-plate waveguide", *IEE Proceedings-H*, vol. 140, no. 5, pp. 387-389, Oct. 1993.
15. J. H. Lee, H. J. Eom, and J. W. Lee, "Scattering and radiation from finite thick slits in parallel-plate waveguide", *IEEE Trans. Antennas Propagat.*, vol. 44, no. 2, pp. 212-216, Feb. 1996.
16. J. K. Park and H. J. Eom, "Radiation from multiple circumferential slots on a coaxial cable", *Microwave Opt. Technol. Lett.*, vol. 26, no. 3, pp. 160-162, Aug. 2000.
17. J. H. Lee, H. J. Eom, and K. H. Jun, "Reflection of a coaxial line radiating into a parallel plate", *IEEE Microwave Guided Wave Lett.*, vol. 6, no. 3, pp. 135-137, March 1996.
18. D. C. Stinson, *Intermediate Mathematics of Electromagnetics.* Englewood Cliffs, NJ: Prentice-Hall, 1976, pp. 227-246.

A

Coordinates and Vector Formulas

A.1 Coordinate Relations

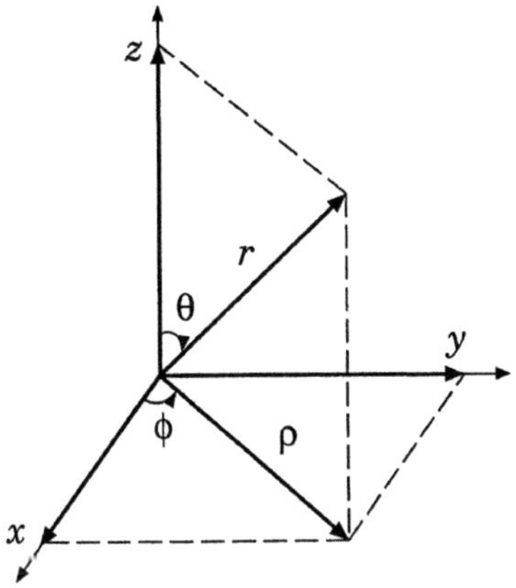

Fig. A.1. Orthogonal coordinate systems.

The relationships among rectangular coordinates (x, y, z), cylindrical coordinates (ρ, ϕ, z), and spherical coordinates (r, θ, ϕ) are

$$x = \rho \cos\phi = r \sin\theta \cos\phi \tag{A.1}$$

$$y = \rho \sin\phi = r \sin\theta \sin\phi \tag{A.2}$$

$$z = r \cos\theta \,. \tag{A.3}$$

The unit vectors for the three orthogonal coordinates are denoted by $(\hat{x}, \hat{y}, \hat{z})$, $(\hat{\rho}, \hat{\phi}, \hat{z})$, and $(\hat{r}, \hat{\theta}, \hat{\phi})$, respectively. The transformations among the unit vectors are given in Tables A.1, A.2, and A.3. For instance, the second column in Table A.2 yields the relation

$$\hat{x} = \hat{r} \sin\theta \cos\phi + \hat{\theta} \cos\theta \cos\phi - \hat{\phi} \sin\phi \,. \tag{A.4}$$

Similarly, the second row in Table A.2 yields

$$\hat{r} = \hat{x} \sin\theta \cos\phi + \hat{y} \sin\theta \sin\phi + \hat{z} \cos\theta \,. \tag{A.5}$$

Table A.1. Rectangular/cylindrical coordinate transformation

	$\hat{x}$	$\hat{y}$	$\hat{z}$
$\hat{\rho}$	$\cos\phi$	$\sin\phi$	0
$\hat{\phi}$	$-\sin\phi$	$\cos\phi$	0
$\hat{z}$	0	0	1

Table A.2. Rectangular/spherical coordinate transformation

	$\hat{x}$	$\hat{y}$	$\hat{z}$
$\hat{r}$	$\sin\theta\cos\phi$	$\sin\theta\sin\phi$	$\cos\theta$
$\hat{\theta}$	$\cos\theta\cos\phi$	$\cos\theta\sin\phi$	$-\sin\theta$
$\hat{\phi}$	$-\sin\phi$	$\cos\phi$	0

Table A.3. Cylindrical/spherical coordinate transformation

	$\hat{\rho}$	$\hat{\phi}$	$\hat{z}$
$\hat{r}$	$\sin\theta$	0	$\cos\theta$
$\hat{\theta}$	$\cos\theta$	0	$-\sin\theta$
$\hat{\phi}$	0	1	0

A.2 Differential Operators

A.2.1 Rectangular Coordinates (x, y, z)

$$\nabla f = \hat{x}\frac{\partial f}{\partial x} + \hat{y}\frac{\partial f}{\partial y} + \hat{z}\frac{\partial f}{\partial z} \tag{A.6}$$

$$\nabla \cdot \overline{F} = \frac{\partial F_x}{\partial x} + \frac{\partial F_y}{\partial y} + \frac{\partial F_z}{\partial z} \tag{A.7}$$

$$\begin{aligned} \nabla \times \overline{F} = \hat{x}&\left(\frac{\partial F_z}{\partial y} - \frac{\partial F_y}{\partial z}\right) \\ +\hat{y}&\left(\frac{\partial F_x}{\partial z} - \frac{\partial F_z}{\partial x}\right) \\ +\hat{z}&\left(\frac{\partial F_y}{\partial x} - \frac{\partial F_x}{\partial y}\right) \end{aligned} \tag{A.8}$$

$$\nabla^2 f = \frac{\partial^2 f}{\partial x^2} + \frac{\partial^2 f}{\partial y^2} + \frac{\partial^2 f}{\partial z^2} \tag{A.9}$$

$$\nabla^2 \overline{F} = \hat{x}\nabla^2 F_x + \hat{y}\nabla^2 F_y + \hat{z}\nabla^2 F_z \tag{A.10}$$

A.2.2 Cylindrical Coordinates (ρ, ϕ, z)

$$\nabla f = \hat{\rho}\frac{\partial f}{\partial \rho} + \hat{\phi}\frac{1}{\rho}\frac{\partial f}{\partial \phi} + \hat{z}\frac{\partial f}{\partial z} \tag{A.11}$$

$$\nabla \cdot \overline{F} = \frac{1}{\rho}\frac{\partial}{\partial \rho}(\rho F_\rho) + \frac{1}{\rho}\frac{\partial F_\phi}{\partial \phi} + \frac{\partial F_z}{\partial z} \tag{A.12}$$

$$\begin{aligned} \nabla \times \overline{F} = \hat{\rho}&\left(\frac{1}{\rho}\frac{\partial F_z}{\partial \phi} - \frac{\partial F_\phi}{\partial z}\right) \\ +\hat{\phi}&\left(\frac{\partial F_\rho}{\partial z} - \frac{\partial F_z}{\partial \rho}\right) \\ +\hat{z}&\left[\frac{1}{\rho}\frac{\partial}{\partial \rho}(\rho F_\phi) - \frac{1}{\rho}\frac{\partial F_\rho}{\partial \phi}\right] \end{aligned} \tag{A.13}$$

$$\nabla^2 f = \frac{1}{\rho}\frac{\partial}{\partial \rho}\left(\rho\frac{\partial f}{\partial \rho}\right) + \frac{1}{\rho^2}\frac{\partial^2 f}{\partial \phi^2} + \frac{\partial^2 f}{\partial z^2} \tag{A.14}$$

$$\nabla^2 \overline{F} = \nabla\left(\nabla \cdot \overline{F}\right) - \nabla \times \nabla \times \overline{F} \tag{A.15}$$

A.2.3 Spherical Coordinates (r, θ, ϕ)

$$\nabla f = \hat{r}\frac{\partial f}{\partial r} + \hat{\theta}\frac{1}{r}\frac{\partial f}{\partial \theta} + \hat{\phi}\frac{1}{r\sin\theta}\frac{\partial f}{\partial \phi} \tag{A.16}$$

$$\nabla \cdot \overline{F} = \frac{1}{r^2}\frac{\partial}{\partial r}(r^2 F_r) + \frac{1}{r\sin\theta}\frac{\partial}{\partial\theta}(\sin\theta F_\theta) + \frac{1}{r\sin\theta}\frac{\partial F_\phi}{\partial\phi} \tag{A.17}$$

$$\begin{aligned}\nabla \times \overline{F} = {} & \frac{\hat{r}}{r\sin\theta}\left[\frac{\partial}{\partial\theta}(F_\phi \sin\theta) - \frac{\partial F_\theta}{\partial\phi}\right] \\ & + \frac{\hat{\theta}}{r}\left[\frac{1}{\sin\theta}\frac{\partial F_r}{\partial\phi} - \frac{\partial}{\partial r}(rF_\phi)\right] \\ & + \frac{\hat{\phi}}{r}\left[\frac{\partial}{\partial r}(rF_\theta) - \frac{\partial F_r}{\partial\theta}\right]\end{aligned} \tag{A.18}$$

$$\begin{aligned}\nabla^2 f = {} & \frac{1}{r^2}\frac{\partial}{\partial r}\left(r^2\frac{\partial f}{\partial r}\right) + \frac{1}{r^2\sin\theta}\frac{\partial}{\partial\theta}\left(\sin\theta\frac{\partial f}{\partial\theta}\right) \\ & + \frac{1}{r^2\sin^2\theta}\frac{\partial^2 f}{\partial\phi^2}\end{aligned} \tag{A.19}$$

$$\nabla^2\overline{F} = \nabla\left(\nabla\cdot\overline{F}\right) - \nabla\times\nabla\times\overline{F} \tag{A.20}$$

A.3 Vector Formulas

$$\overline{F}\cdot\overline{F}^* = \left|\overline{F}\right|^2 \tag{A.21}$$

$$\nabla\cdot\nabla f = \nabla^2 f \tag{A.22}$$

$$\nabla\cdot\left(\nabla\times\overline{F}\right) = 0 \tag{A.23}$$

$$\nabla\times\nabla f = 0 \tag{A.24}$$

$$\nabla\cdot\left(\overline{F}\times\overline{G}\right) = \overline{G}\cdot\nabla\times\overline{F} - \overline{F}\cdot\nabla\times\overline{G} \tag{A.25}$$

$$\nabla\times\nabla\times\overline{F} = \nabla\left(\nabla\cdot\overline{F}\right) - \nabla^2\overline{F} \tag{A.26}$$

B

Bessel Functions

B.1 Bessel Functions and Modified Bessel Functions

Bessel's equation is

$$x^2\frac{d^2y}{dx^2}+x\frac{dy}{dx}+\left(x^2-\nu^2\right)y=0\,. \tag{B.1}$$

When ν is not an integer, the solution is a linear combination of $J_\nu(x)$ and $J_{-\nu}(x)$ as

$$y(x)=c_1J_\nu(x)+c_2J_{-\nu}(x) \tag{B.2}$$

where $J_\nu(x)$ is a Bessel function of the first kind of order ν

$$J_\nu(x)=\sum_{m=0}^{\infty}\frac{(-1)^m\left(\dfrac{x}{2}\right)^{2m+\nu}}{m!(m+\nu)!}\,. \tag{B.3}$$

When ν is an integer n, then $J_{-n}(x)=(-1)^nJ_n(x)$, indicating that $J_{-n}(x)$ and $J_n(x)$ are no longer independent of each other. Therefore, the second independent solution $N_n(x)$ is required to form the complete solution as

$$y(x)=c_1J_n(x)+c_2N_n(x) \tag{B.4}$$

where $N_n(x)$ is a Bessel function of the second kind of order n and

$$\begin{aligned}N_n(x)&=\lim_{\nu\to n}N_\nu(x)\\&=\lim_{\nu\to n}\frac{J_\nu(x)\cos(\nu\pi)-J_{-\nu}(x)}{\sin(\nu\pi)}\,.\end{aligned} \tag{B.5}$$

The graphical representations for $J_n(x)$, $J_n'(x)=\dfrac{dJ_n(x)}{dx}$, and $N_n(x)$ for $n=0,1,2,3$ are shown in Figs. B.1, B.2, and B.3, respectively.

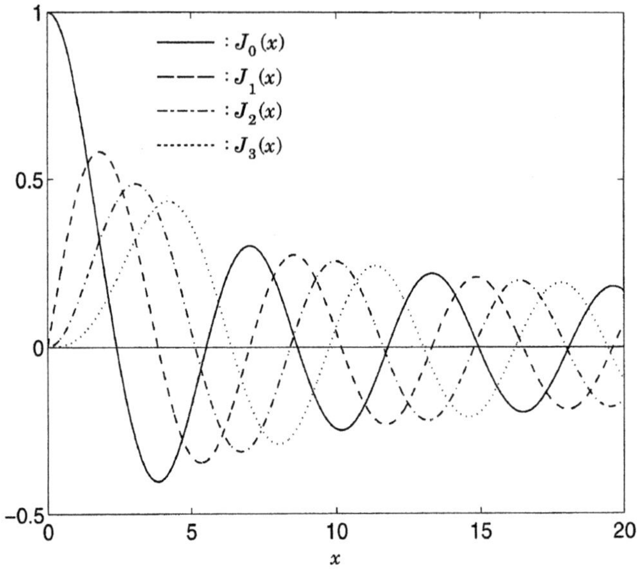

Fig. B.1. Bessel functions of first kind $J_n(x)$, $n = 0, 1, 2, 3$.

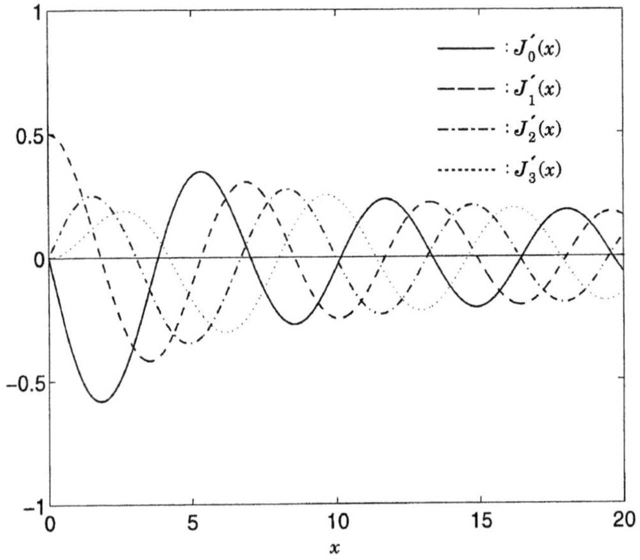

Fig. B.2. Derivatives of Bessel functions of first kind $J_n'(x) = \dfrac{dJ_n(x)}{dx}$, $n = 0, 1, 2, 3$.

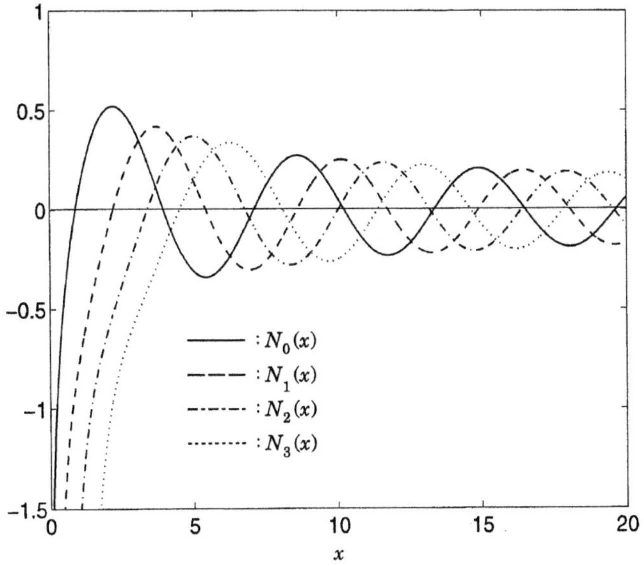

Fig. B.3. Bessel functions of second kind $N_n(x)$, $n = 0, 1, 2, 3$.

B.1.1 Limiting Forms for Small and Large Arguments

For $x \to 0$

$$J_0(x) \sim 1 \tag{B.6}$$

$$N_0(x) \sim \frac{2}{\pi} \ln x \,. \tag{B.7}$$

For $x \to 0$ and $\nu > 0$

$$J_\nu(x) \sim \frac{1}{\nu!} \left(\frac{x}{2}\right)^\nu \tag{B.8}$$

$$N_\nu(x) \sim -\frac{(\nu - 1)!}{\pi} \left(\frac{2}{x}\right)^\nu \,. \tag{B.9}$$

For $x \to \infty$

$$J_\nu(x) \sim \sqrt{\frac{2}{\pi x}} \cos\left(x - \frac{\pi}{4} - \frac{\nu\pi}{2}\right) \tag{B.10}$$

$$N_\nu(x) \sim \sqrt{\frac{2}{\pi x}} \sin\left(x - \frac{\pi}{4} - \frac{\nu\pi}{2}\right) \,. \tag{B.11}$$

B.1.2 Wronskian

The Wronskian of the Bessel functions gives

$$J_\nu(x)N_{\nu+1}(x) - N_\nu(x)J_{\nu+1}(x) = -\frac{2}{\pi x} \tag{B.12}$$

$$J_\nu(x)N'_\nu(x) - N_\nu(x)J'_\nu(x) = \frac{2}{\pi x} . \tag{B.13}$$

B.1.3 Generating Function

The generating function for the Bessel function gives [11, page 973]

$$\begin{aligned} e^{ix\cos\phi} &= \sum_{n=-\infty}^{\infty} i^n J_n(x) e^{in\phi} \\ &= J_0(x) + 2\sum_{n=1}^{\infty} i^n J_n(x)\cos n\phi . \end{aligned} \tag{B.14}$$

B.1.4 Hankel Functions

The Hankel functions of the first kind $H_\nu^{(1)}(x)$ and the second kind $H_\nu^{(2)}(x)$ of order ν are defined as

$$H_\nu^{(1)}(x) = J_\nu(x) + iN_\nu(x) \tag{B.15}$$

$$H_\nu^{(2)}(x) = J_\nu(x) - iN_\nu(x) . \tag{B.16}$$

For $x \to \infty$

$$H_\nu^{(1)}(x) \sim \sqrt{\frac{2}{\pi x}} \exp\left[i\left(x - \frac{\pi}{4} - \frac{\nu\pi}{2}\right)\right] \tag{B.17}$$

$$H_\nu^{(2)}(x) \sim \sqrt{\frac{2}{\pi x}} \exp\left[-i\left(x - \frac{\pi}{4} - \frac{\nu\pi}{2}\right)\right] . \tag{B.18}$$

B.1.5 Recurrence Formulas

$$Z_{\nu-1}(x) + Z_{\nu+1}(x) = \frac{2\nu}{x} Z_\nu(x) \tag{B.19}$$

$$Z_{\nu-1}(x) - Z_{\nu+1}(x) = 2\frac{dZ_\nu(x)}{dx} \tag{B.20}$$

where $Z_\nu(x)$ denotes $J_\nu(x)$, $N_\nu(x)$, $H_\nu^{(1)}(x)$, or $H_\nu^{(2)}(x)$.

B.1.6 Modified Bessel Functions

The modified Bessel functions of the first kind $I_\nu(x)$ and the second kind $K_\nu(x)$ of order ν are defined as

$$I_\nu(x) = i^{-\nu} J_\nu(ix) \tag{B.21}$$

$$K_\nu(x) = \frac{\pi}{2} i^{\nu+1} H_\nu^{(1)}(ix) \,. \tag{B.22}$$

For $x \to \infty$

$$I_\nu(x) \sim \frac{e^x}{\sqrt{2\pi x}} \tag{B.23}$$

$$K_\nu(x) \sim \sqrt{\frac{\pi}{2x}} e^{-x} \,. \tag{B.24}$$

B.2 Spherical Bessel Functions

Consider the differential equation

$$x^2 \frac{d^2 y}{dx^2} + 2x \frac{dy}{dx} + \left[x^2 - n(n+1)\right] y = 0 \,. \tag{B.25}$$

Its solution is a linear combination of the spherical Bessel functions $j_n(x)$, $n_n(x)$, and the spherical Hankel functions $h_n^{(1)}(x)$, $h_n^{(2)}(x)$ of order n. The spherical Bessel function $z_n(x)$ and the Bessel function of fractional order $Z_{n+\frac{1}{2}}(x)$ are related by

$$z_n(x) = \sqrt{\frac{\pi}{2x}} Z_{n+\frac{1}{2}}(x) \tag{B.26}$$

where $z = j, n, h^{(1)}$, and $h^{(2)}$. Some typical forms for the spherical Bessel functions are

$$j_0(x) = \frac{\sin x}{x} \tag{B.27}$$

$$j_1(x) = \frac{\sin x}{x^2} - \frac{\cos x}{x} \tag{B.28}$$

$$n_0(x) = -\frac{\cos x}{x} \tag{B.29}$$

$$n_1(x) = -\frac{\cos x}{x^2} - \frac{\sin x}{x} \,. \tag{B.30}$$

The generating function for the spherical Bessel function gives [11, page 973]

$$e^{ix\cos\phi} = \sqrt{\frac{\pi}{2x}}\sum_{n=0}^{\infty}(2n+1)i^n J_{n+\frac{1}{2}}(x)P_n(\cos\phi)\,. \tag{B.31}$$

Another form of the spherical Bessel function $\hat{Z}_n(x) = xz_n(x)$ is often used. Substituting $z_n(x) = \dfrac{\hat{Z}_n(x)}{x}$ into (B.25) yields

$$x^2\frac{d^2\hat{Z}_n(x)}{dx^2} + \left[x^2 - n(n+1)\right]\hat{Z}_n(x) = 0 \tag{B.32}$$

where

$$\hat{Z}_n(x) = \sqrt{\frac{\pi x}{2}}Z_{n+\frac{1}{2}}(x)\,. \tag{B.33}$$

C

Residue Theorem

Let the function $f(z)$ be analytic in the complex z-plane except for the multiple singularities at $z_1, z_2, \cdots, z_m$ inside the simple closed contour C. The function $f(z)$ is said to be analytic at $z = z_0$ if it is defined and has its derivative at $z = z_0$ and in some small region around z_0. Then, the contour integral along C in the counterclockwise direction is given by

$$\oint_C f(z)\, dz = 2\pi i \sum_{j=1}^{m} \operatorname*{Res}_{z=z_j} f(z) \tag{C.1}$$

where $\operatorname*{Res}_{z=z_j} f(z)$ denotes the residue of $f(z)$ at $z = z_j$. When $f(z)$ has the pole of order $n \geq 1$ at $z = z_j$, the corresponding residue is

$$\operatorname*{Res}_{z=z_j} f(z) = \frac{1}{(n-1)!} \lim_{z \to z_j} \left\{ \frac{d^{n-1}}{dz^{n-1}} \left[(z - z_j)^n f(z) \right] \right\} . \tag{C.2}$$

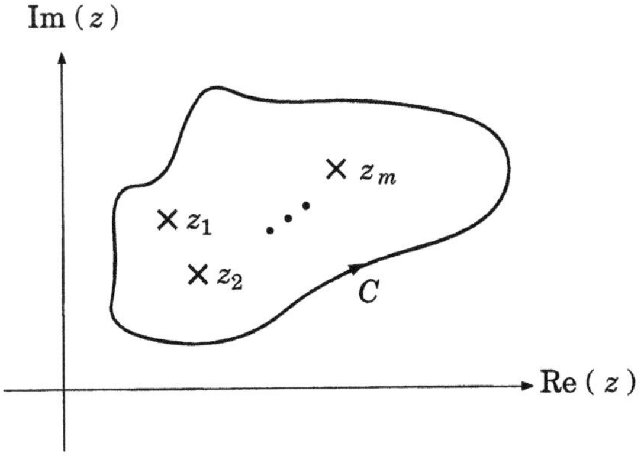

Fig. C.1. Complex z-plane with multiple singularities at $z_1, z_2, \cdots, z_m$.

When $f(z)$ has the simple pole $n = 1$,

$$\operatorname*{Res}_{z=z_j} f(z) = \lim_{z \to z_j} (z - z_j) f(z) . \tag{C.3}$$

Assume that $f(z)$ with the simple pole at $z = z_j$ is written as

$$f(z) = \frac{p(z)}{q(z)} . \tag{C.4}$$

Let us expand $q(z)$ in the Taylor series at $z = z_j$

$$q(z) = (z - z_j) q'(z_j) + \frac{(z - z_j)^2}{2!} q''(z_j) + \cdots \tag{C.5}$$

where the prime denotes differentiation with respect to the argument, such as $q'(z_j) = \left.\frac{dq(z)}{dz}\right|_{z=z_j}$. Therefore

$$\begin{aligned} \operatorname*{Res}_{z=z_j} f(z) &= \lim_{z \to z_j} \frac{p(z)}{\left[q'(z_j) + \frac{(z - z_j)}{2!} q''(z_j) + \cdots \right]} \\ &= \frac{p(z_j)}{q'(z_j)} . \end{aligned} \tag{C.6}$$

D

Associated Legendre Functions

The associated Legendre equation is

$$(1-x^2)\frac{d^2y}{dx^2} - 2x\frac{dy}{dx} + \left[n(n+1) - \frac{m^2}{1-x^2}\right] y = 0 \, . \tag{D.1}$$

The solutions for nonnegative integers n and m are the associated Legendre functions of the first and second kinds, $P_n^m(x)$ and $Q_n^m(x)$, respectively. The solution $Q_n^m(x)$ becomes infinite at $x = \pm 1$. The solution $P_n^m(x)$ is given by

$$P_n^m(x) = (1-x^2)^{\frac{m}{2}} \frac{d^m P_n(x)}{dx^m} \tag{D.2}$$

where $P_n(x)$ $\left[\equiv P_n^0(x)\right]$ are the Legendre polynomials as

$$P_n(x) = \frac{1}{2^n n!}\frac{d^n}{dx^n}(x^2-1)^n \, . \tag{D.3}$$

Some typical functional forms are

$$P_0(x) = 1 \tag{D.4}$$

$$P_1(x) = x \tag{D.5}$$

$$P_2(x) = \frac{1}{2}(3x^2 - 1) \tag{D.6}$$

$$P_3(x) = \frac{1}{2}(5x^3 - 3x) \tag{D.7}$$

$$P_1^1(x) = (1-x^2)^{\frac{1}{2}} \tag{D.8}$$

$$P_2^1(x) = 3x(1-x^2)^{\frac{1}{2}} \tag{D.9}$$

$$P_2^2(x) = 3(1-x^2) \tag{D.10}$$

$$P_3^1(x) = \frac{3}{2}(5x^2-1)(1-x^2)^{\frac{1}{2}} \tag{D.11}$$

$$P_3^2(x) = 15x(1-x^2) \tag{D.12}$$

$$P_3^3(x) = 15(1-x^2)^{\frac{3}{2}} \, . \tag{D.13}$$

E

Transforms and Series

- The Fourier transform of $f(x)$ is defined as

$$\tilde{f}(\zeta) = \int_{-\infty}^{\infty} f(x) e^{-i\zeta x} \, dx \, . \tag{E.1}$$

The inverse Fourier transform of $\tilde{f}(\zeta)$ is

$$f(x) = \frac{1}{2\pi} \int_{-\infty}^{\infty} \tilde{f}(\zeta) e^{i\zeta x} \, d\zeta \, . \tag{E.2}$$

The functions $f(x)$ and $\tilde{f}(\zeta)$ are called the Fourier transform pair.

- The Fourier sine transform pair is given by

$$\tilde{f}(\zeta) = \int_{0}^{\infty} f(x) \sin \zeta x \, dx \tag{E.3}$$

$$f(x) = \frac{2}{\pi} \int_{0}^{\infty} \tilde{f}(\zeta) \sin \zeta x \, d\zeta \, . \tag{E.4}$$

- The Fourier cosine transform pair is given by

$$\tilde{f}(\zeta) = \int_{0}^{\infty} f(x) \cos \zeta x \, dx \tag{E.5}$$

$$f(x) = \frac{2}{\pi} \int_{0}^{\infty} \tilde{f}(\zeta) \cos \zeta x \, d\zeta \, . \tag{E.6}$$

- The Hankel transform pair is given by

$$\tilde{f}(\zeta) = \int_{0}^{\infty} f(x) J_n(\zeta x) x \, dx \tag{E.7}$$

$$f(x) = \int_{0}^{\infty} \tilde{f}(\zeta) J_n(\zeta x) \zeta \, d\zeta \, . \tag{E.8}$$

- The Fourier series representation for the function $f(x)$ for the interval $-L < x < L$ is given by

$$f(x) = \sum_{n=-\infty}^{\infty} c_n \exp\left(\frac{in\pi x}{L}\right) \tag{E.9}$$

$$c_n = \frac{1}{2L}\int_{-L}^{L} f(x) \exp\left(-\frac{in\pi x}{L}\right) dx \,. \tag{E.10}$$

- The spherical harmonics representation for the function $f(\theta, \phi)$ for the interval $0 < \theta < \pi$ and $0 < \phi < 2\pi$ is given by

$$f(\theta, \phi) = \sum_{n=0}^{\infty} \sum_{m=-n}^{n} c_{nm} P_n^m(\cos\theta) e^{im\phi} \tag{E.11}$$

$$c_{nm} = \frac{2n+1}{4\pi}\frac{(n-m)!}{(n+m)!}$$

$$\cdot \int_0^{2\pi} d\phi \int_0^{\pi} d\theta \sin\theta f(\theta, \phi) P_n^m(\cos\theta) e^{-im\phi} \,. \tag{E.12}$$

Index

Printing: Mercedes-Druck, Berlin
Binding: Stein + Lehmann, Berlin

GPSR Compliance

The European Union's (EU) General Product Safety Regulation (GPSR) is a set of rules that requires consumer products to be safe and our obligations to ensure this.

If you have any concerns about our products, you can contact us on ProductSafety@springernature.com

In case Publisher is established outside the EU, the EU authorized representative is:

Springer Nature Customer Service Center GmbH
Europaplatz 3
69115 Heidelberg, Germany

Batch number: 10703970

Printed by Printforce, the Netherlands